Telling the Evolutionary Time: Molecular Clocks and the Fossil Record

The Systematics Association Special Volume Series
Series Editor

Alan Warren
Department of Zoology, The Natural History Museum,
Cromwell Road, London SW7 5BD, UK.

The Systematics Association promotes all aspects of systematic biology by organising conferences and workshops on key themes in systematics, publishing books, and awarding modest grants in support of systematics research. Membership of the Association is open to internationally-based professionals and amateurs with an interest in any branch of biology including palaeobiology. Members are entitled to attend conferences at discounted rates, to apply for grants and to receive the newsletters and mailed information; they also receive a generous discount on the purchase of all volumes produced by the Association.

The first of the Systematics Association's publications, *The New Systematics* (1940), was a classic work edited by its then president Sir Julian Huxley, that set out the problems facing general biologists in deciding which kinds of data would most effectively progress systematics. Since then, more than 70 volumes have been published, often in rapidly expanding areas of science where a modern synthesis is required.

The *modus operandi* of the Association is to encourage leading researchers to organise symposia that result in a multi-authored volume. In 1997 the Association organised the first of its international Biennial Conferences. This and subsequent Biennial Conferences, which are designed to provide fora for systematists of all kinds, included themed symposia that resulted in further publications. The Association also publishes volumes that are not specifically linked to meetings and encourages new publications in a broad range of systematic topics.

Anyone wishing to learn more about the Systematics Association and its publications should refer to our website at www.systass.org.

Forthcoming titles in the series:

Milestones in Systematics
D.M. Williams and P.L. Forey

Other Systematics Association publications are listed after the index for this volume.

The Palaeontological Association

The Palaeontological Association is a learned society and a charity registered in England and Wales (No 276369). It exists in order to promote research in palaeontology and its allied sciences. It publishes the flagship journal *Palaeontology* six times a year and other publications from time to time. Details of these and other activities can be viewed at www.palass.org. Membership and sales enquiries should be adderssed to the Executive Officer: palass@palass.org.

The Systematics Association Special Volume Series 66

Telling the Evolutionary Time: Molecular Clocks and the Fossil Record

Edited by
Philip C.J. Donoghue
and M. Paul Smith

Taylor & Francis
Taylor & Francis Group

LONDON AND NEW YORK

First published 2003 by Taylor & Francis
11 New Fetter Lane, London EC4P 4EE

Simultaneously published in the USA and Canada
by Taylor & Francis Inc,
29 West 35th Street, New York, NY 10001

Taylor & Francis is an imprint of the Taylor & Francis Group

Typeset in 10/12pt Sabon by Graphicraft Limited, Hong Kong
Printed and bound in Great Britain by TJ International Ltd, Padstow, Cornwall

Every effort has been made to ensure that the advice and information in this book is true
and accurate at the time of going to press. However, neither the publisher nor the authors
can accept any legal responsibility or liability for any errors or omissions that may be
made. In the case of drug administration, any medical procedure or the use of technical
equipment mentioned within this book, you are strongly advised to consult the
manufacturer's guidelines.

British Library Cataloguing in Publication Data
A catalogue record for this book is available from the British Library

Library of Congress Cataloging in Publication Data
A catalog record has been requested

ISBN 0-415-27524-5

Contents

Contributors

Francisco J. Ayala, Department of Ecology and Evolutionary Biology, University of California, Irvine, California 92697–2525, USA (fjayala@uci.edu).

Michael J. Benton, Department of Earth Sciences, University of Bristol, Queens Road, Bristol BS8 1RJ, UK (mike.benton@bristol.ac.uk).

Cédric Berney, Department of Zoology and Animal Biology, University of Geneva, 154, route de Malagnou, 1224-Chêne Bougeries, Switzerland (cedric.berney@zoo.unige.ch).

Graham E. Budd, Department of Earth Sciences, Historical Geology and Palaeontology, University of Uppsala, Uppsala, Sweden SE-752 36 (graham.budd@pal.uu.se).

Mark W. Chase, Molecular Systematics Section, Jodrell Laboratory, Royal Botanic Gardens, Kew, Richmond, Surrey TW9 3DS, UK.

Michael I. Coates, Department of Organismal Biology and Anatomy, University of Chicago, 1027 East 57th Street, Chicago, Illinois 60637–1508, USA (mcoates@midway.uchicago.edu).

Philip C.J. Donoghue, Department of Earth Sciences, School of Geography, Earth & Environmental Sciences, University of Birmingham, Birmingham B15 2TT, UK. Present address: Department of Earth Sciences, University of Bristol, Queens Road, Bristol BS8 1RJ (phil.donoghue@bristol.ac.uk).

Gareth J. Dyke, Department of Zoology, University College Dublin, Belfield Dublin 4, Ireland (gareth.dyke@ucd.ie).

Richard A. Fortey, Department of Palaeontology, The Natural History Museum, London SW7 5BD, UK (raf@nhm.ac.uk).

S. Blair Hedges, NASA Astrobiology Institute and Department of Biology, 208 Mueller Laboratory, The Pennsylvania State University, University Park, PA 16802, USA (sbh1@psu.edu).

Jennifer Jackson, Ancient Biomolecules Centre, Zoology Department, South Parks Road, University of Oxford, Oxford OX1 3PS, UK (jennifer.jackson@linacre.oxford.ac.uk).

Sören Jensen, Department of Earth Sciences, University of California, Riverside, CA 92521, USA. Present address: Area de Paleontologia, Facultad de Ciencias, Universidad de Extremadura, 06071 Badajoz, Spain (soren@unex.es).

Christopher R.C. Paul, Department of Earth Sciences, University of Liverpool, Liverpool L69 3GP, UK (crcp@liv.ac.uk).

Jan Pawlowski, Department of Zoology and Animal Biology, University of Geneva, 154, route de Malagnou, 1224-Chêne Bougeries, Switzerland (Jan.Pawlowski@zoo.unige.ch).

Francisco Rodríguez-Trelles, Unidad de Medicina Molecular INGO, Hospital Clínico Universitario, Universidad de Santiago de Compostela, 15706 Santiago de Compostela, Spain (ftrelles@correo.cesga.es).

Marcello Ruta, Department of Organismal Biology and Anatomy, University of Chicago, 1027 East 57th Street, Chicago, Illinois 60637–1508, USA (mruta@ midway.uchicago.edu).

Ivan J. Sansom, Department of Earth Sciences, School of Geography, Earth & Environmental Sciences, University of Birmingham, Birmingham B15 2TT, UK (i.j.sansom@bham.ac.uk).

Vincent Savolainen, Molecular Systematics Section, Jodrell Laboratory, Royal Botanic Gardens, Kew, Richmond, Surrey TW9 3DS, UK.

M. Paul Smith, Lapworth Museum of Geology, School of Geography, Earth & Environmental Sciences, University of Birmingham, Birmingham B15 2TT, UK (m.p.smith@bham.ac.uk).

Jan Strugnell, Ancient Biomolecules Centre, Zoology Department, South Parks Road, University of Oxford, Oxford OX1 3PS, UK (jan.strugnell@merton.ox.ac.uk).

Rosa Tarrío, Misión Biológica de Galicia, CSIC, Apartado 28, 36080, Pontevedra, Spain.

Charles H. Wellman, Department of Animal and Plant Sciences, University of Sheffield, Alfred Denny Building, Western Bank, Sheffield S10 2TN, UK (c.wellman@sheffield.ac.uk).

Niklas Wikström, Department of Botany, The Natural History Museum, London SW7 5BD, UK. Present address: Department of Systematic Botany, Evolutionary Biology Centre, Uppsala University, Norbyvägen 18D, Sweden (niklas.wikstrom@ ebc.uu.se).

Introduction
Molecular clocks and the fossil record – towards consilience?

Molecular clocks or the fossil record: which approach tells evolutionary time? The papers presented in this volume stem from a joint Palaeontological Association/Systematics Association symposium held at the Third Biennial Meeting of the Systematics Association, at Imperial College, London on 5 September 2001, which brought together palaeontologists and molecular biologists with the aim of addressing the disparity between molecular and palaeontological perspectives on evolutionary time.

Resolving the timing and tempo of major evolutionary radiations represents one of the most prominent and controversial debates currently underway in evolutionary biology. Although this temporal dimension has traditionally been the preserve of palaeontology, with a burgeoning molecular database that is progressively more representative of the tree of life, attempts to establish a molecular timescale for evolutionary history have become ever more sophisticated. The so-called 'molecular clock' works, at least at a simplistic level, by calibrating molecular distance between a pair of organisms to time using a reliable fossil estimate. By extrapolating this scaling it is possible to estimate the time of divergence between other taxa, for which the record is perhaps unreliable, or simply as a test of fossil-based estimates. Because of the unlikelihood of the fossilization of the earliest representatives of taxa, even the apparently reliable fossil records of taxa used in calibrating the clock can provide only a conservative estimate of the true divergence date, and so it is to be expected that molecular clock estimates should also be conservative.

However, this is not the case; molecular clocks almost always provide divergence estimates that are not just older, but considerably so (e.g. Hedges, Chapter 2), and sometimes as much as double the age of fossil-based estimates. By implication the fossil record is not just incomplete, but represents only half of evolutionary history, and the latter half at that.

No one would defend total completeness of the fossil record, but it is the degree to which evolutionary history is unrepresented by the fossil record that is in question (Fortey *et al.*, Chapter 3; Benton, Chapter 4). Indeed, the gross disparity between molecular and fossil-based estimates has led to a period of introspection amongst the palaeontological community, and the development of a variety of methods for the evaluation of the quality of the fossil record. These include attempts to assess confidence in the fossil record as a reflection of the true time of origin of a particular clade (e.g. Benton, Chapter 4; Donoghue *et al.*, Chapter 10). These methods are becoming progressively more realistic and, inevitably, more complex in their application (e.g. Tavaré *et al.* 2002), but they do provide a means of testing the timing of evolutionary events within the bounds of statistical confidence. Methods have also been devised to assess congruence between the order of stratigraphic appearance of taxa in the fossil record and the sequence of evolutionary branching as inferred from phylogenetic hypotheses constructed independently of temporal data. These represent powerful independent tests of the fossil record, although they are not without their attendant problems (Paul, Chapter 5). All of these techniques can be used to identify and qualify the veracity of potential calibration points for molecular clock analyses.

The fossil record has also become better understood through significant advances in systematics and classification with, for instance, the gradual elimination of paraphyletic groups. But perhaps the greatest advances have been made through the better understanding, application and realization of the implications of total- and crown-group concepts (Jefferies 1979; Budd & Jensen, Chapter 9; Donoghue *et al.*, Chapter 10; Dyke, Chapter 12). Hitherto enigmatic fossil taxa that have resisted classification because they exhibit only a subset of the characters necessary for inclusion in groups defined on the basis of their living members (crown-groups) have subsequently been recognized as reflecting interim stages in the assembly of crown-group body plans. The augmentation of such fossil taxa to their nearest living crown-group constitutes the more universal total-group, and the difference between total- and crown-groups discriminates the paraphyletic stem-group. The recognition of stem-groups is important not only because it provides an understanding of the steps through which clade divergence proceeded, but also because stem taxa are frequently older than their fossil crown-group relatives and therefore lead frequently to significant extensions to the inferred time of divergence of one crown-group from another. The discrimination of stem- from crown-group taxa has also proved to be profitable because it provides more appropriate time estimates for clade divergence in comparison with those derived from molecular clock analyses which are, rather ironically, usually framed within the context of total-groups.

Reanalysis and revision of the fossil database has raised the possibility for rapprochement between fossil- and molecular-based approaches. Generally, however, optimism has not been rewarded. In short, two main camps have arisen amongst practising palaeontologists, delimited by (a) those who accept the molecular clock estimates, concluding that the fossil record is corrupted by its reliance upon negative evidence (Fortey *et al.*, Chapter 3) and, (b) those who find that the fossil record exhibits no evidence for a significant deterioration in quality of the fossil record through time, and conclude that molecular clock estimates are spurious, for one reason or another (Benton, Chapter 4; Wellman, Chapter 7; Budd & Jensen, Chapter 9). Indeed, if there is any agreement, it is that pictures of evolutionary history widely accepted less than ten years ago are incorrect.

Molecular clocks, in turn, are not without their own attendant problems (Rodríguez-Trelles *et al.*, Chapter 1; Fortey *et al.*, Chapter 3). Although the molecular clock originally found support in Kimura's neutral theory of molecular evolution, observed rates of substitution are generally much higher than expected. In attempting to allow for this, various supplementary hypotheses to the neutral theory have been invoked, the validity of which are explored by Rodríguez-Trelles *et al.* (Chapter 1). However, the theoretical basis underlying most recent analyses is more empirical, relying instead upon a so-called 'law of large numbers' (Rodríguez-Trelles *et al.*, Chapter 1;

e.g. Hedges, Chapter 2). This assumes that because the evolutionary process is time-dependent and because the number of individual clocks (=genes) is vast, it is expected that their results will converge on average values reflecting the time elapsed since the divergence of species. In practice, datasets are always filtered for sequences that exhibit deviation from clock-like behaviour. The implications of such sequences for molecular evolution are not readily apparent and represent an interesting and important subject for study in their own right (Pawlowski & Berney, Chapter 6).

However, sources of much of the controversy surrounding the mechanics of molecular clock analyses lie elsewhere. These include phylogenetic uncertainty, the unreliability of calibration points, neglect of confidence intervals both on calibration points and on clock estimates, and problems with accurately relating sequence changes to rate variation (Fortey *et al.*, Chapter 3). These are further exacerbated by the fact that such potential errors are cumulative and progressively distant clock analyses will be subject to gross absolute errors.

In addition to these specific errors, molecular clock estimates, palaeontological estimates, and attempts to correlate the two, are all hostage to a geological timescale that is under continual revision and, thus, the choice of timescale is non-trivial. Dating techniques are advancing at a considerable pace, with large increases in precision, and a timescale that is a decade old is likely to have been supplanted by more accurate ones – although the pace of advance is uneven in different parts of the column. In addition, the accuracy of available dates varies throughout the geological column, with poorly dated intervals not restricted to the oldest periods of geological history. Unless all of these sources of error are taken into account, in addition to attempts to correlate fossil occurrences to the geological timescale, and those errors attendant to molecular clocks themselves, errors will propagate, potentially beyond the age of the events being estimated.

This volume presents analyses by experts in the mechanics and application of molecular clocks as well as analyses of the fossil records of protists (Pawlowski & Berney, Chapter 6), plants (Wellman, Chapter 7; Wikström *et al.*, Chapter 8), the divergence of complex animals (Fortey *et al.*, Chapter 4; Budd & Jensen, Chapter 9), chordates (Donoghue *et al.*, Chapter 10), tetrapods (Ruta & Coates, Chapter 11), and modern birds (Dyke, Chapter 12). Although there is evidence for creeping increases in palaeontological estimates, as might be expected, there remain serious disagreements over the time of origin of major groups such as land plants and animals. Interestingly, disagreement does not follow disciplinary boundaries suggesting that conference between molecular biologists and palaeontologists heralds a realizable consilience in the future.

In concluding, we would like to thank all of the contributors to the symposium and the volume for their enthusiasm and willingness to look anew upon their old datasets and cherished assumptions. We wish to express our gratitude also to the reviewers for their hard work and commitment in the face of anonymity, and to Mandy Donoghue for assisting with production of this volume. Finally, we wish to acknowledge the support of the Palaeontological Association and Systematics Association for co-sponsoring the symposium.

Philip Donoghue and Paul Smith

References

Jefferies, R.P.S. (1979) 'The origin of chordates: a methodological essay', M.R. Horse (ed.) *The Origin of Major Invertebrate Groups*, London: systematics Association, pp. 443–7.

Tavaré, S., Marshall, C.R., Will, O., Soligo, C. and Martin, R.D. (2002) 'Using the fossil record to estimate the age of the last common ancestor of extant primates', *Nature*, 416: 726–9.

Chapter 1

Molecular clocks: whence and whither?

Francisco Rodríguez-Trelles, Rosa Tarrío and Francisco J. Ayala

ABSTRACT

The neutrality theory of molecular evolution predicts that the rate of molecular evolution is constant over time, and thus that there is a molecular clock that can be used for timing evolutionary events. Experimental data have shown that the variance of the rate of evolution is generally larger than expected according to the neutrality theory. This raises the question of how reliable the molecular clock is or, indeed, whether there is a molecular clock. We have carried out an extensive investigation of nine proteins in organisms belonging to the three multicellular kingdoms, namely ADH, AMD, DDC, GPDH, G6PD, PGD, SOD, TPI, and XDH. We observe that the nine proteins evolve erratically through time and across lineages. The observations are inconsistent with the neutrality theory and also with various subsidiary hypotheses proposed to account for the overdispersion of the molecular clock.

Introduction: the hypothesis of a molecular clock

Biological evolution is a time-dependent process, by and large unidirectional. Some degree of correlation is, therefore, expected between the biological differentiation of two organisms and the time elapsed since their separation, by the comparison between an organism and its ancestor, or between two organisms sharing a common ancestor. The correlation, however, need not be exact, if only because organisms evolve in response to the vagaries of environmental change in time and space. It is well known that some organisms have evolved fast morphologically, at least with respect to some traits, whereas others have changed but little over millions of years (see, e.g. Dobzhansky *et al.* 1977, pp. 327–31). Zuckerkandl and Pauling (1962, 1965; see also Margoliash, 1963) proposed that the time-change correlation might be more approximately precise if change were measured in the protein and nucleic acid components of organisms, indeed that there might be a molecular clock of evolution.

The hypothesis of the molecular clock was advanced on the grounds that most amino acid substitutions in a protein (or nucleotides in a gene) occur between functionally equivalent residues, so that their replacement along evolving lineages would be determined by mutation rate and time elapsed, rather than by natural selection (Zuckerkandl and Pauling 1965). Natural selection is rather fickle, subject to the vagaries of environmental change and organism interactions, whereas mutation rate for a given gene is likely to remain constant through time and across lineages. The number of amino acid replacements (or nucleotide substitutions) between species would, then, reflect the time elapsed since their last common ancestor. The time of remote events, as well as the degree of relationship among contemporary lineages could, thus, be determined on the basis of amino acid (or nucleotide) differences. A notable feature of the hypothesis of the molecular evolutionary clock is multiplicity: every one of the thousands of proteins or genes of an organism is an independent clock, each ticking at a different rate but all measuring the same events (Ayala 1986; Gillespie 1991; Li 1997).

Early investigations showed that the evolution of the globins in vertebrates conformed fairly well to the clock hypothesis, which allowed reconstruction of the history of globin gene duplications (Zuckerkandl and Pauling 1965). Fitch and Margoliash (1967) would soon provide a 'genetic distance' method that was effectively used for reconstruction of the history of 20 organisms, from yeast to moth to human, based on the amino acid sequence of a small protein, cytochrome c. A theoretical foundation for the clock was provided by Kimura (1968), who developed a 'neutral theory of molecular evolution', with great mathematical simplicity; notably, the theory states that the rate of substitution of adaptively equivalent ('neutral') alleles, k, is precisely the rate of mutation, u, of neutral alleles, $k = u$. The neutrality theory predicts that molecular evolution behaves like a stochastic clock, such as radioactive decay, with the properties of a Poisson distribution, in which the mean, M, and variance, V, are expected to be identical, so that $V/M = 1$. The 'index of dispersion', measuring the deviation of this ratio from the expected value of 1, is a way to test whether observations fit the theory.

Experimental data have shown that often the rate of molecular evolution is 'overdispersed', that is, that the index of dispersion is often significantly greater than 1 (Gillespie 1991; Li 1997). Deviations from rate constancy occur between lineages, for example between rodents and mammals, as well as at different times along a given lineage, both factors having significant effects (Langley and Fitch 1974). Consequently, several modifications of the neutral theory have been proposed, seeking to account for the excess variance of the molecular clock. It has been proposed, for example, that most protein evolution involves 'slightly deleterious' replacements rather than strictly neutral ones (Ohta 1973); or that certain 'biological properties', such as the effectiveness of the error-correcting polymerases, vary among organisms, so that mutation rates for a given gene vary from one organism to another (Kimura and Ohta 1972; Kimura 1980, 1983; Gillespie 1991; Li and Graur 1997). A 'population size' hypothesis proposes that organisms with larger effective population size have a slower rate of evolution than organisms with smaller population size, because the time required to fix new mutations increases with population size (Ohta 1972; Kimura 1983). Another supplementary hypothesis invokes a generation-time effect. Protein evolution has been extensively investigated in primates and rodents with the common observation that the number of replacements is greater in the rodents (Kohne 1970; Li *et al.* 1996). In plants, the overall rate at the *rbc*L locus is more than five times greater in annual grasses than in palms, which have much longer generations (Gaut *et al.* 1992). These rate differences could be accounted for, according to the generation-effect hypothesis, by assuming that the time-rate of evolution depends on the number of germ-line replications per year, which is several times greater for the short-generation rodents and grasses than for the long-generation primates and palms. The rationale of the assumption is that the larger the number of replication cycles, the greater the number of mutational errors that will occur.

From a theoretical, as well as operational, perspective, these and other supplementary hypotheses have the discomforting consequence that they involve additional empirical parameters, often not easy to estimate. It is of great epistemological significance that the original proposal of the neutral theory is (i) highly predictive and therefore, (ii) eminently testable. The supplementary hypotheses lead, nevertheless, to certain predictions that can be tested. The 'generation-time', 'population size', and 'biological properties' hypotheses uniformly predict that rate variations observed between lineages or at different times will equally affect (in direction and magnitude) all genes of any particular organism, since these attributes are common to all genes of the same species. The 'slightly deleterious' hypothesis predicts that the rate of evolution will be inversely related to population size, and thus reduces to the 'population size' hypothesis (Ohta 1973).

In this chapter, we present an analysis of nine genes undertaken as a test of the four supplementary hypotheses, as well as of the neutrality theory, the more general or 'null' hypothesis underlying the molecular clock hypothesis. We have, in the past, reported results for three of these genes, showing that they exhibit overdispersed patterns of molecular evolution that are incompatible with the proposed supplementary hypotheses (Rodríguez-Trelles *et al.* 2001a, and references therein). The additional tests reported here lead to the same conclusion. We surmise that inferences about the timing of past events (and about phylogenetic relationships among species) based on molecular evolution are subject to sources of error not altogether disparate from inferences based on anatomy, embryology, or other phenotypic characteristics. Nevertheless, molecular investigations have two obvious advantages over phenotypic traits, in degree if not completely in kind; namely, that the *number* of 'traits' is very large, that is, every one of the thousands of genes in the make-up of each organism, and that differences can be more precisely *quantified*, measured as they are in terms of distinct units, such as amino acids or nucleotides. There are many evolutionary issues concerning both timing and phylogenetic relationships between species for which molecular sequence data provide the best, if not the only, dependable evidence. The large-scale reconstruction of the 'universal tree' of life is a case in point: the phylogenetic relationships among Archaea and bacterial prokaryotes and between them and the eukaryotes have best been determined with DNA sequences encoding ribosomal RNA genes. The multiplicity of genes opens up the possibility of combining data for numerous genes in assessing the timing of particular evolutionary events, or the phylogeny of species. Because of the time-dependence of the evolutionary process, the multiplicity of independent results would probably tend to converge (by the so-called 'law of large numbers') on average values reflecting, with reasonable accuracy, the time elapsed since the divergence of species.

The nine genes and their protein products

We have investigated the following nine nuclear genes: (1) *alcohol dehydrogenase* (*Adh*; E.C.1.1.1.1), (2) *aromatic-L-amino acid decarboxylase* (*Ddc*; E.C.4.1.1.28), and its paralogue (3) *α-methyl-dopa* (*Amd*; E.C.4.1.1.-), (4) *glycerol-3-phosphate dehydrogenase* (*Gpdh*; E.C.1.1.1.8), (5) *glucose-6-phosphate dehydrogenase* (*G6pd*;

Table 1.1 Likelihood-ratio test carried out on the amino acid data sequences for various genes

Dataset	Locus	Number of species	Constant rate between amino acids Poisson : JTT-F (19)	Constant rate among sites JTT-F : JTT-F+dG (1)	
			$-2\log\Lambda$	$-2\log\Lambda$	α
DIPTERA	Adh	22	624.3	191.5	0.86
	Amd	17	746.5	404.7	0.21
	Ddc	18	367.2	290.0	0.16
	Gpdh	17	139.4	74.5	0.06
	Sod	21	186.0	179.2	0.22
	Xdh	20	1533.8	650.4	0.45
GLOBAL	Gpdh	30	667.2	366.0	0.82
	G6pd	16	1190.1	541.8	0.66
	Pgd	14	506.6	164.9	0.81
	Sod	61	851.4	707.2	0.57
	Tpi	26	351.4	204.8	0.77
	Xdh	34	2083.4	1135.5	0.84

The null hypothesis (H_0) is compared with a hypothesis (H_1) that removes the assumption indicated on top. Log-likelihood values are obtained assuming the topology in Figure 1.1. The probability of obtaining the observed value of the likelihood ratio statistic ($-2\log\Lambda$) if H_0 were true, with degrees of freedom (*d.f.*) indicated, is $<10^{-6}$ for all tests. Poisson: Poisson model; JTT-F: Yang *et al.* (1998) model; JTT-F+dG assuming discrete gamma distributed rates at sites. The α values are obtained with the JTT-F+dG model.

E.C.1.1.1.49), (6) *phosphogluconate dehydrogenase* (*Pgd*; E.C.1.1.1.44), (7) *super-oxide dismutase* (*Sod*; E.C.1.15.1.1), (8) *triosephosphate isomerase* (*tpi*; E.C.5.3.1.1), and (9) *xanthine dehydrogenase* (*Xdh*; E.C.1.1.1.204) (Table 1.1 and Figure 1.1). Six genes (1–4, 7, 9) have been analysed in 34 species of Diptera, comprising representatives of the families Drosophilidae and Tephritidae ('Diptera' dataset in Table 1.1; see also Figure 1.1) and the last six (genes 4–9) in 95 species that include representatives of the three multicellular eukaryote kingdoms, i.e. animals, plants, and fungi ('Global' dataset, Table 1.1; see Figure 1.1).

The enzymes encoded by the nine genes (i.e. ADH, AMD, DDC, GPDH, G6PD, PGD, SOD, TPI, and XDH) are globular soluble proteins generally composed of two identical subunits with approximate molecular masses of 28, 54, 54, 40, 60, 53, 15, 25, and 145 KDa, respectively. The encoded enzymes are oxidoreductases, except for DDC and AMD, which are carboxylases, and TPI, which is an isomerase.

ADH is a member of the insect-type, or 'short-chain' dehydrogenase/reductase family (SDR). In *Drosophila melanogaster* the *Adh* coding sequence is interrupted by two introns. The protein consists of two domains for binding the coenzyme (NAD) and the substrate (Benyajati *et al.* 1981). The crystal structure and reaction mechanisms of ADH have been elucidated in *Scaptodrosophila lebanonensis* (Benach *et al.* 1998). The enzyme plays a key role in adaptation to substrates undergoing alcoholic fermentation. It has been intensively investigated in *Drosophila* from a variety of perspectives (Powell 1997).

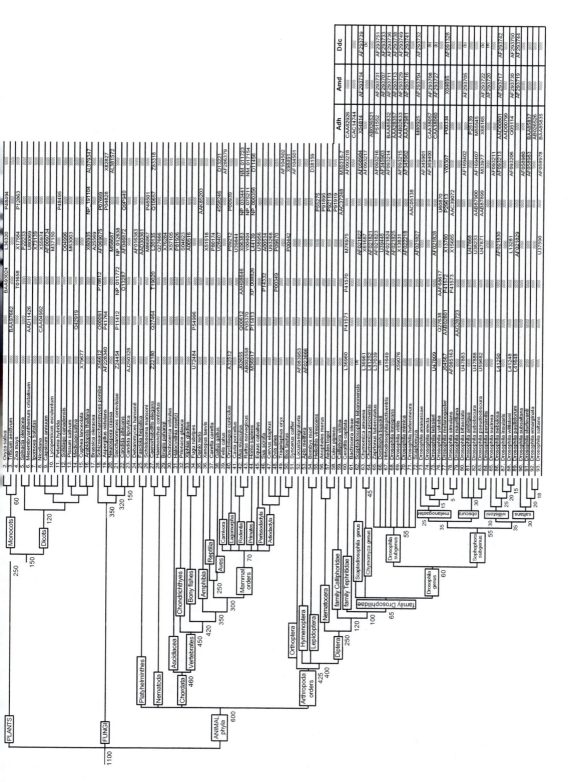

Figure 1.1 Tree topology for the species used in this study. Labels and numbers on the branches represent taxonomic categories and divergence times, respectively. *Nicotiana* and *Scaptomyza* species are *N. tabacum* for *G6pd* and *N. plumbaginifolia* for *Sod*; *S. adusta* for *Amd*, *Ddc*, and *Sod*; and *S. albovittata* for *Adh*. a: *Gpdh* sequences from Kwiatowski and Ayala (1999); b: *Ddc* sequences from Tatarenkov *et al.* (1999).

AMD and DDC are paralogues that arose as a result of a gene duplication, holding structural and functional relationships. They code for two decarboxylases involved in morphological differentiation, being essential for the sclerotization and melanization of the newly moulted cuticle of Diptera. In addition, DDC is required for the production of the neurotransmitters dopamine and serotonin (Wright 1996, and references therein). DDC is conserved between *Drosophila* and humans and is expressed in the central nervous system (CNS) as well as in the peripheral nervous system of insects and mammals (Wang *et al.* 1996; Wright 1996). In *D. melanogaster* there are two isoforms of the enzyme, which result from two alternative splicing pathways in neural and epidermal tissue (O'Keefe *et al.* 1995). The two genes combined have proven utility in determining previously unresolved phylogenetic relationships between *Drosophila* subgenera and species (Tatarenkov and Ayala 2001; Tatarenkov *et al.* 2001).

GPDH, the nicotinamide–adenine dinucleotide (NAD)-dependent cytoplasmic glycerol-3-phosphate dehydrogenase, plays a crucial role in insect flight metabolism because of its keystone position in the glycerophosphate cycle, which provides energy for flight in the thoracic muscles of *Drosophila* (O'Brien and MacIntyre 1978). In *Drosophila melanogaster* the *Gpdh* gene is located on chromosome 2 (O'Brien and MacIntyre 1972) and consists of eight coding exons (Bewley *et al.* 1989; von Kalm *et al.* 1989). It produces three isozymes by differential splicing of the last three exons (Cook *et al.* 1988). The GPDH polypeptide can be divided into two main domains, the NAD-binding domain and the catalytic domain. The NAD-binding domain (which in the rabbit is encompassed by the first 118 amino acids) is more highly conserved than the catalytic domain (Bewley *et al.* 1989). We have previously investigated the evolution of GPDH in Diptera (Kwiatowski *et al.* 1997) and across the three multicellular kingdoms (Rodríguez-Trelles *et al.* 2001a).

G6PD and PGD are prototype housekeeping enzymes, present in bacteria and all eukaryotic cell types. They catalyse the first and the last steps in the pentose shunt. Because the first step is rate limiting for the pathway, G6PD regulates the production of NADPH, critical for lipid synthesis and detoxification, and ribose 5-phosphate required for nucleotide and nucleic acid synthesis. The enzymes exhibit two domains for binding the coenzyme and substrate. The *G6pd* locus is X-linked in *D. melanogaster* (denoted *Zw* gene) and mammals. The active human G6PD exists in a dimer↔tetramer equilibrium depending on pH and ionic strength (reviewed in Shannon *et al.* 2000). Unlike other PGDs, the enzyme of *Schizosaccharomyces pombe* is tetrameric (Tsai and Chen 1998). Deficiency of G6PD and/or PGD causes chronic haemolytic anaemia, a common human enzymopathy.

The superoxide dismutases are abundant enzymes in aerobic organisms, with highly specific superoxide dismutation activity that protects the cell against harm from free oxygen radicals (Fridovich 1986). These enzymes have active centres that contain either iron or manganese, or both copper and zinc (Fridovich 1986). The Cu Zn superoxide dismutase (SOD) is a well-studied protein, found in eukaryotes but also in some bacteria (Steinman 1988). The population genetics and evolution of SOD have been, for three decades, the subject of numerous investigations in our laboratory (e.g. Ayala *et al.* 1971, 1974; Ayala 1972; Lee *et al.* 1981; Peng *et al.* 1986; Ayala 1997; Rodríguez-Trelles *et al.* 2001a).

TPI is a glycolytic enzyme essential for efficient energy production. It catalyses the interconversion of dihydroxyacetone phosphate and D-glyceraldehyde 3-phosphate, coupling the branches between triglyceride metabolism and glycolysis. TPI is metabolically adjacent to GPDH.

XDH is a complex metalloflavoprotein that plays an important role in nucleic acid degradation in all organisms: it catalyses the oxidation of hypoxanthine to xanthine and xanthine to uric acid with concomitant reduction of NAD to NADH. XDH protects the cell against damage induced by free oxygen radicals through the antioxidant action of uric acid (Xu *et al.* 1994). The chief physiological function of the enzyme changes, none the less, from one organism to another: its primary role is purine metabolism in mammals and chicken, but pteridine metabolism in *Drosophila* (review in Chovnic *et al.* 1977). The *Xdh* locus is widely expressed in human tissues (Xu *et al.* 1994); in *Drosophila melanogaster* (*rosy* locus) and *Bombyx mori* it is transcribed in the fat body, midgut, and Malpighian tubules, and in *D. melanogaster* some part of the protein is transported to the eyes. In higher plants, XDH takes part in ureide biosynthesis through *de novo* synthesis of purines from glutamine (Sagi *et al.* 1998). In mammals, but not in chicken and *Drosophila*, the enzyme can be converted to the oxidase form *xanthine oxidase* (XO; Hille and Nishino 1995). Defective *xanthine dehydrogenase* causes xanthinuria in humans; the enzyme is a target of drugs against gout and hyperuricaemia, and has been associated with blood pressure regulation (Enroth *et al.* 2000). We have previously investigated the rate of molecular evolution of *Xdh* in *Drosophila* as well as other species, including mammals, fungi, and plants (Rodríguez-Trelles *et al.* 2001b; see also Rodríguez-Trelles *et al.* 2001a).

Sources and models

The 95 species studied and GenBank accession numbers are given in Figure 1.1 (#8 and #72 in Figure 1.1 refer each to two species; respectively, *Nicotiana tabacum* and *N. plumbaginifolia*, and *Scaptomyza adusta* and *S. albovittata*). For GPDH and SOD we use the protein alignments from Ayala *et al.* (1996), and Fitch and Ayala (1994), slightly modified to fit additional sequences newly deposited in GenBank. For XDH we use the alignment from Rodríguez-Trelles *et al.* (2001b). ADH, AMD, DDC, G6PD, PGD, and TPI alignments were generated using ClustalX with default options. After removal of the gaps, the actual ADH, AMD, DDC, GPDH, G6PD, PGD, SOD, TPI, and XDH alignments consisted of 236, 344, 298, 241, 367, 208, 107, 78, and 599 amino acid residues, respectively. We treat *Dorsilopha*, *Hirtodrosophila*, and *Zaprionus* as *Drosophila* subgenera, following Tatarenkov *et al.* (1999), and Tarrío *et al.* (2001), but *Scaptodrosophila* as a genus, according to Grimaldi (1990), Tatarenkov *et al.* (1999), and Tarrío *et al.* (2001).

We follow a model-based maximum likelihood framework of statistical inference. We first model the substitution processes of the genes using a tree topology that is approximately correct; with the models so identified we proceed to generate maximum likelihood distances between pairs of amino acid sequences. As a tree topology for model fitting, we use the hypothesis shown in Figure 1.1. Relationships which are not well established (e.g. the branching order of animal phyla) are set as polytomies. Use of other reasonable topologies does not change parameter estimates (see also Yang 1994).

The amino acid substitution models used in this study (Table 1.1) are all special forms of the model of Yang *et al.* (1998), which is based on the empirical matrix of Jones *et al.* (1992), with amino acid frequencies set as free parameters (referred to as

JTT-F). Substitution-rate variation from site to site is accommodated in the substitution models using the discrete-gamma approximation of Yang (1996a) with eight equally probable categories of rates to approximate the continuous gamma distribution (referred to as dG models). The transition probability matrices of models, and details about parameter estimation are given in Yang (2000).

Likelihood ratio tests are applied to test several hypotheses of interest. For a given tree topology (e.g. that shown in Figure 1.1), a model (H_1) containing p free parameters and with log-likelihood L_1 fits the data significantly better than a nested sub-model (H_0) with $q = p - n$ restrictions and likelihood L_0, if the deviance $D = -2\log\wedge = -2(\log L_1 - \log L_0)$ falls in the rejection region of a χ^2 distribution with n degrees of freedom (Yang 1996b). We use several starting values in the iterations to guard against the possible existence of multiple local optima. These analyses are conducted with the CODEML programs from the PAML version 3.0b package (Yang 2000).

Evolution of six genes in Diptera

Table 1.1 shows the log-likelihood ratio statistic values for models of protein evolution assuming the tree topology shown in Figure 1.1. The best description of the substitution process of ADH, AMD, DDC, GPDH, SOD, and XDH is provided by the JTT-F+dG model, which treats amino acid frequencies as free parameters and allows variable replacement rates among sites. The discrete gamma distribution that better accommodates the variation of the replacement rate from site to site along GPDH is extremely L-shaped ($\alpha = 0.06$; i.e. $\alpha \ll 1$), reflecting that most (216/241; i.e. ≈ 90 per cent) of the aligned residues are conserved in dipterans; the number of conserved residues is 95 per cent when comparisons are confined to the genus *Drosophila*. Among-site rate variation is also (but less) extreme in DDC ($\alpha = 0.16$), AMD ($\alpha = 0.21$), and SOD ($\alpha = 0.22$), moderate in XDH ($\alpha = 0.45$), and lowest in ADH ($\alpha = 0.86$), indicating that it is the least constrained (note, however, that after removal of the two tephritid sequences from the ADH alignment the value of α decreases to 0.53, meaning that low conservancy of ADH in dipterans is to a great extent due to the dramatic divergence of this protein between drosophilids and tephritids).

The substitution models and estimates of the among-site rate variation obtained above are used for calculating amino acid distances between pairs of sequences. The results are summarized in Table 1.2 and Figure 1.2. For any given gene, except XDH, the rate of evolution varies from one level to another. For some genes, the rate is generally fastest for comparisons between species from different families (Drosophilidae and Tephritidae). The rate of evolution is faster for comparisons between drosophilid genera (*Di*) than between species of the *Drosophila* genus (*Da*) for SOD and GPDH, whereas the opposite is the case for AMD; the two rates are fairly similar for the other three genes. Specific comments about individual genes follow.

ADH

The average rate of amino acid replacement ($\times 10^{-10}$/site per year) in this protein for comparisons between species of the *Drosophila* genus is 32 (Figure 1.2), about the same as between genera (33, Figure 1.2 and Table 1.2). However, the rate is 41.9

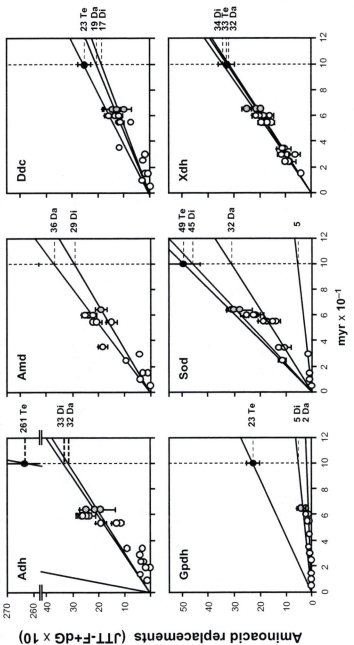

Figure 1.2 Rates of amino acid replacement for ADH, AMD, DDC, GPDH, SOD, and XDH in dipterans. The time unit (abscissa) is 10 million years (myr). White circles indicate comparisons made between *Drosophila* species, grey circles between the drosophilid genera, and black circles between tephritids and the drosophilids. The rates on the right are for replacements $\times 10^{-10}$ per site per year. *Da* is the rate for comparisons between species of the *Drosophila* genus, *Di* for comparisons between drosophilid genera, and *Te* between tephritids and the drosophilids. The average values in Figure 1.2 are from Table 1.2 for *Di* (row 4) and *Te* (row 5). For *Da* the values used are 32, 36, 19, 2, 32, and 32, respectively for ADH, AMD, DDC, GPDH, SOD, and XDH. Averages (with their standard errors) are calculated to minimize the impact of the phylogenetic structure of the sequence data shown in Figure 1.1. Thus, for example, for the *Drosophila melanogaster* species-group the average amino acid distance from XDH (0.0855 ± 0.0243) is the arithmetic mean of the pair-wise distances *D. ananassae* to *D. melanogaster* (0.0971) and *D. ananassae* to *D. erecta* (0.0723). The rates are obtained by linear regression, with the intercept constrained to be the origin. The rate 5 on the right of the SOD graph corresponds to comparisons between species within the *melanogaster* and *obscura* groups. Note the broken ordinate on the ADH graph to accommodate the regression line for the comparison between tephritids and the drosophilids.

Table 1.2 Normalized rates of evolution of ADH, AMD, DDC, GPDH, SOD, and XDH for increasingly remote lineages of Diptera

Comparison	myr	Rate of amino acid replacement					
		ADH	AMD	DDC	GPDH	SOD	XDH
1. Within *Drosophila* groups	25–30	10.6–25.1	13.8–36.2	6–11.5	0.0–2.0	4.3–44.8	20.9–38.7
2. Between *Drosophila* groups	55 ± 10	22.2	38.1	19.3	1.7	30.7	32.6
3. Between *Drosophila* subgenera	60 ± 10	41.9	38.1	20.3	2.7	38.2	31.6
4. Between drosophilid genera	65 ± 10	33.1	28.6	17.3	5.2	45.3	34.2
5. Between dipteran families	100 ± 20	260.8	–	23.3	22.6	49.4	32.8

The species compared are listed in Figure 1.1. The plus/minus values for myr are crude estimates of error. Rate values are expressed in units of 10^{-10} per site per year. The rates are estimated using the α values obtained from the dipteran dataset given in Table 1.1.

when *Drosophila* species from different subgenera are compared (Table 1.2, row 3), and decreases to 22.2 and 15.9, when the species compared are from different groups of the same subgenus, or from the same group, respectively (see rows 2 and 1, Table 1.2). The rate increases to 260.8×10^{-10}/site per year, when drosophilid species are compared with the tephritids *Ceratitis* or *Bactrocera*. Thus, the rate for comparisons between different families is ≈16 times greater than the average rate within *Drosophila* groups (15.9×10^{-10}/site per year). Tephritids and drosophilids diverge about 11 times faster for ADH than for DDC (23.3×10^{-10}/site per year), and eight and five times faster than for XDH (32.8×10^{-10}/site per year), and SOD (49.4×10^{-10}/site per year), respectively. For comparisons between *Drosophila* species at various taxonomic levels, the rates of divergence are fairly similar for these four genes. Besides the extensive differentiation of ADH between drosophilids and tephritids, ADH-based trees cluster the tephritids closer to the more distantly related Calyptrata sarcophagids than to the drosophilids, which are, like the tephritids, Acalyptrata (Brogna et al. 2001).

Two scenarios, which are not mutually exclusive, could account for the ostensibly fast rate of ADH divergence between Drosophilidae and Tephritidae; namely an accelerated rate of evolution in the last common ancestor to the Drosophilidae and/or paralogy resulting from a duplication prior to the divergence of Drosophilidae and Tephritidae. Evidence for the first scenario comes from the observation that ADH evolves about four times faster in tephritids (53.1×10^{-10}/site per year between *Ceratitis* and *Bactrocera*, assuming that the genera diverged at 35 Ma, Beverly and Wilson 1984) than in the *saltans* and *willistoni* groups (12.5×10^{-10}/site per year), reflecting that the evolutionary rate of ADH can in effect fluctuate dramatically. In favour of the paralogy scenario is the observation that *Adh* has undergone multiple duplication events during evolution (see Brogna *et al.* 2001). However, even if the *Adh* genes of drosophilids and tephritids are paralogous, they differentiated at an unusually rapid rate. The age of the duplicates (estimated assuming that the fastest, most conservative rate of 53.1×10^{-10}/site per year between *Ceratitis* and *Bactrocera* was the pervasive evolutionary rate after the duplication) would be 491 Ma, which precedes the origin of insects.

AMD

The average rate within *Drosophila* species groups is 22.91×10^{-10}/site per year. But the *willistoni* group rate is 36.2, twice as fast as the *melanogaster* and *obscura* group rate, which is 16.3. In turn, the rate between *Drosophila* subgenera is 38.1×10^{-10}/site per year, but between drosophilid genera reduces to 28.6×10^{-10}/site per year. Comparisons between the two families cannot be made, because no AMD sequences are available for the Tephritidae.

DDC

The average rates of this gene are quite uniform in dipterans. The rate ranges from 19.3×10^{-10}/site per year between *Drosophila* groups, to 23.3×10^{-10}/site per year between dipteran families (Table 1.2). However, the rates between species of the same group are considerably lower, ranging from six to 11.5 (see row 1, Table 1.2). Note in Table 1.2 that the fairly equitable rates of *Ddc* contrast with the greater variation observed in the closely linked paralogue *Amd*.

GPDH

The rate of replacement is $\leq 2 \times 10^{-10}$/site per year between *Drosophila* species; 2.5 times greater ($\sim 5 \times 10^{-10}$) between species of different genera; and more than 10 times greater ($\sim 23 \times 10^{-10}$) between species of different families.

SOD

The rate increases as progressively more distantly related dipterans are compared, but the increase is much less than for GPDH. The rate between *Drosophila* species is $\sim 20 \times 10^{-10}$, twice as fast when the taxonomic window is enlarged to drosophilids ($\sim 40 \times 10^{-10}$), and 2.5 times faster when drosophilids are compared with *Ceratitis* ($\sim 50 \times 10^{-10}$). The average rate of replacement between *Drosophila* species in SOD, however, conceals important differences among species groups. Thus, the mean rate between species of the *obscura* group (4.3×10^{-10}) is about half the rate in the *melanogaster* group (9.4) or between the two *Chymomyza* species (8.8), and 10 times slower than the rate between species of the *willistoni* group (44.8).

XDH

The rate of amino acid replacement is fairly regular. The rates are 32.6×10^{-10}/site per year between *Drosophila* groups, 31.6 between *Drosophila* subgenera, and 34.2 between drosophilid genera, acceptable as sample variation of the same stochastic clock. The average of these three rates is $\approx 32.8 \times 10^{-10}$, similar to the rate between dipteran families (see Table 1.2 and Figure 1.2).

Global rates of evolution of six genes

The best description of the amino acid substitution process of GPDH, G6PD, PGD, SOD, TPI, and XDH is provided by the JTT-F+dG model (see Table 1.1). However, the among-site rate variation along GPDH, SOD, and XDH is much less than for the dipterans (i.e. α values are larger; a similar comparison of α values between dipterans and global is not feasible for G6PD, PGD, and TPI, because the available dipteran sequences from these three proteins are too few). Amino acid rate variation from site to site decreases least for SOD, with $\alpha = 0.57$ (0.22 for dipterans), and most for GPDH, $\alpha = 0.82$ (0.06 for dipterans), nearly equal to that of XDH, $\alpha = 0.84$ (0.45 for dipterans). The increase in α from dipterans to global is expected because the proportion of invariable positions decreases with the enlarged timescale (i.e. 1000 myr versus 100 myr since the split of the three dipteran families); but also because the variable positions of one lineage are not the same as those of another (i.e. the proteins evolve in non-stationary fashion with regard to the among-site rate variation). Fungi show significantly larger α (0.94 ± 0.20) than dipterans (0.06 ± 0.02) and mammals (0.21 ± 0.14) for GPDH, and than dipterans (0.22 ± 0.05) for SOD; and dipterans show a greater α (0.45 ± 0.04) than mammals (0.34 ± 0.09) for XDH (normal deviate tests with standard errors computed by the curvature method of Yang 2000).

The results of the pairwise distance analysis conducted with the models and estimates of the among-site rate variation obtained above (see Table 1.1) are summarized in Table 1.3 and Figure 1.3. The rates change erratically among genes from one level of taxonomic comparison to another. The GPDH rate of amino acid replacement is $\leq 2.0 \times 10^{-10}$/site per year between *Drosophila* species, but 40.0 between fungi, and ~13 between animal phyla or between kingdoms. For G6PD, the rate between *Drosophila* species is 44.1, but ~12 between animal phyla or between kingdoms. The rate is substantially larger between animal phyla than between kingdoms for SOD and XDH (19.2 versus 12.6 and 19.2 versus 11.5, respectively), but very nearly the same for GPDH and G6PD (as noted), and TPI. Fungi exhibit much faster rates of evolution than most other comparisons for GPDH, G6PD, and TPI, but slower than most for PGD and XDH. For the three levels of comparison highlighted in Figure 1.3, we see that *Drosophila* exhibits the slowest rate for GPDH, the highest for G6PD, PGD, and XDH, and intermediate for SOD and TPI. Specific comments about individual genes follow.

GPDH

We have noted earlier (Ayala *et al.* 1996; Ayala 1997), that the rate of amino acid replacement is highly erratic, very slow within the genus *Drosophila* ($\leq 2.0 \times 10^{-10}$/site per year) much greater for comparisons between drosophilid genera (4.4×10^{-10}) and still greater for comparisons between dipteran families (9.31×10^{-10}). This heterogeneity conceals even more disparate actual rates, which become manifest when we take into account that these rates apply to largely overlapping lineages. Thus, the rate of 4.4×10^{-10} for comparisons between *Chymomyza* and *Drosophila* (row 4 in Table 1.3) needs to be decomposed into a rate of $\leq 2.0 \times 10^{-10}$ for most of the overlapping period of *Drosophila* evolution and a rate $> 10 \times 10^{-10}$ for the 25 million years between

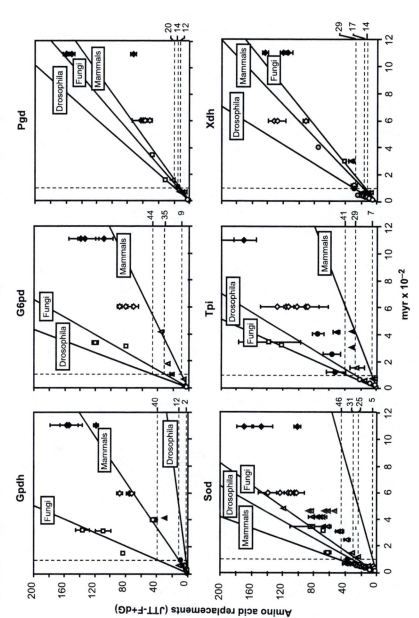

Figure 1.3 Global rates of amino acid replacement for GPDH, G6PD, PGD, SOD, TPI, and XDH. The time unit (abscissa) is 100 million years. The rates on the right are for replacements × 10⁻¹⁰ per site per year. These rates correspond to comparisons between *Drosophila* subgenera, mammal orders, or fungi (rows 3, 5, and 7 in Table 1.3). The comparisons between *Drosophila* subgenera and mammal orders correspond to roughly contemporary time lapses (60 Ma and 70 Ma, respectively). Other points in the figure are for other comparisons, such as between kingdoms (1100 Ma) or animal phyla (600 Ma; see also Table 1.3). For SOD the rate of 5 on the right is for comparisons between species within the *melanogaster* and *obscura* groups (see also Figure 1.2).

Table 1.3 Normalized rates of evolution of GPDH, G6PD, PGD, SOD, TPI, and XDH for increasingly remote lineages

Comparison	myr	Amino acid replacements per 100 myr						
		GPDH	G6PD	PGD	SOD	TPI	XDH	Average
1. Within Drosophila groups	25–30	0.0–1.9	44.1[a]	19.7[a]	4.8–40.6	0–8.8	20.3–36.7	25.0
2. Between Drosophila groups	55 ± 10	1.5	—	—	25.7	26.1	30.4	25.9
3. Between Drosophila subgenera	60 ± 10	2.0	—	—	30.7	35.5	29.2	28.9
4. Between drosophilid genera	65 ± 10	4.4	—	—	34.9	—	31.7	27.1
5. Between mammalian orders	70 ± 10	11.6	8.5	12.4	46.0	6.8	17.1	15.2
6. Between dipteran families	100 ± 20	9.3	21.2	16.1	33.7	43.5[b]	28.9	23.6
7. Between Fungi	300 ± 50	40.0	35.3	13.8	24.9	40.5	13.7	24.8
8. Between animal phyla	600 ± 100	13.2	13.4	9.7	19.2	17.8	19.2	15.7
9. Between kingdoms	1100 ± 200	13.0	11.7	11.7	12.6	19.9	11.5	12.3

The species compared are listed in Figure 1.1. The plus/minus values are crude estimates of error for myr. Rate values are expressed in units of 10^{-10} per site per year. Averages across loci are obtained by weighing the rate of each gene by the length of its sequence (i.e. 0.15, 0.23, 0.13, 0.07, 0.05, and 0.37, corresponding to 241, 367, 208, 107, 78, and 599 residues of GPDH, G6PD, 6PGDH, SOD, TPI, and XDH, respectively); averages for rows two, three, and four are taken across the proteins in Table 1.2 (i.e. ADH, AMD, DDC, GPDH, SOD, and XDH). The rates are estimated using the α values obtained from the global dataset given in Table 1.1.

[a] For the comparison between the *melanogaster* group species in Figure 1.1.
[b] For the comparison between *Drosophila* and *Calliphora* assuming that the two dipteran families diverged at 120 Ma.

the divergence of the two genera and the time of divergence between the *Drosophila* subgenera and between the *Chymomyza* species (Ayala 1997). The range of rates of evolution is ostensibly greater for GPDH (from ≤ 2 to ~40) than for any other gene in this survey.

G6PD

The rate within *Drosophila* groups (*melanogaster* group) is 44.1×10^{-10}/site per year (row 1, Table 1.3). This rate is about twice as great as the rate between dipteran families (21.2×10^{-10}/site per year; row 6), and four times greater than the rate between animal phyla (13.4×10^{-10}/site per year; row 8) or multicellular kingdoms (11.7×10^{-10}/site per year; row 9). Note that *Drosophila* (*melanogaster* group) evolves ≈10 times faster (44.1×10^{-10}/site per year) than rodents (4.6×10^{-10}/site per year between mouse and rat; not shown in Table 1.3), and ≈5 times faster than mammals (8.5×10^{-10}/site per year; row 5).

PGD

The rate of amino acid replacement of PGD generally slows down with increasing divergence time. Rate differences are, however, less conspicuous than for G6PD. Thus, the average rate within *Drosophila* groups (*melanogaster* group) is 19.7×10^{-10}/site per year, just slightly faster than the rate between dipteran families (16.1×10^{-10}/site per year), but about twice as great as the rate between animal phyla (9.7×10^{-10}/site per year) or multicellular kingdoms (11.7×10^{-10}/site per year).

SOD

We have noted earlier the baffling contrast between SOD and GPDH. When animals are compared with plants or fungi, the two enzymes ostensibly evolve at similar rates (12.6 and 13.0), but for comparisons between *Drosophila* species, SOD evolves as much as 30–40 times faster than GPDH (Ayala 1997; Rodríguez-Trelles *et al.* 2001a).

TPI

The rate within *Drosophila* groups is 4.4×10^{-10}/site per year (averaged across the *melanogaster* and *obscura* groups). This rate becomes six times greater between *Drosophila* groups (26.1×10^{-10}/site per year), eight times greater between *Drosophila* subgenera (35.5×10^{-10}), and 10 times greater between brachyceran families (43.5×10^{-10}; between *Drosophila* and *Calliphora*). Compared with this last-named value, the rate decelerates by a factor of two when comparisons are made between the more diverged brachycerans and nematocerans (23.4×10^{-10}/site per year; obtained assuming that brachycerans and nematocerans split at 250 Ma). Assuming that *Anopheles* diverged from *Aedes* and *Culex* at 160 Ma, the rate between these two lineages is only 11.0×10^{-10}/site per year, much slower than between *Drosophilidae*. Notice also the mammalian rate of 6.8×10^{-10}, which is several times slower than the rate of the contemporaneously evolving *Drosophila* subgenera (35.5×10^{-10}).

XDH

When comparisons are made between different kingdoms, this enzyme evolves at a rate (11.5×10^{-10}/site per year) no greater than any other enzyme. Yet, XDH evolves very fast in the Drosophilidae. As noted for other enzymes, the XDH rate variations displayed in Table 1.3 mask a much greater variation between lineages at different times because the rates given apply to largely overlapping lineages. Thus, the average number of replacements between birds and mammals is 10.5×10^{-10}/site per year. If we accept that the lineage of mammals has evolved at an average rate of 17.1 (the mammal rate in Table 1.3) since they separated from birds, to attain an average of 10.5 between mammals and birds, the bird lineage must have evolved at a rate of only 3.9, ~8 times slower than the *Drosophila* rate. Additional rate discrepancies have been pointed out by Rodríguez-Trelles *et al.* (2001a).

The estimated rates of amino acid replacement shown in Table 1.2 (i.e. between dipteran families and successively lower categories) differ from their correlates in Table 1.3 because the rates assume different degrees of among-site rate variation: the rates in Table 1.2 use α values obtained from dipterans, which are substantially smaller than the values used in Table 1.3, derived from the global dataset (see above). Which of the two sets of α values is more nearly correct is not easily decided (Nei *et al.* 2001). In the absence of sampling bias, for a substitution process stationary with respect to among-site rate variation (i.e. sites retain the same relative rates of change throughout the tree), the estimates obtained from closely related species should be closer to the true value of α (Zang and Gu 1998). If the process is non-stationary, however, using closely related species to estimate α can be misleading, because different lineages can have disparate α values. In the present case, it is apparent that the GPDH α value obtained from Diptera is very low because this gene is extremely conserved in *Drosophila*; therefore, the rate of GPDH amino acid replacement between *Ceratitis* and the drosophilids obtained with this value of α is likely to be unduly large (i.e. the variation introduced by *Ceratitis* is outweighed).

We have unveiled disparate differences in evolutionary rates among and within lineages, which are inconsistent across genes. Doubtless, these differences reflect biological processes, as illustrated by GPDH. In *Drosophila*, GPDH is subjected to constraints which considerably restrict the number of sites that can accept amino acid replacements and the particular amino acid replacement that can occur at each site (Ayala 1997). The question is, nevertheless, whether observed rate differences are detected by standard statistical tests. We have conducted likelihood ratio tests of the molecular clock hypothesis on the global datasets. Strictly speaking this comparison is valid only if the likelihood values are calculated using the true topology, and caution is needed when the phylogeny is uncertain (Yang *et al.* 1995). This problem is attenuated taking into account that relationships which are not well established (e.g. the branching order of animal phyla) are set as polytomies in the tree of Figure 1.1. The JTT-F+dG model is used to calculate the likelihood values with and without the clock assumption. We focus on GPDH, SOD, and XDH because they are more extensively represented in our study than the other proteins. The JTT-F+dG clock model is rejected only for the case of XDH ($-2log\Lambda = 42.7$, $P < 10^{-6}$, 33 *d.f.*). For GPDH and SOD, relaxation of the clock assumption does not improve significantly the likelihood ($-2log\Lambda = 42.7$, $P > 0.01$, 29 *d.f.*; and $-2log\Lambda = 69.3$, $P > 0.01$, 60 *d.f.*, respectively). As we have seen (see Figures 1.2, 1.3) XDH is the most clock-like of the three proteins. Yet, its conspicuously greater length (599 residues versus 241 and 107, for XDH versus GPDH and SOD, respectively) yields the molecular clock test for XDH more sensitive to departures from the rate constancy assumption.

Whither the clock?

The theoretical foundation originally proposed for the clock, namely the neutrality theory of molecular evolution, is untenable. The variance of molecular rates of evolution has contributed much to invalidating the theory. Supplementary hypotheses have been proposed, such as those enunciated in the introduction of this chapter. These hypotheses invoke parameters or processes that might be ascertained, at least in favourable cases, and thus lead to predictive inferences. The tests that we have designed rely on the prediction made by several hypotheses that all genes of a given lineage will be equally affected, because they postulate attributes that are equally shared by all genes of a species, such as population size, generation time, polymerases, or some other (defined or not) biological characteristic of the species. The evidence brought forward in this chapter makes these remediating hypotheses untenable. The rate variation pattern is erratic. Figures 2 and 3 make it graphically obvious that some lineages, such as Fungi (see Figure 1.3), evolve fastest for some genes (GPDH and TPI), slowest for other genes (SOD and XDH), and intermediately for the remaining two genes (G6PD and GPD). These vagaries may be a consequence of natural selection, whether in response to the fickleness of the biotic and physical external environment, or to complex interactions within or between the organisms of the species. But, at the present state of knowledge, there seems to be no way to make predictions as to how molecular evolutionary rates would be impacted, so that we could derive precise inferences about phylogenetic relationships or the timing of past events. One might expect, for example, that functionally related proteins might evolve in similar patterns across lineages. But our results do not favour this conclusion. G6PD is metabolically adjacent to PGD, and so are GPDH and TPI, and all four enzymes are involved in central cell metabolism, but the members of each pair do not exhibit consistent patterns of molecular evolution across lineages, much less all four enzymes (see Figure 1.3).

There are, nevertheless, genes and proteins for which the molecular clock seems to hold with fair accuracy, at least for certain groups of organisms and/or time intervals (Kimura 1983; Nei 1987; Li 1997). Even in these cases caution should be exercised before accepting observed rate variation as insignificant. The molecular clock tests used to identify and exclude sequences that violate the rate-constancy assumption have only limited statistical power (Dobzhansky *et al.* 1977; Scherer 1989; Robinson *et al.* 1998; Bromham *et al.* 2000). Dramatic evolutionary rate differences among lineages can pass undetected by conventional molecular clock tests for most common alignment lengths. The possibility of deriving extremely erroneous conclusions on the basis of statistically insignificant rate variation is illustrated by GPDH and SOD.

Be that as it may, there are important evolutionary questions, such as the configuration of a 'universal tree' of life (Woese *et al.* 1990), for which gene and protein sequences may provide the best, if not the only dependable information. Our own recommendation is rather pat: (1) Use molecular sequence data not as definitive,

Table 1.4 Normalized rates of evolution across the loci, normalized to the rates of *Drosophila* subgenera and mammal orders, and corresponding estimates of divergence times

Comparison	Normalized rates		Clock estimates (Ma)	
	(1)	*(2)*	*(1)*	*(2)*
1. *Drosophila* subgenera	1.0	1.9	60	111
2. Drosophilid genera	0.9	1.6	60	110
3. Mammal orders	0.5	1.0	38	70
4. Dipteran families	0.8	1.6	179	330
5. Fungi	0.9	1.6	276	509
6. Animal phyla	0.5	1.0	337	621
7. Kingdoms	0.4	0.8	485	893

The normalized rates are derived from the averages in Table 1.3: (1) normalized to the *Drosophila*-subgenera average; (2) normalized to the mammal-orders average. The clock estimates assume that the divergence times are 60 Ma for the *Drosophila* subgenera and 70 Ma for the mammal orders.

as sometimes is done (e.g. Wray *et al.* 1996; see Ayala *et al.* 1999), but as one more source of information when other evidence is available. (2) Pay attention to careful choice of the sequences. Close approximation to the molecular clock premise should be a necessary condition. Given the limited power of available tests, however, acceptance of this second premise seems safe only for long and fast evolving (yet alignable) sequences. (3) Combine data for as many genes as feasible, so that average values may converge towards a good correlation between amount of change and time elapsed – the so-called 'law of large numbers'.

Table 1.4 shows two sets of evolutionary rates and clock estimates derived from these rates. The evolutionary rates are based on the average rates given in Table 1.3 (last column) and are normalized to the average values for comparisons between (1) *Drosophila* subgenera and (2) mammal orders. Time estimates derived from (1) underestimate the time of divergence of mammals, animal phyla and kingdoms. This underestimation occurs because of the relatively fast rate of evolution of the *Drosophila* lineages for most genes (see Figure 1.3 and Table 1.4), with the notable exception of GPDH, which is not sufficient to overcome the effect of the other genes. Time estimates derived from (2) would seem more nearly accurate for the animal phyla and kingdoms, but overestimate the time of divergence of the *Drosophila* subgenera. In any case, the time estimates derived from the average rates are, as expected, generally more accurate than those that would be obtained with individual genes, using the separate rates displayed in Table 1.3 (see also the estimates of Rodríguez-Trelles *et al.* 2001a, Table 1.2).

Acknowledgements

F.R-T. and R.T. have received support from contracts Ramón y Cajal and Doctor 13P, respectively, from the Ministerio de Ciencia y Technología (Spain). Research supported by NIH grant GM42397 to F.J.A.

References

Ayala, F.J. (1972) 'Frequency-dependent mating advantage in *Drosophila*', *Behavior Genetics*, 2: 85–91.

—— (1986) 'On the virtues and pitfalls of the molecular evolutionary clock', *Journal of Heredity*, 77: 226–35.

—— (1997) 'Vagaries of the molecular clock', *Proceedings of the National Academy of Sciences, USA*, 94: 7776–83.

Ayala, F.J., Powell, J.R. and Dobzhansky, Th. (1971) 'Polymorphisms in continental and island populations of *Drosophila willistoni*', *Proceedings of the National Academy of Sciences, USA*, 68: 2480–3.

Ayala, F.J., Tracey, M.L., Barr, L.G., McDonald, J.F. and Pérez-Salas, S. (1974) 'Genetic variation in natural populations of five *Drosophila* species and the hypothesis of selective neutrality of protein polymorphisms', *Genetics*, 77: 343–84.

Ayala, F.J., Barrio, E. and Kwiatowski, J. (1996) 'Molecular clock or erratic evolution? A tale of two genes', *Proceedings of the National Academy of Sciences, USA*, 93: 11729–34.

Ayala, F. Jose, Rzhetsky, A. and Ayala, F.J. (1999) 'Molecular clocks and the origin of animals', in S.P. Wasser (ed.) *Evolutionary Theory and Processes: Modern Perspectives*, The Netherlands: Kluwer Academic Publishers, pp. 151–69.

Benach, J., Atrian, S., Gonzalez-Duarte, R. and Ladenstein, R. (1998) 'The refined crystal structure of *Drosophila lebanonensis* alcohol dehydrogenase at 1.9 Å resolution', *Journal of Molecular Biology*, 282: 383–99.

Benyajati, C., Place, A.R., Powers, D.A. and Sofer, W. (1981) 'Alcohol dehydrogenase gene of *Drosophila melanogaster*: relationship of intervening sequences to functional domains in the protein', *Proceedings of the National Academy of Sciences, USA*, 78: 2717–21.

Beverly, S.M. and Wilson, A.C. (1984) 'Molecular evolution in *Drosophila* and the higher Diptera II. A time scale for fly evolution', *Journal of Molecular Evolution*, 21: 1–13.

Bewley, G.C., Cook, J.L., Kusakabe, S., Mukai, T., Rigby, D.L. and Chambers, G.K. (1989) 'Sequence, structure and evoluton of the gene coding for *sn*-glycerol-3-phosphate dehydrogenase in *Drosophila melanogaster*', *Nucleic Acids Research*, 17: 8553–67.

Brogna, S., Benos, P.V., Gasperi, G. and Savakis, C. (2001) 'The *Drosophila* alcohol dehydrogenase gene may have evolved independently of the functionally homologous medfly, olive fly, and flesh fly genes', *Molecular Biology and Evolution*, 18: 322–9.

Bromham, L., Penny, D., Rambaut, A. and Hendy, M.D. (2000) 'The power of relative rates tests depends on the data', *Journal of Molecular Evolution*, 50: 296–301.

Chovnick, A., Gelbart, W. and McCarron, M. (1977) 'Organization of the *rosy* locus in *Drosophila melanogaster*', *Cell*, 11: 1–10.

Cook, J.L., Bewley, G.C. and Schaffer, J.B. (1988) '*Drosophila sn*-glycerol-3-phosphate dehydrogenase isozymes are generated by alternate pathways of RNA processing resulting in different carboxyl-terminal amino acid sequences', *Journal of Biological Chemistry*, 263: 10858–64.

Dobzhansky, Th., Ayala, F.J., Stebbins, G.L. and Valentine, J.W. (1977) *Evolution*, San Francisco: Freeman.

Enroth, C., Eger, B.T., Okamoto, K., Nishino, T., Nishino, T. and Pai, E.F. (2000) 'Cystal structures of bovine milk xanthine dehydrogenase and xanthine oxidase: Structure-based mechanism of conversion', *Proceedings of the National Academy of Sciences, USA*, 97: 10723–8.

Fitch, W.M. and Ayala, F.J. (1994) 'The superoxide molecular clock revisited', *Proceedings of the National Academy of Sciences, USA*, 91: 6802–7.

Fitch, W.M. and Margoliash, E. (1967) 'Construction of phylogenetic trees', *Science*, 155: 279–84.

Fridovich, I. (1986) 'Superoxide dismutases', *Advances in Enzymology*, 58: 61–97.

Gaut, B.S., Muse, S.V., Clark, W.F. and Clegg, M.T. (1992) 'Relative rates of nucleotide substitution at the *rbc*L locus of monocotyledonous plants', *Journal of Molecular Evolution*, 35: 292–303.

Gillespie, J.H. (1991) *The Causes of Molecular Evolution*, New York: Oxford University Press.

Grimaldi, D.A. (1990) 'A phylogenetic revised classification of genera in the *Drosophilidae* (Diptera)', *Bulletin of the American Museum of Natural History*, 197: 1–139.

Hille, R. and Nishino, T. (1995) 'Xanthine oxidase and xanthine dehydrogenase. *FASEB Journal*, 9: 995–1003.

Jones, D.T, Taylor, W.R. and Thornton, J.M. (1992) 'The rapid generation of mutation data matrices from protein sequences', *Computer Applications in the Biosciences*, 8: 275–82.

Kimura, M. (1968) 'Evolutionary rate at the molecular level', *Nature*, 217: 624–6.

—— (1980) 'A simple method for estimating evolutionary rate of base substitution through comparative studies of nucleotide sequences', *Journal of Molecular Evolution*, 16: 111–20.

—— (1983) *The Neutral Theory of Molecular Evolution*, Cambridge: Cambridge University Press.

Kimura, M. and Ohta, T. (1972) 'Population genetics, molecular biometry, and evolution', *Proceedings of the Sixth Berkeley Symposium on Mathematical Statistics and Probabilities*, Vol. 5, pp. 43–68.

Kohne, D.E. (1970) 'Evolution of higher-organism DNA', *Quarterly Review of Biophysics*, 33: 327–75.

Kwiatowski, J. and Ayala, F. (1999) 'Phylogeny of *Drosophila* and related genera: Conflict between molecular and anatomical analyses', *Molecular Phylogenetics and Evolution*, 13: 319–28.

Kwiatowski, J., Krawczyk, M., Jaworski, M., Skarecky, D. and Ayala, F.J. (1997) 'Erratic evolution of glycerol-3-phosphate dehydrogenase in *Drosophila Chymomyza*, and *Ceratitis*', *Journal of Molecular Evolution*, 44: 9–22.

Langley, C.H. and Fitch, W.M. (1974) 'An examination of the constancy of the rate of molecular evolution', *Journal of Molecular Evolution*, 3: 161–77.

Lee, Y.M., Misra, H.P. and Ayala, F.J. (1981) 'Superoxide dismutase in *Drosophila melanogaster*: Biochemical and structural characteristics of allozyme variants', *Proceedings of the National Academy of Sciences, USA*, 78: 7052–5.

Li, W.-H. (1997) *Molecular Evolution*, Sunderland, MA: Sinauer.

Li, W.-H. and Graur, D. (1997) *Fundamentals of Molecular Evolution*, Sunderland, MA: Sinauer.

Li, W.-H., Ellsworth, D.L., Kruchkal, J.K., Chang, B.H.-J. and Hewett-Emmett, D. (1996) 'Rates of nucleotide substitution in primates and rodents and the generation time effect hypothesis', *Molecular Phylogenetics and Evolution*, 5: 182–7.

Margoliash, E. (1963) 'Primary structure and evolution of cytochrome *c*', *Proceedings of the National Academy of Sciences, USA*, 50: 672–9.

Nei, M. (1987) *Molecular Evolutionary Genetics*, New York: Columbia University Press.

Nei, M., Xu, P. and Glasko, M. (2001) 'Estimation of divergence times from multiprotein sequences for a few mammalian species and a few distantly related organisms', *Proceedings of the National Academy of Sciences, USA*, 98: 2497–502.

O'Brien, S.J. and MacIntyre, R.J. (1972) 'The α-glycerophosphate cycle in *Drosophila melanogaster*. II. Genetic aspects', *Genetics*, 71: 127–38.

—— (1978) 'Genetics and biochemistry of enzymes and specific proteins of *Drosophila*', in M. Ashburner and T.R.F. Wright (eds) *The Genetics and Biology of Drosophila*, Vol. 2a, New York: Academic Press, pp. 396–552.

Ohta, T. (1972) 'Population size and rate of evolution', *Journal of Molecular Evolution*, 1: 304–14.

—— (1973) 'Slightly deleterious mutant substitutions in evolution', *Nature*, 246: 96–8.

O'Keefe, S., Schouls, M. and Hodgetts, R. (1995) 'Epidermal cell-specific quantitation of DOPA decarboxylase mRNA in *Drosophila* by competitive RT-PCR: an effect of Broad-Complex mutants', *Developmental Genetics*, 16: 77–84.

Peng, T., Moya, A. and Ayala, F.J. (1986) 'Irradiation-resistance conferred by superoxide dismutase: Possible adaptive role of a natural polymorphism in *Drosophila melanogaster*', *Proceedings of the National Academy of Sciences, USA*, 83: 684–7.

Powell, J.R. (1997) *Progress and Prospects in Evolutionary Biology*, New York: Oxford University Press.

Robinson, M., Gouy, M., Gautier, C. and Mouchirod, D. (1998) 'Sensitivity of relative rates tests to taxonomic sampling', *Molecular Biology and Evolution*, 15: 1091–8.

Rodríguez-Trelles, F., Tarrío, R. and Ayala, F.J. (2001a) 'Erratic overdispersion of three molecular clocks: GPDH, SOD, and XDH', *Proceedings of the National Academy of Sciences, USA*, 98: 11405–10.

—— (2001b) 'Xanthine dehydrogenase (XDH): episodic evolution of a neutral protein?', *Journal of Molecular Evolution*, 53: 485–95.

Sagi, M., Omarov, R.T. and Lips, S.H. (1998) 'The Mo-hydroxylases xanthine dehydrogenase and aldehyde oxidase in ryegrass as affected by nitrogen and salinity', *Plant Science*, 135: 125–35.

Scherer, S. (1989) 'The relative-rate test of the molecular clock hypothesis: a note of caution', *Molecular Biology and Evolution*, 6: 436–41.

Shannon, W.N., Gover, S., Lam, V.M.S. and Adams, M.J. (2000) 'Human glucose-6-phosphate dehydrogenase: the crystal structure reveals a structural NADP+ molecule and provides insights into enzyme deficiency', *Structure*, 8: 293–303.

Steinman, H.M. (1988) 'Bacterial superoxide dismutases', *Basic Life Sciences*, 49: 641–6.

Tarrío, R., Rodríguez-Trelles, R. and Ayala, F.J. (2001) 'Shared nucleotide composition biases among species and their impact on phylogenetic reconstructions of the Drosophilidae', *Molecular Biology and Evolution*, 18: 1464–73.

Tatarenkov, A. and Ayala, F.J. (2001) 'Phylogenetic relationships among species groups of the *virilis-repleta* radiation of *Drosophila*', *Molecular Phylogenetics and Evolution*, 21: 327–31.

Tatarenkov, A., Kwiatowski, J., Skarecky, D., Barrio, E. and Ayala, F.J. (1999) 'On the evolution of dopa decarboxylase (Ddc) and *Drosophila* systematics', *Journal of Molecular Evolution*, 48: 445–62.

Tatarenkov, A., Zurovcova, M. and Ayala, F.J. (2001) '*Ddc* and *amd* sequences resolve phylogenetic relationships of *Drosophila*', *Molecular Phylogenetics and Evolution*, 20: 311–25.

Tsai, C.S. and Chen, Q. (1998) 'Purification and kinetic characterization of 6-phosphogluconate dehydrogenase from *Schizosaccharomyces pombe*', *Biochemistry and Cell Biology*, 76: 637–44.

von Kalm, L., Weaver, J., DeMarco, J., MacIntyre, R.J. and Sullivan, D.T. (1989) 'Structural characterization of the α-glycerol-3-phosphate dehydrogenase-encoding gene of *Drosophila melanogaster*', *Proceedings of the National Academy of Sciences, USA*, 86: 5020–4.

Wang, D., Marsh, J.L. and Ayala, F.J. (1996) 'Evolutionary changes in the expression pattern of a developmentally essential gene in three *Drosophila* species', *Proceedings of the National Academy of Sciences, USA*, 93: 7103–7.

Woese, C.R., Kandler, O. and Wheelis, M.L. (1990) 'Toward a natural system of organisms: Proposal for the domains Archaea, Bacteria, and Eucarya', *Proceedings of the National Academy of Sciences, USA*, 87: 4576–9.

Wray, G.A., Levinton, J.S. and Shapiro, L.H. (1996) 'Molecular evidence for deep Precambrian divergences among metazoan phyla', *Science*, 274: 568–73.

Wright, T.R.F. (1996) 'Phenotypic analysis of the Dopa decarboxylase gene cluster mutants in *Drosophila melanogaster*', *Journal of Heredity*, 87: 175–90.

Xu, P., Huecksteadt, T.P., Harrison, R. and Hoidal, J.R. (1994) 'Molecular cloning, tissue expression of human xanthine dehydrogenase', *Biochemical and Biophysical Research Communications*, 199: 998–1004.

Yang, Z. (1994) 'Estimating the pattern of nucleotide substitution', *Journal of Molecular Evolution*, 39: 105–11.

—— (1996a) 'The among-site rate variation and its impact on phylogenetic analyses', *Trends in Ecology and Evolution*, 11: 367–72.

—— (1996b) 'Maximum likelihood models for combined analyses of multiple sequence data', *Journal of Molecular Evolution*, 42: 587–96.

—— (2000) 'PAML: a program package for phylogenetic analysis by maximum likelihood' *Computer Applications in the Biosciences*, 13: 555–6.

Yang, Z., Lauder, I.J. and Lin, H.J. (1995) 'Molecular evolution of the Hepatitis B virus genome. *Journal of Molecular Evolution*, 41: 587–96.

Yang, Z., Nielsen, R. and Hasegawa, M. (1998) 'Models of amino acid substitution and applications to mitochondrial DNA evolution', *Molecular Biology and Evolution*, 15: 1600–11.

Zang, J. and Gu, X. (1998) 'Correlation between the substitution rate and rate variation among sites in protein evolution', *Genetics*, 149: 1615–25.

Zuckerkandl, E. and Pauling, L. (1962) 'Molecular disease, evolution and genic heterogeneity', in M. Kasha and B. Pullman (eds) *Horizons in Biochemistry*, New York: Academic Press, pp. 189–225.

—— (1965) 'Evolutionary divergence and convergence in proteins', in V. Bryson and H.J. Vogel (eds) *Evolving Genes and Proteins*, New York: Academic Press, pp. 97–166.

Chapter 2

Molecular clocks and a biological trigger for Neoproterozoic Snowball Earth events and the Cambrian explosion

S. Blair Hedges

ABSTRACT

Two major events occurred in the history of the Earth and its biota during the late Precambrian and Cambrian (750–500 Ma; million years ago): a sequence of global glaciations (Snowball Earth events) followed by the sudden appearance in the fossil record of approximately half of the living animal phyla (the 'Cambrian explosion'). This has fuelled speculation that the two events were associated, perhaps through the generation of biological diversity during periods of isolation in glacial refugia, or in the period subsequent to isolation. However, recent molecular clock analyses have suggested that fungi and plants colonized land in the late Precambrian, considerably earlier than indicated by the fossil record, raising the possibility of a different connection between Snowball Earth events and the Cambrian explosion. These new data suggest a biological rather than geological trigger for the global glaciations, through increased rates of weathering by land fungi (e.g. lichens) and plants, and burial of decay-resistant carbon compounds of early land plants. The weathering and carbon burial would have lowered levels of carbon dioxide, possibly leading to the Snowball Earth events. A biological trigger can also explain the cyclic nature of the glaciations if they reflect cycles of extinction and recovery. Moreover, the fungal photobionts and early land plants may have generated sufficient oxygen in the latest Precambrian and Early Cambrian for animals to evolve larger body sizes and hard parts, explaining the Cambrian explosion. This model can be tested by increased precision of molecular clock estimates of divergence times and by searching for biomarkers and fossils of land fungi and plants in rocks of this time period.

Introduction

The last decade has seen major advances in our knowledge of Earth and biotic history in the late Precambrian, especially the Mesoproterozoic (1600–1000 Ma) and Neoproterozoic (1000–545 Ma). These have come from discoveries in geology and geochemistry, palaeontology, molecular evolution, and developmental biology. These discoveries and their resulting models continue to be debated but are changing our perceptions of the late Precambrian biosphere. For example, the Earth may have passed through several cycles of global glaciations during the period 750–580 Ma, each of which may have been characterized by complete freezing of all of the oceans for 10 myr or longer (Hoffman *et al.* 1998). At the same time, molecular clocks have suggested that major groups of complex multicellular organisms such as plants, animals, and fungi were present during, if not before, these global glaciations (Wray *et al.* 1996; Feng *et al.* 1997; Wang *et al.* 1999; Heckman *et al.* 2001). Fossils of complex organisms (Wood *et al.* 2002), metazoan embryos (Li *et al.* 1998; Xiao *et al.* 1998), and trace fossils (Rasmussen *et al.* 2002) have been found considerably earlier than expected, lending some support to molecular clock estimates.

On the one hand, these new revelations appear contradictory: an earlier history of complex life in an environment that was much harsher. On the other hand, the contradiction disappears if one is causally connected to the other. A connection that is explored in this chapter is the suggestion that the early colonization of land by fungi and plants became a biological trigger for the Snowball Earth events and the Cambrian explosion (Heckman *et al.* 2001). Firstly, I will review the evidence from molecular clocks for the early diversification of complex life (plants, animals, and fungi), and recent fossil evidence. This will be followed by a discussion of the data supporting Neoproterozoic global glaciations and a proposed geological trigger. Finally, I will discuss the biological trigger model for the initiation of Snowball Earth events and the Cambrian explosion.

Molecular clocks and the early diversification of animals, fungi, and plants

From their conceptual inception three decades ago, molecular clocks have consistently found early divergences for selected animal phyla (Brown *et al.* 1972; Runnegar 1982a,b; Runnegar 1986; Doolittle *et al.* 1996; Wray *et al.* 1996; Feng *et al.* 1997; Bromham *et al.* 1998; Wang *et al.* 1999). In most of these studies, relatively small numbers of genes or proteins were used, and there has been some discussion concerning methodology (Ayala *et al.* 1998; Gu 1998). However, in one case a relatively large number of proteins (50) was used and the vertebrate–arthropod divergence time was estimated at approximately 1000 Ma (Wang *et al.* 1999). The split between cephalochordates (amphioxus) and vertebrates was dated at approximately 750 Ma using nine proteins (Hedges 2001), suggesting that even relatively closely related groups of animals might have deep divergence times. However, most animal phyla have yet to be included in molecular clock analyses because of the paucity of protein sequence data.

The nuclear small subunit ribosomal RNA gene has been used to date the divergence of major groups of fungi, and some splits have been found to be older than 800 Ma (Berbee and Taylor 1993, 2001). However, a relatively young date (965 Ma) for the divergence of fungi and animals was used as a calibration point in those studies, derived from another molecular clock study (Doolittle *et al.* 1996). Subsequently, that split has been dated at 1200 Ma (Feng *et al.* 1997) and 1576 Ma (Wang *et al.* 1999), and the use of those dates as calibrations would proportionately extend the fungal divergence times (Berbee and Taylor 2001) deeper into the Neoproterozoic. In a more recent study, divergences among nine lineages of fungi were dated using 111 proteins and all were found to be Precambrian, with most, including lineages associated with land plants, diverging in the interval 800–1200 Ma (Heckman *et al.* 2001). In that study, the divergence of chlorophytan green algae and seed plants (41 proteins), and between land and vascular plants (50 proteins), were found to be approximately 1100 Ma and 700 Ma, respectively, suggesting an interval during which land plants arose

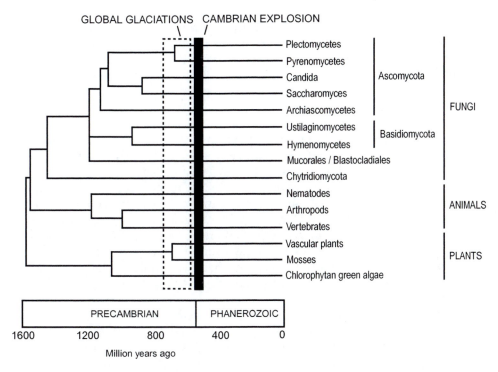

Figure 2.1 A time-calibrated evolutionary tree of fungi, animals, and plants showing times of divergence of selected lineages based on molecular clock studies (Wang *et al.* 1999; Heckman *et al.* 2001; Hedges 2001).

(Figure 2.1). Most of these molecular clock estimates have associated standard errors derived from differences among single protein estimates (Wang *et al.* 1999; Heckman *et al.* 2001). As with other fields of science, different estimates may be obtained with different methods and therefore the error in the estimate does not refer to the difference from the true value (always unknown) but rather the error as estimated under the specific conditions of analysis. In that sense, it is no different from the interpretation of error estimates of phylogenies (Nei and Kumar 2000).

In nearly all of these studies, the molecular clocks have been calibrated ultimately with robust fossil-based estimates of vertebrate divergences or secondary calibrations (molecular time estimates) derived from those primary fossil calibrations. The most commonly employed vertebrate fossil calibration point is the divergence of the mammalian and avian lineages at 310 Ma (Benton 2000). The robustness of this calibration point has been justified elsewhere (Kumar and Hedges 1998; Wang *et al.* 1999) and includes the following factors: (i) the fossils are exceptionally well-preserved, (ii) the earliest representatives of the two lineages are similar in morphology, suggesting that the palaeontological record for the divergence date is not a significant underestimate, (iii) no older remains of either lineage have been found since the mid-1800s, and (iv) earlier branches in the vertebrate tree constrain this divergence from being significantly older. A claim that this divergence should date more appropriately at 288 Ma (Lee 1999) is not supported by others (Carroll 1997; Benton 2000); even if true this would reduce time estimates by only 7 per cent. If anything, calibrations are likely to be underestimates of the true divergence, but the amount of underestimation is unknown and therefore calibrations are typically presented without errors. If the error in a calibration time is known, such as from variation in the age of the fossils, it could be used as a propagated error in computing each single-gene time estimate (van Tuinen and Hedges 2001). In such instances, the among-gene standard error of the overall estimate would encompass the calibration error.

What are the potential biases in molecular clock analysis that could explain such old time estimates for plants, animals, and fungi? It is widely known that relative rate tests do not detect all rate heterogeneity. Therefore, it is obvious that any undetected heterogeneity could bias the resulting time estimate (Bromham *et al.* 2000). However, in large studies involving many genes, there is no reason to expect a directional bias in the overall time estimate. None the less, we tested this in a study of 658 proteins in vertebrates (Kumar and Hedges 1998) by increasing the stringency of the rate test far beyond the 5 per cent level, effectively removing nearly all rate heterogeneity. Although many more proteins and comparisons were rejected, there was no effect on the overall time estimates indicating that there was no directionality to the rate variation. In specific cases involving species with branches that are consistently short (or long) in many gene trees, some directional bias might be expected, and this would be evident if the stringency of the rate test were increased as described above. However, time may be estimated even in those cases showing rate differences by using lineage specific and variable rate methods (Sanderson 1997; Schubart *et al.* 1998; Thorne *et al.* 1998). The power of the rate test is higher in longer sequences and therefore short sequences distinguished by only one or a few differences should be avoided.

It has been claimed that the well-known statistical bias resulting from averaging ratios might cause an overestimation of time in multiprotein studies, favouring a sequence concatenation approach (Nei *et al.* 2001). Although theoretically correct, its effect on time estimation has been shown to be minimal, probably because large extrapolations are typically avoided and modes are used rather than means (Heckman *et al.* 2001; Hedges *et al.* 2001). In contrast, concatenation may prevent detection of contaminant proteins (paralogues) that are normally detected in the multiprotein approach (Kumar and Hedges 1998). Although some paralogues are easily detected in individual gene trees, especially if different sequences of the same species appear in the tree (clearly indicating the result of gene duplication), other cases of paralogy are not easily detected by such simple inspection. For example, if some gene loss has occurred and multiple sequences of the same species are not present, detection of paralogy might require additional sequences. However, such a gene (without use of additional sequences) would be likely to be an outlier in a multiprotein clock analysis because the branching event being dated would be an earlier event (gene duplication) and not the speciation event in question. Another limitation of the concatenation approach is that, in effect, it gives the fastest evolving proteins the highest weight (i.e. because they contain the highest proportion of variable sites) whereas those proteins may produce the greatest distance estimation errors.

Yet another statistical bias has been attributed to the multiprotein approach (Rodríguez-Trelles *et al.* 2002). This may be a problem with short proteins having low rates of change and estimations involving large extrapolations. In such instances, the substitutions between closely related sequences (e.g. the calibration) may be underestimated, resulting in an overestimate of divergence time for a distant node. However, simulations (Rodríguez-Trelles *et al.* 2002) have shown that the bias is minimal (~1–2 per cent) for proteins of typical length and rate of change, and, in practice, authors have been aware of this potential bias and have avoided it (Kumar and Hedges 1998; Wang *et al.* 1999). Also, most distributions are not right skewed, and modes (rather than means) have been used for those non-normal distributions (Kumar and Hedges 1998). Some authors have objected to the use of secondary calibration points based on molecular clock estimates as not being 'independent' (Smith and Peterson 2002). However, independence is not a requirement for calibration. Such secondary calibrations are simply used to acquire more proteins or genes for comparison and thus increase the precision of the time estimates.

Another criticism of molecular clocks is that rates may have changed in many lineages concurrently, such as during an adaptive radiation, resulting in consistently biased time estimates (Gingerich 1986; Foote *et al.* 1998; Benton 1999; Conway Morris 2000). For this criticism to be valid the rate change would have to take place to exactly the same degree in the calibration lineages (unlikely) or else the rate test would detect the differences. However, even in such a case, inconsistencies would arise between the fossil record and molecular time estimates that would reveal the distortion in the timescale. As has been pointed out previously (Kumar and Hedges 1998; Easteal 1999), time estimates before and after the Late Cretaceous 'gap' in the fossil record of birds and mammals are largely consistent, suggesting that a widespread increase in molecular rate of change at the Cretaceous–Tertiary boundary was not responsible for the older divergence time estimates (Hedges *et al.* 1996; Kumar and Hedges 1998). In the case of the deep Precambrian divergence times, there are fewer fossil constraints to rule out a uniform rate change, but the occurrence of fossil red algae at 1200 Ma (Butterfield 2000) constrains the plant–animal–fungus divergence to an even earlier date, given that red algae are part of the plant lineage (Moreira *et al.* 2000). Thus, it would not be possible to compress the ~1000 Ma divergences between animal phyla (Wang *et al.* 1999) and fungal lineages (Heckman *et al.* 2001) by 50 per cent, up to the Precambrian–Cambrian boundary, without creating inconsistencies between the molecular and fossil divergence times for plants versus animals and fungi. Moreover, there is no known molecular mechanism to explain such rate acceleration. Typical amino acid substitutions in the housekeeping genes that are employed in these clock studies are considered effectively neutral and are unlikely to be associated with the major morphological changes that take place in adaptive radiations.

Fossil evidence

No widely accepted fossils of animals, land plants, or fungi have yet been collected from deep in the Proterozoic that would corroborate the 1 Ga divergence time estimates calculated in molecular clock studies. None the less, fossils of all three groups have been collected in recent years that have significantly extended their times of origin. In the case of animals, body fossils have been found as early as 555 Ma (Martin *et al.* 2000), embryos to ~570 Ma (Li *et al.* 1998; Xiao *et al.* 1998), and radially symmetrical impressions of possible metazoans at 600–610 Ma (Martin *et al.* 2000). The most convincing evidence for the existence of metazoans prior to this are the

1200 Ma trace fossils of vermiform organisms from rocks in southwestern Australia (Rasmussen *et al.* 2002).

Until recently, the oldest known fossil remains of fungi were from the Rhynie Chert (400 Ma) but, with the discovery of glomalean fungi from Ordovician shallow marine sediments, this has been extended by 60 myr to 460 Ma (Redecker *et al.* 2000). The classification of many Ediacaran organisms remains controversial because they do not resemble animals (Seilacher 1994) and one interpretation is that at least some were marine lichens (Retallack 1994). This was suggested after analysis and consideration of differential compression in animals (e.g. jellyfish) versus lichens. The Ediacaran organisms apparently were more rigid than animals and that durability may have been conferred by chitin (as in lichens and other fungi). Additional support for the interpretation as lichens is the large size of Ediacaran organisms at a time when oxygen levels were probably low (Retallack 1994).

The oldest land plants are also known from the Ordovician (Gray and Shear 1992; Wellman and Gray 2000). Although widespread evidence of aerially dispersed land plant spores might be expected in Precambrian strata if land plants were present then, this need not be the case. For example, if Precambrian land plants were restricted in distribution, their spores might not be globally distributed. Also, the spores of the earliest land plants may not have fossilized as well as later spores, and the habitats where they occurred may be under-represented in the exposed Precambrian strata.

Some palaeontologists have argued that the absence of fossil evidence for animals much earlier than the late Neoproterozoic (600–700 Ma) is because they had not yet evolved (Valentine *et al.* 1999; Conway Morris 2000). However, other palaeontologists have entertained the possibility that molecular clock estimates indicating older divergences between animal lineages may be correct (Runnegar 1982b; Xiao *et al.* 1998; Runnegar 2000; Rasmussen *et al.* 2002). Various explanations have been proposed for the absence of metazoan fossils from this earlier period, although the most commonly cited reason is that animals were smaller and soft-bodied (Runnegar 1982a,b; Bengtson and Lipps 1992; Lipps *et al.* 1992; Bengtson 1994; Fedonkin 1994; Weiguo 1994; Davidson *et al.* 1995; Fortey *et al.* 1996; Cooper and Fortey 1998). In fact, there is evidence from trace fossils of a size increase in bilaterian animals and for the acquisition of hard parts at the Proterozoic–Phanerozoic boundary (Bengtson and Farmer 1992; Lipps *et al.* 1992; Valentine *et al.* 1999). Nearly one-third of animal phyla believed to have arisen in the Cambrian, based on phylogenetic relationships, have virtually no fossil record (Valentine *et al.* 1999). All of those phyla are small in size and most are soft-bodied, essentially confirming that such traits can render animals 'invisible' in the fossil record, and lending plausibility to the hypothesis of a long and cryptic history of animal evolution prior to the Cambrian explosion.

Neoproterozoic Snowball Earth events

Several global glaciations (Snowball Earth events) occurred during the Neoproterozoic, from 750–570 Ma (Kirschvink 1992; Kaufman *et al.* 1997; Hoffman *et al.* 1998; Kennedy *et al.* 1998; Walter *et al.* 2000). The number of glaciations continues to be debated, but there is a consensus that there were at least two major episodes, the Sturtian (700 Ma) and the Marinoan (600 Ma); there may have been an additional, smaller, glaciation at 570 Ma (Walter *et al.* 2000). Temporal constraints are poor during this time period and other glacial episodes may be identified in the future. These Snowball Earth events have been identified primarily by carbon isotopic excursions and from glacial deposits, and they follow the same pattern at localities on different continents (Kaufman *et al.* 1997; Hoffman *et al.* 1998; Walter *et al.* 2000). This and other evidence has led to the following model. First, carbon dioxide levels in the atmosphere declined, causing ice sheets to expand below 30 degrees latitude, triggering a runaway albedo affect, reflecting more solar energy back out to space; this lowered temperatures further and caused all oceans to freeze over (Snowball Earth). Normal volcanic activity continued to contribute carbon dioxide to the atmosphere and, after approximately 10 million years, this was sufficient to warm the Earth and melt the ice. An extreme greenhouse period followed the Snowball Earth state, perhaps for hundreds to thousands of years during which large volumes of limestone were created. After an interlude of millions of years, the cycle began again. A 'soft' Snowball Earth model has been proposed which allows a zone of ice-free equatorial oceans (Hyde *et al.* 2000).

Palaeogeography has been implicated as the trigger for the Neoproterozoic Snowball Earth events (Kirschvink 1992; Hoffman *et al.* 1998). At the time of the Sturtian glaciation, the supercontinent Rodinia straddled the equator and was breaking apart, according to some reconstructions (Meert and Powell 2001). The equatorial position of the continents may have had two affects. First, the greater fraction of the equatorial region having higher albedo (continents) rather than lower albedo (oceans) may have contributed to a lower overall global temperature (Kirschvink 1992). In addition, these tropical landmasses would have weathered more rapidly, lowering carbon dioxide levels more than usual because there were no polar landmasses to provide a buffer (Hoffman *et al.* 1998). The buffer normally works by shutting down continental weathering at an early stage, through freezing of high latitude continental areas, as temperatures begin to drop. This allowed temperatures to increase through build-up of carbon dioxide. Second, during the Snowball Earth events, the absence of that buffer allowed weathering to continue until the polar oceans froze and the ice caps extended to equatorial regions (Hoffman *et al.* 1998). This proposed geological trigger is only speculative, and palaeogeography in the Precambrian is not well known. It has been suggested that the increased solar luminosity in the Phanerozoic, coupled with less efficient carbon burial, may explain why no Snowball Earth events have occurred since 570 Ma (Hoffman *et al.* 1998).

The extreme conditions associated with Snowball Earth events, including a postglacial greenhouse period, would have posed considerable hardships for any life forms, especially eukaryotes. Nevertheless, the fossil record confirms that several groups of eukaryotic algae survived through this period, and members of the animal and fungal lineages must have survived as well if the animal–fungal divergence occurred prior to 750 Ma as is generally believed. If molecular clock estimates are correct, major lineages of animals and fungi survived the Snowball Earth events. Deep sea vents would have provided one possible refuge, and vents and rift zones near the surface, such as in modern Iceland, possibly provided refuge for terrestrial organisms. Continental thermal springs would have been less likely refugia because the water required to charge them may not have been available during these periods of low precipitation. Connections between the Snowball Earth events and animal evolution have been suggested (Kaufman *et al.* 1997), specifically through genetic bottlenecks leading to diversification before and after the last snowball event (Hoffman *et al.* 1998) or through cycles of allopatric speciation, in refugia, of many lineages that had evolved prior to the glaciations (Hedges 2001).

A biological trigger

I have proposed elsewhere that the presence of fungi and plants on land in the Precambrian, as inferred from molecular clocks, became a biological trigger for Snowball Earth events and the Cambrian explosion of animal diversity (Figure 2.2; Heckman *et al.* 2001). Fungi and plants would have increased rates of weathering and carbon burial, lowering levels of carbon dioxide and global temperatures. At the

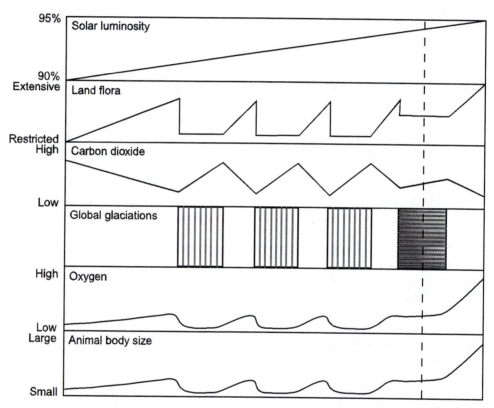

Figure 2.2 Proposed 'Biological Trigger' model for Neoproterozoic Snowball Earth events and the Cambrian explosion (see text). The panels show changes in different components of Earth and biotic history in the interval 1000–500 Ma, but the timescale is not linear because the 10 myr periods of glaciation are expanded to illustrate the model. The first panel (solar luminosity) shows the approximate increase during that period (Kump *et al.* 1999) relative to the present (100 per cent). Other panels show only relative, diagrammatic changes implied by the model and are not quantitative. The land flora includes fungi (e.g. lichens) and primitive land plants. Three Snowball Earth events (vertical lines) are illustrated followed by a pseudo-snowball (horizontal lines) at the Precambrian–Cambrian boundary. Animal body size is largely hypothetical (although size is known to increase in the Early Cambrian), assuming a direct relationship to oxygen levels. The vertical dashed line represents the Precambrian–Cambrian boundary.

same time, the oxygen produced by lichen photobionts and plants would have increased levels of this gas, possibly permitting animals to increase in size and facilitating the development of hard parts.

The fact that rates of weathering can be enhanced by the presence of organisms on land is well established, and lichens alone may increase rates by 10–100-fold (Schwartzman and Volk 1989; Schwartzman 1999). In general, biologically enhanced weathering during the last two billion years has been claimed to have lowered global temperatures, allowing complex life to develop (Schwartzman 1999). In particular, this mechanism for temperature change has been suggested as a possible cause of the Neoproterozoic glaciations (Carver and Vardavas 1994; Retallack 1994). However, this has been largely speculative because, prior to recent molecular clock studies, there have been no fossils or other evidence of terrestrial eukaryotes during the Precambrian (Horodyski and Knauth 1994; Kenrick and Crane 1997; Redecker et al. 2000).

The molecular time estimates for the diversification of major groups of fungi at around 800–1200 Ma, and the origin of land plants at 700–1100 Ma, raised the possibility that these organisms may have triggered the Snowball Earth events (Heckman et al. 2001). At first, the appearance of land fungi prior to land plants may come as a surprise because fungi are heterotrophs that acquire their nutrients from absorption, and most species today are involved in symbiosis with land plants. However, lichens represent an ancient ecological form for fungi (Taylor et al. 1995), and their symbiotic partners (green algae and cyanobacteria) were present in the Proterozoic. They can withstand severe environmental stresses and live in extreme habitats where neither fungi nor algae could live alone (Ahmadjian and Hale 1973; Gray and Shear 1992; Selosse and LeTacon 1998). Today, lichens form a rock and soil crust flora in harsh terrestrial environments, sometimes in combination with primitive plants (mosses and liverworts) and cyanobacteria. The earlier appearance of fungi (lichens), and associated weathering of rock, would have provided soil for the later colonization of land by plants. Nematodes and tardigrades, which themselves could be classified as extremophile eukaryotes, sometimes are found in symbiosis with lichens and the soil crust. The divergence of nematodes from other animals has been calculated using molecular clocks at approximately 1200 Ma (Wang et al. 1999) raising the possibility that nematodes were among the first animals on land. It is possible that terrestrial eukaryotes (fungi, plants, and animals) formed a similar biological crust on exposed land in the Proterozoic.

Besides the increased rates of weathering produced by these terrestrial eukaryotes, the burial of decay-resistant carbon of early land plants would also have contributed to a lowering of temperatures. Both vascular and bryophytic plants have lignins or lignin-like compounds that are decay resistant (Kroken et al. 1996) and make up a significant fraction of terrestrially derived carbon deposited in marine sediments (Berner 1999). The burial of such carbon would have lowered levels of carbon dioxide and global temperature. Some prokaryotes, such as cyanobacteria, are adapted to terrestrial life and were probably the first colonists on land (Horodyski and Knauth 1994). The biological trigger model described here might also have occurred simply by the weathering, carbon burial, and oxygen production of these organisms alone, or by marine prokaryotes and eukaryotes. However, that does not explain why the Neoproterozoic Snowball Earth events began at 700 Ma, because organisms (e.g.

cyanobacteria) were present much earlier. An involvement of fungi and land plants in the biological trigger is more likely because their early evolutionary history (Heckman *et al.* 2001) more closely corresponds to the timing of the Neoproterozoic glaciations and Cambrian explosion. In addition, lichens have an enhanced ability to weather the terrestrial environment and land plants uniquely have decay-resistant carbon compounds.

A biological trigger is also a better explanation for the cyclic nature of the Neoproterozoic glaciations than a geological trigger. During each glaciation, most life on land would have disappeared, but the recovery period that followed would have resulted in an increase in productivity and weathering and a concomitant decline in levels of carbon dioxide and global temperature, leading once again to a Snowball Earth episode (Figure 2.2). Carbon isotope excursions reflect this repeated pattern of biotic collapse followed by a recovery period and then another collapse (Kaufman *et al.* 1997; Hoffman *et al.* 1998).

The same mechanism may explain a Neoproterozoic rise in oxygen and the Cambrian explosion of animal diversity. Great attention has been paid to understanding changes in oxygen levels through geological time, but no consensus has been reached aside from a general (though not universal) agreement that there was an initial rise to approximately 1 per cent of present atmospheric levels (PAL), at around 2300 Ma, and a second major increase in the Neoproterozoic (Holland 1994; Canfield and Teske 1996; Ohmoto 1997; Kasting 2001). The second increase has been implicated as a possible explanation for the Cambrian explosion because it would have permitted animals to become larger in size and form skeletons that would more readily reveal their existence in the fossil record (Bengtson and Farmer 1992; Knoll 1992; Bengtson 1994; Knoll 1994).

The carbon isotope record reveals a sharply negative $\delta^{13}C_{carbonate}$ anomaly immediately prior to the Precambrian–Cambrian boundary (545 Ma) that in some ways resembles those negative excursions associated with earlier glacial events, yet there is no evidence that a glaciation occurred at this time (Kaufman *et al.* 1997; Knoll and Carroll 1999). The fact that the Cambrian explosion occurred immediately following this 'pseudo-snowball' event is curious and suggests a possible connection. A geological explanation for this particular carbon isotope excursion is that it is due to the release of methane from oceanic clathrates that were destabilized by combined sea level fall and global warming resulting from volcanic release of carbon dioxide (Walter *et al.* 2000). Alternatively, I suggest that it may have been an extension of the biological trigger model discussed above. Under this model, carbon dioxide levels lowered sufficiently (from biological activity on land) to reduce temperature and productivity (resulting in the carbon isotope excursion) but not enough to cause widespread glaciations or a full Snowball Earth event. As a result, a sufficient diversity of land plants and/or lichens may have survived the episode (in contrast to a full Snowball Earth where most would have perished), permitting a rapid and extensive recovery that generated a major pulse of oxygen. This may go some way towards explaining the Cambrian explosion of animals that took place over the subsequent 40 million years.

This model can be tested by using larger numbers of proteins and taxa, when they become available, to increase the precision of molecular time estimates. In addition, searches should be made for biomarkers or fossils of fungi and land plants from the late Precambrian and Cambrian. Such searches may require the same methods that have been used to uncover the earliest plant and fungal fossils from the Phanerozoic, such as the examination of shallow marine (nearshore) sediments with acid baths (Gray and Shear 1992; Redecker *et al.* 2000). If the pseudo-snowball event at the Precambrian–Cambrian boundary led to a land plant diversification and expansion, evidence of land plant spores should be recovered from Cambrian sediments.

Acknowledgements

This research was supported by the NASA Astrobiology Institute and National Science Foundation.

References

Ahmadjian, V. and Hale, M.E. (1973) *The Lichens*, New York: Academic Press.

Ayala, F.J., Rzhetsky, A. and Ayala, F.J. (1998) 'Origin of the metazoan phyla: molecular clocks confirm paleontological estimates', *Proceedings of the National Academy of Sciences, USA*, 95: 606–11.

Bengtson, S. (1994) 'The advent of animal skeletons', in S. Bengston (ed.) *Early life on Earth*, New York: Columbia University Press, pp. 412–25.

Bengtson, S. and Farmer, J.D. (1992) 'The evolution of metazoan body plans', in W.J. Schopf and C. Klein (eds) *The Proterozoic Biosphere*, Cambridge: Cambridge University Press, pp. 443–6.

Bengtson, S. and Lipps, J.R. (1992) 'The Proterozoic–Early Cambrian evolution of metaphytes and metazoans', in J.W. Schopf and C. Klein (eds) *The Proterozoic Biosphere*, Cambridge: Cambridge University Press, pp. 427–8.

Benton, M.J. (1999) 'Early origins of modern birds and mammals: molecules vs. morphology', *BioEssays*, 21: 1043–51.

—— (2000) *Vertebrate palaeontology*, Oxford: Blackwell Science.

Berbee, M.L. and Taylor, J.W. (1993) 'Dating the evolutionary radiations of the true fungi', *Canadian Journal of Botany*, 71: 1114–27.

—— (2001) 'Fungal molecular evolution: gene trees and geologic time', in D.J. McLaughlin and E. McLaughlin (eds) *The Mycota Vol VIIB, Systematics and Evolution*, New York: Springer-Verlag, pp. 229–46.

Berner, R.A. (1999) 'Atmospheric oxygen over Phanerozoic time', *Proceedings of the National Academy of Sciences, USA*, 96: 10955–7.

Bromham, L., Rambaut, A., Fortey, R., Cooper, A. and Penny, D. (1998) 'Testing the Cambrian explosion hypothesis by using a molecular dating technique', *Proceedings of the National Academy of Sciences, USA*, 95: 12386–9.

Bromham, L., Penny, D., Rambaut, A. and Hendy, M.D. (2000) 'The power of relative rate tests depends on the data', *Journal of Molecular Evolution*, 50: 296–301.

Brown, R.H., Richardson, M., Boulter, D., Ramshaw, J.A.M. and Jeffries, R.P.S. (1972) 'The amino acid sequence of cytochrome c from *Helix aspera* Müeller (Garden Snail)', *Biochemical Journal*, 128: 971–4.

Butterfield, N.J. (2000) '*Bangiomorpha pubescens* n. gen., n. sp.: implications for the evolution of sex, multicellularity, and the Mesoproterozoic–Neoproterozoic radiation of eukaryotes', *Paleobiology*, 26: 386–404.

Canfield, D.E. and Teske, A. (1996) 'Late Proterozoic rise in atmospheric oxygen concentration inferred from phylogenetic and sulphur-isotope studies', *Nature*, 382: 127–32.

Carroll, R.L. (1997) *Patterns and Processes of Vertebrate Evolution*, Cambridge: Cambridge University Press.

Carver, J.H. and Vardavas, I.M. (1994) 'Precambrian glaciations and the evolution of the atmosphere', *Annales Geophysicae*, 12: 674–82.

Conway Morris, S. (2000) 'The Cambrian "explosion": slow-fuse or megatonnage?', *Proceedings of the National Academy of Sciences, USA*, 97: 4426–9.

Cooper, A. and Fortey, R.A. (1998) 'Evolutionary explosions and the phylogenetic fuse', *Trends in Ecology and Evolution*, 13: 151–6.

Davidson, E.H., Peterson, K.J. and Cameron, R.A. (1995) 'Origin of bilaterian body plans: evolution of developmental regulatory mechanisms', *Science*, 270: 1319–25.

Doolittle, R.F., Feng, D.-F., Tsang, S., Cho, G. and Little, E. (1996) 'Determining divergence times of the major kingdoms of living organisms with a protein clock', *Science*, 271: 470–7.

Easteal, S. (1999) 'Molecular evidence for the early divergence of placental mammals', *BioEssays*, 21: 1052–8.

Fedonkin, M.A. (1994) 'Vendian body fossils and trace fossils', in Bengtson, S. (ed.) *Early Life on Earth*, New York: Columbia University Press, pp. 370–88.

Feng, D.-F., Cho, G. and Doolittle, R.F. (1997) 'Determining divergence times with a protein clock: update and reevaluation', *Proceedings of the National Academy of Sciences, USA*, 94: 13028–33.

Foote, M., Hunter, J.P., Janis, C.M. and Sepkoski, J.J. Jr (1998) 'Evolutionary and preservational constraints on origins of biologic groups: divergence times of eutherian mammals', *Science*, 283: 1310–14.

Fortey, R.A., Briggs, D.E.G. and Wills, M.A. (1996) 'The Cambrian evolutionary "explosion": decoupling cladogenesis from morphological disparity', *Biological Journal of the Linnean Society*, 57: 13–33.

Gingerich, P.D. (1986) 'Temporal scaling of molecular evolution in primates and other mammals', *Molecular Biology and Evolution*, 3: 205–21.

Gray, J. and Shear, W. (1992) 'Early life on land', *American Scientist*, 80: 444–56.

Gu, X. (1998) 'Early metazoan divergence was about 830 million years ago', *Journal of Molecular Evolution*, 47: 369–71.

Heckman, D.S., Geiser, D.M., Eidell, B.R., Stauffer, R.L., Kardos, N.L. and Hedges, S.B. (2001) 'Molecular evidence for the early colonization of land by fungi and plants', *Science*, 293: 1129–33.

Hedges, S.B. (2001) 'Molecular evidence for the early history of living vertebrates', in P.E. Ahlberg (ed.) *Major Events in Early Vertebrate Evolution: Palaeontology, Phylogeny, Genetics and Development*, London: Taylor & Francis, pp. 119–34.

Hedges, S.B., Parker, P.H., Sibley, C.G. and Kumar, S. (1996) 'Continental breakup and the ordinal diversification of birds and mammals', *Nature*, 381: 226–9.

Hedges, S.B., Chen, H., Kumar, S., Wang, D.-Y., Thompson, A.S. and Watanabe, H. (2001) 'A genomic timescale for the origin of eukaryotes', *BMC Evolutionary Biology*, 1: 4.

Hoffman, P.F., Kaufman, A.J., Halverson, G.P. and Schrag, D.P. (1998) 'A Neoproterozoic snowball Earth', *Science*, 281: 1342–6.

Holland, H.D. (1994) 'Early Proterozoic atmosphere change', in S. Bengtson (ed.) *Early Life on Earth*, New York: Columbia University Press, pp. 237–44.

Horodyski, R.J. and Knauth, L.P. (1994) 'Life on land in the Precambrian', *Science*, 263: 494–8.

Hyde, W.T., Crowley, T.J., Baum, S.K. and Peltier, W.R. (2000) 'Neoproterozoic "snowball Earth" simulations with a coupled climate/ice-sheet model', *Nature*, 405: 425–9.

Kasting, J.F. (2001) 'The rise of atmospheric oxygen', *Science*, 293: 819–20.

Kaufman, A.J., Knoll, A.H. and Narbonne, G.M. (1997) 'Isotopes, ice ages, and terminal Proterozoic Earth History', *Proceedings of the National Academy of Sciences, USA*, 94: 6600–5.

Kennedy, M.J., Runnegar, B.N., Prave, A.R., Hoffmann, K.-H. and Arthur, M.A. (1998) 'Two or four Neoproterozoic glaciations?', *Geology*, 26: 1059–63.

Kenrick, P. and Crane, P.R. (1997) 'The origin and early evolution of plants on land', *Nature*, 389: 33–9.

Kirschvink, J.L. (1992) 'Late Proterozoic low-latitude global glaciation: the Snowball Earth', in J.W. Schopf and C. Klein (eds) *The Proterozoic Biosphere*, Cambridge: Cambridge University Press, pp. 51–2.

Knoll, A.H. (1992) 'The early evolution of eukaryotes: a geological perspective', *Science*, 256: 622–7.

—— (1994) 'Neoproterozoic evolution and environmental change', in S. Bengtson (ed.) *Early Life on Earth*, New York: Columbia University Press, pp. 439–49.

Knoll, A.H. and Carroll, S.B. (1999) 'Early animal evolution: emerging views from comparative biology and geology', *Science*, 284: 2129–37.

Kroken, S., Graham, L. and Cook, M. (1996) 'Occurrence and evolutionary significance of resistant cell walls in charophytes and bryophytes', *American Journal of Botany*, 83: 1241–54.

Kumar, S. and Hedges, S.B. (1998) 'A molecular timescale for vertebrate evolution', *Nature*, 392: 917–20.

Kump, L.R., Kasting, J.F. and Crane, R.G. (1999) *The Earth System*, Upper Saddle River, New Jersey: Prentice-Hall.

Lee, M.S.Y. (1999) 'Molecular clock calibrations and metazoan divergence times', *Journal of Molecular Evolution*, 49: 385–91.

Li, C.-W., Chen, J.-Y. and Hua, T.-E. (1998) 'Precambrian sponges with cellular structures', *Science*, 279: 879–82.

Lipps, J.H., Bengtson, S. and Farmer, J.D. (1992) 'The Precambrian–Cambrian evolutionary transition', in J.W. Schopf and C. Klein (eds) *The Proterozoic Biosphere*, Cambridge: Cambridge University Press, pp. 453–7.

Martin, M.W., Grazhdankin, D.V., Bowring, S.A., Evans, D.A.D. and Fedonkin, M.A. (2000) 'Age of Neoproterozoic bilaterian body and trace fossils, White Sea, Russia: implications for metazoan evolution', *Science*, 288: 841–5.

Meert, J.G. and Powell, C.M. (2001) 'Assembly and break-up of Rodinia: introduction to the special volume', *Precambrian Research*, 110: 1–8.

Moreira, D., LeGuyader, H. and Philippe, H. (2000) 'The origin of red algae and the evolution of chloroplasts', *Nature*, 405: 69–72.

Nei, M. and Kumar, S. (2000) *Molecular Evolution and Phylogenetics*, New York: Oxford University Press.

Nei, M., Xu, P. and Glazko, G. (2001) 'Estimation of divergence times from multiprotein sequences for a few mammalian species and several distantly related organisms', *Proceedings of the National Academy of Sciences, USA*, 98: 2497–502.

Ohmoto, H. (1997) 'When did the Earth's atmosphere become oxic?', *The Geochemical News*, 93: 12–26.

Rasmussen, B., Bengston, S., Fletcher, I.R. and McNaughton, N.J. (2002) 'Discoidal impressions and trace-like fossils more than 1200 million years old', *Science*, 296: 1112–15.

Redecker, D., Kodner, R. and Graham, L.E. (2000) 'Glomalean fungi from the Ordovician', *Science*, 289: 1920–1.

Retallack, G.J. (1994) 'Were the Ediacaran fossils lichens', *Paleobiology*, 20: 523–44.

Rodríguez-Trelles, F., Tarrío, R. and Ayala, F.J. (2002) 'A methodological bias toward overestimation of molecular evolutionary timescales', *Proceedings of the National Academy of Sciences, USA*, 99: 8112–15.

Runnegar, B. (1982a) 'The Cambrian explosion: animals or fossils?', *Journal of the Geological Society of Australia*, 29: 395–411.

—— (1982b) 'A molecular-clock date for the origin of the animal phyla', *Lethaia*, 15: 199–205.

—— (1986) 'Molecular palaeontology', *Palaeontology*, 29: 1–24.

Runnegar, B. (2000) 'Loophole for the snowball Earth', *Nature*, 405: 403–4.

Sanderson, M.J. (1997) 'A nonparametric approach to estimating divergence times in the absence of rate constancy', *Molecular Biology and Evolution*, 14: 1218–31.

Schubart, C.D., Diesel, R. and Hedges, S.B. (1998) 'Rapid evolution to terrestrial life in Jamaican crabs', *Nature*, 393: 363–5.

Schwartzman, D.W. (1999) *Life, Temperature, and the Earth*, New York: Columbia University Press.

Schwartzman, D. and Volk, T. (1989) 'Biotic enhancement of weathering and the habitability of Earth', *Nature*, 340: 457–60.

Seilacher, A. (1994) 'Early multicellular life: late Proterozoic fossils and the Cambrian explosion', in S. Bengtson (ed.) *Early Life on Earth*, New York: Columbia University Press, pp. 389–400.

Selosse, M.-A. and LeTacon, F. (1998) 'The land flora: a phototroph-fungus partnership', *Trends in Ecology and Evolution*, 13: 15–20.

Smith, A.B. and Peterson, K.J. (2002) 'Dating the time of origin of major clades: molecular clocks and the fossil record', *Annual Review of Earth and Planetary Sciences*, 30: 65–88.

Taylor, T.N., Hass, H., Remy, W. and Kerp, H. (1995) 'The oldest fossil lichen', *Nature*, 378: 244.

Thorne, J.L., Kishino, H. and Painter, I.S. (1998) 'Estimating the rate of evolution of the rate of molecular evolution', *Molecular Biology and Evolution*, 15: 1647–57.

Valentine, J.W., Jablonski, D. and Erwin, D.H. (1999) 'Fossils, molecules and embryos: new perspectives on the Cambrian explosion', *Development*, 126: 851–9.

van Tuinen, M. and Hedges, S.B. (2001) 'Calibration of avian molecular clocks', *Molecular Biology and Evolution*, 18: 206–13.

Walter, M.R., Veevers, J.J., Calver, C.R., Gorjan, P. and Hill, A.C. (2000) 'Dating the 840–544 Ma Neoproterozoic interval by isotopes of strontium, carbon, and sulfur in seawater, and some interpretative models', *Precambrian Research*, 100: 371–433.

Wang, D.Y.-C., Kumar, S. and Hedges, S.B. (1999) 'Divergence time estimates for the early history of animal phyla and the origin of plants, animals, and fungi', *Proceedings of the Royal Society, London,* B266: 163–71.

Weiguo, S. (1994) 'Early multicellular fossils', in S. Bengtson (ed.) *Early Life on Earth*, New York: Columbia University Press, pp. 358–69.

Wellman, C.H. and Gray, J. (2000) 'The microfossil record of early land plants', *Philosophical Transactions of the Royal Society, London*, B355: 717–32.

Wood, R.A., Grotzinger, J.P. and Dickson, J.A.D. (2002) 'Proterozoic modular biomineralized metazoan from the Nama Group, Namibia', *Science*, 296: 2383–6.

Wray, G.A., Levinton, J.S. and Shapiro, L.H. (1996) 'Molecular evidence for deep Precambrian divergences among metazoan phyla', *Science*, 274: 568–73.

Xiao, S., Zhang, Y. and Knoll, A.H. (1998) 'Three-dimensional preservation of algae and animal embryos in a Neoproterozoic phosphorite', *Nature*, 391: 553–8.

Chapter 3

Phylogenetic fuses and evolutionary 'explosions': conflicting evidence and critical tests

Richard A. Fortey, Jennifer Jackson and Jan Strugnell

ABSTRACT

Evolutionary radiations are often considered to have been times of rapid origination of major clades as represented by their sudden appearance in the fossil record. The Cambrian evolutionary 'explosion' is considered as an example in this chapter. Whether or not this 'explosive' phase was preceded by a prolonged period (phyloge-netic 'fuse') of more cryptic evolution poorly represented by fossils can, in principle, be tested by molecular estimates of divergence times. However, these methods them-selves have several limitations, not all of which have been acknowledged in previous work. We critically examine some of these limitations concerning phylogenetic uncertainty, the unreliability of calibration points, neglect of confidence intervals, and problems with accurately relating sequence changes to rate variation. Many of these problems are cumulative with time and hence add to the difficulties in accurately determining deep divergences, as within the Precambrian. The techniques considered preferable are summarized.

Introduction

How faithfully does the fossil record reflect the major events in evolution? Where fossils are abundant the answer might seem to be self-evident, but even in this case there may be factors at work that introduce the possibility of spurious patterns. For example, Smith *et al.* (2001) have shown that the volume of rock exposed for a given age influences the taxonomic richness recovered from the same time slice. Hence fluctuating diversity curves may reflect little more than sedimentation/outcrop extent, which is in turn related to former relative sea level. The greatest mass extinctions are real phenomena, but minor extinction events and 'radiations' should be viewed with more circumspection. If the fossil record is putatively poor there is all the more reason to view negative evidence with a sceptical eye before claiming major patterns. We know that marine copepods are among the most abundant elements in the marine realm, yet their fossil occurrences are negligible. Nobody would conclude from this that such a great crustacean group only achieved its current importance post-Pleistocene. Yet assumptions about the completeness of the fossil record are often implicit in hypotheses concerning major evolutionary events. We examine some of these assumptions here, and point to ways in which molecular phylogenies and 'clocks' may be used to test them. The latter come with another battery of qualifications.

We are particularly concerned with what has been termed the Cambrian evolutionary 'explosion'. Even here, we find confusion in what data might or might not support the idea of an 'explosion'. For example, Ayala *et al.* (1998) in a highly critical essay estimated the protostome–deuterostome split at 670 Ma and the echinoderm–chordate split at 600 Ma, claiming that these supported palaeontological estimates. In fact, these estimates allow for a >130 myr phylogenetic fuse before the base of the Cambrian.

Evolutionary radiations

Major evolutionary radiations are apparently rapid bursts of evolution – often prompted by 'ecological release' – recorded by the short-term appearance of many morphologies (and hence high-level taxa) in the fossil record. Familiar though they are, in some examples a question that remains is whether the radiation coincided with the actual appearance of all of the major clades, or whether some or all required an antecedent period wherein their relevant synapomorphies were acquired one by one. This is what Cooper and Fortey (1998) termed the 'phylogenetic fuse' and Jablonski and Bottjer (1990) the 'macroevolutionary lag'. During this formative period of time it has been suggested that the groups in question may have been rare, and/or small, or unlikely to fossilize for some other reason. On the contrary, some analyses of the fossil record (Sole *et al.* 1999; Benton *et al.* 2000) have suggested that it is adequate to record true ranges, and that rarity is not a problem. None the less it remains difficult to understand the significance of an 'absence' below a radiation in a way that is not the same as an 'absence' above an alleged extinction, since only in the latter case do we *know* the previous existence of a clade, and can hence statistically assess the increasing improbability of its survival above its last record. The taphonomic rules may have been different during 'fuse' phases.

It is not unreasonable to invoke an early innovative phase involving organisms of small population size, and/or small absolute size (Martin and Palumbi 1993). It has been claimed that evolutionary rates are higher in small taxa, not least because reproductive rates correlate negatively with size. Morphological change may also be accelerated among highly vicariant and isolated populations. The problem is that direct fossil evidence of such a phase is, in principle, likely to be rare, and hence impossible to distinguish from other reasons for absence. This is why independent testing of divergence times from molecular evidence becomes so important.

However, it is possible to test the 'fuse' idea by looking at a converse example: where the early representatives of a major radiation are *known* to be large animals. The best case is probably terrestrialization and the origin of tetrapods. The Devonian complex of animals in this transition (for a recent summary, see Ahlberg and Johansen 1998; Ahlberg 2001) including the well-known *Ichthyostega* are about one metre long and robust. Although their occurrences may be localized, they are also easily fossilizable by virtue of the habitat in which they lived. They also record a selection of what might loosely be described as 'hopeful monsters'. So far as we are aware, there is no suggestion that there was a prior history of major clades in the 'fish'–tetrapod transition, and the sister taxa of early tetrapods are selectable from near contemporaries in the osteolepiforms. In short, if there is no suggestion of small and cryptic ancestors, there is no suggestion either of a 'phylogenetic fuse'.

The examples of evolutionary radiation where molecular 'clock' estimates indicate such a prior history also tend to be those where the fossil examples show an apparently sudden appearance of several clades. Worked examples include the radiations that occurred after the extinction of the dinosaurs at the Cretaceous–Tertiary (K–T) boundary. The modern birds undoubtedly proliferated in the Tertiary, and some of the major clades appeared apparently suddenly. Molecular studies indicate that they descended from a common ancestor well within the Cretaceous – a phylogenetic fuse is implied (Cooper and Penny 1997). On the contrary, some avian palaeontologists dispute this (Feduccia 1995), as the record of such birds is unknown, and such a scenario is considered by them unnecessary. The mammals can be interpreted analogously, although 'fuses' have been disputed by mammologists (Alroy 1999). However, in this case every year brings the apparent discovery of ancestors of clades within the Cretaceous. For example, Archibald *et al.* (2001) have proposed rabbit/rodent relatives from the Cretaceous rocks of Uzbekistan, while Hooker (2001) has described evidence that the apparently highly-derived bat clade was already present pre-Tertiary. However, fossils of these animals remain rare, but it does give more confidence in the fuse schema. Another good example of recently discovered fossils challenging our acceptance of the origin of groups was the report of a cirrate octopod from the Upper Carboniferous Mazon Creek fauna (Kluessendorf and Doyle 2000). Prior to its discovery the earliest known octobrachid was from the Jurassic.

The origin of land plants is discussed elsewhere in this volume (Wellman, Chapter 7). The appearance of plant body fossils in the Silurian suggests that, again, a 'fuse' of perhaps 50 million years is involved. In this case, however, there is evidence of the time involved from the presence of air-dispersed cryptospore fossils in the absence of body fossils, which occur back to the mid-Ordovician. However, there is an analogy with the mammalian case in that some researchers were initially reluctant to accept a Silurian age for the 'anomalously' advanced plant *Baragwanathia* from Australia (Rickards 2000). The acceptance of the 'fuse' removes the anomaly, and increases the expectation of discoveries to come.

Cambrian evolutionary 'explosion'

The most contentious case of all is the so-called Cambrian evolutionary 'explosion' (see reviews in Zhuravlev and Riding 2001) at about 540 Ma, which is the main focus of this discussion. The questions involved were explored by Fortey *et al.* (1996) and have since been articulated many times (e.g. Fortey *et al.* 1997; Conway Morris 2000). To summarize briefly, the appearance of clades in the Cambrian is sudden. As research on early faunas has continued, more and more putative Early Cambrian representatives of major groups have appeared, for example vertebrates have been added to the list (Shu *et al.* 1999). Even at their first appearance some (e.g. trilobites and other arthropods) are both taxonomically and biogeographically differentiated, implying an earlier history without a fossil record. Trees of various groups (e.g. molluscs, arthropods, brachiopods) have now been described, from palaeontological, morphological, and molecular bases, all of which imply nodes that predate the appearance of various major groups (usually classes) and hint at a phylogenetic history antedating the first record of fossils. Connections at a still deeper level (e.g. between phyla) are implicated. Nobody questions that the appearance of mineralized tissue was a Cambrian

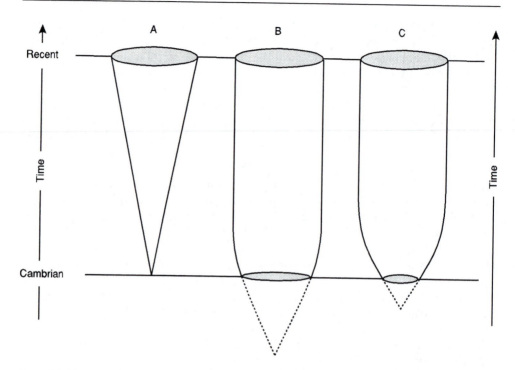

Figure 3.1 Three models depicting the evolution of clade diversity among the Metazoa through time. A, Traditional model. Metazoa originated simply and became more complex throughout the Phanerozoic. B, Fortey *et al.* (1996). Evidence from phylogeny, palaeobiogeography, trace fossils and molecular clocks imply extensive Precambrian cladogenesis prior to the appearance of the first body fossils. C, Budd and Jensen (2000). Most Cambrian stem-lineage forms lacked some of the defining features of modern phyla, suggesting that they were considerably less disparate than the latter. Cladogenesis producing the stem-lineages of modern phyla probably occurred in the latest Proterozoic but no earlier.

phenomenon, and some of the discussion has centred on arguments such as 'can you have a brachiopod without a mineralized shell?', but *a propos* of arthropods, molluscs, and chordates the answer clearly is 'yes, you can'. The persistent paradox remains that fossils of these earlier animals stay obstinately undiscovered. This in turn leads to two principal possibilities: (i) that the fossil record is relatively complete and that it indicates greatly accelerated rates (presumably both morphological and molecular) at about the Precambrian–Cambrian boundary (Budd and Jensen 2000), or (ii) fossils of putative 'ancestors' remain undiscovered for taphonomic reasons, or because they were soft bodied and of small size – and that there was, indeed, a phylogenetic fuse (Figure 3.1). Under this scenario, rates do not have to be drastically elevated, and the question of acquisition of shells and skeletons becomes decoupled from that of the origin of clades. Molecular tests of rate constancy and divergence times can, in principle, distinguish between these alternatives.

Problems with interpreting fossil evidence of the Cambrian evolutionary 'explosion'

Duration of Early Cambrian

The formalization of the base of the Cambrian at the base of the Placentian Stage in Avalonia (= Nemakit Daldynian of Siberia; 545 Ma) has altered the perception of the radiation in terms of stratigraphical nomenclature. First, new isotope ages have shown that the Early Cambrian under this definition is considerably longer than was previously believed (Landing *et al.* 1998). The earliest stage may be as much as 10 myr. Second, the appearance of most of the major clades with which we are concerned is at the succeeding Tommotian (or the next, Atdabanian) Stage. The fauna of the Nemakit Daldynian is a relatively sparse array of enigmatic small shelly fossils, while the Placentian has yielded mostly trace fossils. With regard to the phyla that concern us here, the earliest Cambrian Stage should be added to the 'fuse' time. Hence the former use of the Precambrian–Cambrian boundary as if it were synonymous with the time of 'explosion' requires a certain modification. This is, however, a matter of definition and not of science. It should be noted that one of the earliest trace fossils in the Placentian (*Rusophycus avalonensis*) is of 'trilobite' type, and assuredly made by a normal-sized schizoramous arthropod, whether or not it was a 'soft-bodied trilobite'. This sets the timing of this part of the arthropod 'fuse' at the (formal) base of the Cambrian.

Definitions of phyla

Cambrian fossil faunas include a variety of more or less plesiomorphic animals, often exhibiting in addition peculiar autapomorphies (e.g. Edgecombe 1998). Not surprisingly, they frequently offer challenges in cladistic analysis, as reflected in unresolved polytomies, or shifting 'basal' positions (Wills and Fortey 2000). Budd and Jensen (2000) regard the difficulties of attempting to include these problematic fossils in a classification based upon Recent organisms (for which inexhaustible phylogenetic data are obtainable in principle) as impractical and probably ultimately unsolvable. They prefer to define major animal clades as crown-groups embraced by the most basal living representative and its descendants. Thus, arthropodan Chelicerata would presumably include the common ancestor of the primitive *Limulus* and all its sister taxa. Fossil taxa falling outside this definition would find their place (or not) on the stem-group, without formal status. The more usual view would attempt to place stem taxa within the *total*-group, whose inclusiveness would be defined from the basal node separating Chelicerata from its sister clade (for example, Crustacea). The former approach has the advantage that it should be possible to define major groups with a battery of synapomorphies (say, n). Cambrian taxa will tend to fall on the stem, and, except in those cases where there are Early Cambrian representatives lying inboard of living animals, this will displace phylum level taxa upwards. In Budd and Jensen's view there would be no 'explosion of phyla' in the Lower Cambrian – because on their definition phyla appear in younger strata. They effectively reduce the length of the phylogenetic 'fuse'. The problems with this approach are exemplified by the discovery of a fossil with $n - 1$ synapomorphies – it seems perverse to 'define out' such a fossil

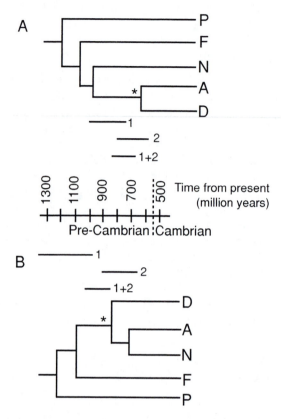

Figure 3.2 Alternative topologies which have been used as consensus trees in divergence time estimation papers. A, shown in Wang *et al.* (1998), Nei *et al.* (2001) and Hausdorf (2000). B, shown in Wray *et al.* (1996) and Lynch (1999). Key: D = Deuterostomata, A = Arthropoda, N = Nematoda, F = Fungi, P = Plantae.

from other animals it resembles in significant features merely because such an animal did not win the lottery of survival (Wills and Fortey 2000). The argument is well displayed in the discussion concerning a recently discovered ostracode-like Early Cambrian fossil (see Siveter *et al.* 2001, and replies). Regardless of phylum definitions, there are striking morphological differences between Early Cambrian arthropods in the Chengjiang fauna of China (Bergstrom and Hou 1998) sufficient to place them on different stem-lineages, which must themselves have split earlier: the problem of deep divergences will not simply go away. Molecular estimates of divergences (e.g. between chelicerates and crustaceans) will be based upon the split between the total-groups, regardless of definitions of convenience.

Relationships between major groups

More disturbing, perhaps, is the lack of consensus on phylogenetic relationships using either morphological or molecular data, between some of the major groups involved in the Cambrian 'explosion' (Figure 3.2). Considering arthropods, for example, despite previous enormous databases and total evidence analyses (e.g. Wheeler 1998), a recent edition of *Nature* featured two papers back to back (Giribet *et al.* 2001; Hwang *et al.* 2001) which claimed myriapods as a sister group of 'pancrustacea' and chelicerates, respectively. This kind of phylogenetic uncertainty directly affects the estimates of upper and lower bounds on divergence times, as differing opinions on the affinities of fossil taxa can lead to very different divergence times (Marshall 1990). Similar debates exist as to the sister group of Brachiopoda, or relationships within the Mollusca. At a deeper level, the concept of a clade Ecdysozoa (Aguinaldo *et al.* 1997) to embrace all moulting animals has been disputed by others (Erwin 2001).

Unresolved nodes between major taxa involved in the Cambrian 'explosion' have been used in support of the evolutionary 'explosion' hypothesis (Field *et al.* 1988). Philippe *et al.* (1994) reported a method to estimate the number of nucleotides that would be required to resolve uncertain nodes confidently. They showed that an 'experimentally unfeasible' number of nucleotides is required to resolve many of these uncertain nodes. Although this method provides an indication of the amount of resolution we can obtain from a gene, its utility is limited by inconsistent substitution rates between genes and lineages. This analysis was based on 18S rRNA sequences. The lack of tree resolution is likely to be at least partially due to both very short and very long internal branches being present, which can cause long branch attraction (LBA) (Philippe and Germot 2000). LBA can cause the false placement of distantly related taxa together in a phylogeny (Philippe and Adoutte 1998; Philippe and Germot 2000) and may thus obscure or 'blur' phylogenetic signal. Causes of LBA include differing G + C contents, rate heterogeneity and increased proportions of variable positions and other biases within ingroup taxa in comparison with those taxa in crown-groups (Philippe and Germot 2000). LBA can apply to both nucleotide and amino acid data. It is important to eliminate LBA as much as possible when tree building, by testing for nucleotide and amino acid skew (Foster and Hickey 1999; Penny *et al.* 1999). Where inequality exists, corrections can sometimes be applied such as RY coding (Phillips *et al.* 2001) and non-stationary modelling (Galtier and Gouy 1998). Furthermore, the addition of sequence data from a greater number of species breaks up long branches, thus reducing LBA (Hillis *et al.* 1996).

Molecular estimates of divergence times can be confounded by phylogenetic uncertainty. The placement of both nematode and molluscan clades has not been clearly resolved despite analysis involving many datasets (Aguinaldo *et al.* 1997; Adoutte *et al.* 2000) (see Figures 3.2, 3.5). In cases where the true phylogeny is unresolved, estimates of divergence time and their confidence intervals should be calculated not only for the consensus tree (Cutler 2000) but also for those trees not significantly worse fitting. This allows the calculation of a total interval estimate for all likely phylogenies rather than only the constrained tree which is assumed to be the correct one.

As new sequence data become available, particularly from slowly evolving genes, it is expected that the resolution of the metazoan phylogeny will improve. Although data from protein-coding nuclear genes are likely to provide an improvement over mitochondrial genes and nuclear rRNA in resolving deep divergences and providing rate estimates, they add a level of complication due to gene duplications and gene loss (Nichols 2001). Gene duplication can introduce marked differences between

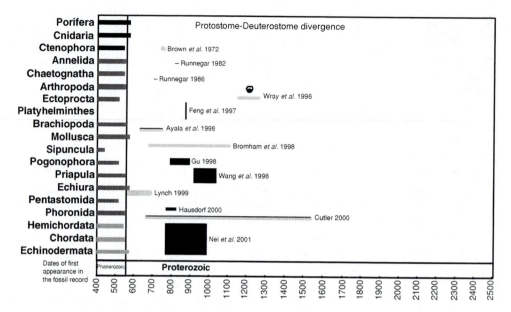

Figure 3.3 The confidence intervals on times of divergence between protostome and deuterostome phyla as measured by recent studies. Estimates are sorted in chronological order. Dots indicate point estimates. Line thickness indicates number of genes analysed. Black lines, grey lines and the ribosome image ● indicate nuclear, mitochondrial and 18S genes used in the analyses, respectively.

gene phylogenies and species phylogenies, particularly in the vertebrates where many genes actually occur as large gene families (Page 2000). Attempts to infer phylogeny from nuclear genes must take gene duplications into account and ensure as far as possible that the sequences under comparison are orthologous rather than paralogous. For example, one criterion for ensuring that sequences are orthologous is to check that the phylogeny of the sequences being used matches the known phylogeny of the species (Nei *et al.* 2001). However, this may become too lax a criterion when the number of sequences is low. Alternatively, 'blasting' a sequence in Genbank (Altschul *et al.* 1997) to find genes with high sequence similarity can provide information about the existence of orthologues in the species under scrutiny.

Molecular evidence: criticisms and critical tests

All molecular estimates of protostome–deuterostome divergence times fall in the Precambrian with the majority of studies ranging from 800–1000 Ma (Figure 3.3). The 'individual protein' (IP) approach has been taken in a number of metazoan divergence studies (e.g. Kumar 1996; Ayala *et al.* 1998; Gu 1998). With this approach, individual divergence time estimates are made for each protein, and the mean of these estimates is often presented as the final estimate. Nei *et al.* (2001) suggested that this approach has an upward bias, and that distances based on concatenated sequences may help reduce this bias.

Pairwise distance methods have commonly been used to estimate divergence times (e.g. Kumar 1996; Ayala *et al.* 1998; Gu 1998). Later studies have used a maximum

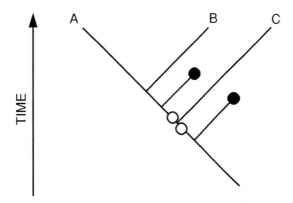

Figure 3.4 The ideal case is shown by fossils ○ wherein well-dated fossils fall just before and after a branching event enabling tight constraint on the true divergence time. More realistically, fossils may be sister taxa, (shown as ●) and will instead provide an underestimate of the true divergence date.

likelihood (ML) framework for branch length estimation (Bromham *et al.* 1998; Cutler 2000; Hausdorf 2000; Yoder and Yang 2000).

Calibration points

The calibration of genetic distance to produce an estimate of molecular change per unit time is arguably the most serious difficulty with divergence time calculations. This is likely to be particularly acute when deep divergences, such as the basal Cambrian are considered. The incomplete nature of the fossil record means that it is unlikely that the very earliest specimens following a divergence will be preserved. Fossils used to date divergences are seldom claimed as direct ancestors of the taxa in question, but must always be regarded as members of a plesion (Figure 3.4). Such fossil dates will tend to underestimate the true date of the divergence, thus causing evolutionary rates to be overestimated. This in turn causes molecular date estimates to be too young (Rambaut and Bromham 1998).

Additionally, since allelic divergence generally predates species divergence, appearance of any taxon-defining morphological–taxonomic features (including those seen in the 'earliest' fossils) will post-date actual genetic divergence by a significant period of time (Smith and Peterson 2002). Thus, when these earliest fossils are used as calibration points in divergence time estimation (even assuming a perfect molecular clock) they will still underestimate genetic divergence times.

Despite such problems associated with calibration from the fossil record, a surprisingly large number of studies use only one or two calibration points (Hedges *et al.* 1996; Kumar 1996; Gu 1998; Wang *et al.* 1998). The argument for using few calibration point estimates is that the majority of fossil-based divergence times are underestimates. If supposedly 'accurate' calibration points are used instead (those points very close to the true divergence date), it is argued that this will provide a more realistic estimate than calibration based on an average of many calibration points (Wang *et al.* 1998). In most studies a fossil calibration point of 310 Ma, an estimate of the mammal–bird split, is used (Hedges *et al.* 1996; Kumar 1996; Gu 1998; Wang *et al.* 1998; Nei *et al.* 2001). Despite such widespread use, Lee (1999) argued that 310 Ma is an over-estimate and stated that the first tetrapods to be confidently assigned to either the mammal or bird lineage are approximately 288 million years old. Using this date reduces inferred dates calibrated by the mammal–bird divergence to 93 per cent of their reported value (Lee 1999). Yoder and Yang (2000) obtained conflicting divergence estimates within the mammalian clade using different calibration points. Strikingly, Lee (1999) pointed out that the second calibration point of the primate–rodent divergence of 100 Ma, used by Hedges *et al.* (1996) and Gu (1998) is a molecular clock estimate which is also based on the 310 Ma mammal–bird divergence estimate. This is clearly not a second, independent, calibration point.

It should be noted that while these errors would all tend to cause underestimation of divergence, the opposite is often seen to occur in metazoan divergence analysis, with molecular dates pushing back fossil-based dates. It has been proposed that the smaller body sizes and faster generation times of many Cambrian taxa may correlate with relatively fast rates of evolution (Martin and Palumbi 1993), compared with more recent taxa. When calibration points taken from recent taxa with larger body sizes and/or generation times (e.g. the synapsid–diapsid split) are used to estimate divergence times, they will cause an overestimation of the true date of divergence.

Rather than using fossils to 'assign' dates to nodes within a phylogeny, a number of authors (Sanderson 1997; Rambaut and Bromham 1998; Cutler 2000) have allowed fossils to provide constraints on divergence estimates, using a ML-based approach. This is a more realistic approach as we can never be certain when two lineages diverged. Furthermore, the addition of fossils can only improve estimates when using this method.

If possible, calibration point estimates should not be applied to distantly related reference taxa. This is because rates of molecular evolution are well known to differ markedly between groups (Hillis *et al.* 1996) and relative rate tests cannot detect changes shared across long branches. However, calibration points can only be as good as the fossil record from which they are drawn and are often unavailable or unreliable for the group in question (especially when small and soft bodied). Accurately dated, well-characterized fossils from just above and below the lineage divergence to be dated are ideal, although admittedly also rare (e.g. Doolittle *et al.* 1996; Cooper and Penny 1997; Sanderson 1997; Bromham *et al.* 1998). If applied well, the use of multiple calibration points can provide good constrained estimates of absolute rates of molecular evolution (Marshall 1990).

Another problem with using multiple calibration points is the availability of calibrating sequences for less frequently studied taxa. Often, the only sequences available for many genes are those of human, mouse, and chicken (Wang *et al.* 1998) although it is likely that sequence data for a wider range of taxa will become rapidly available in the coming years.

Confidence intervals

Divergence times of lineages cannot be dated precisely (Figures 3.3, 3.5). Confidence intervals are an essential part of any molecular dating technique and grow in size as errors are introduced to these analyses at a number of levels. Statements such as 'the divergence time between *Drosophila* and vertebrates was about 830 million years ago'

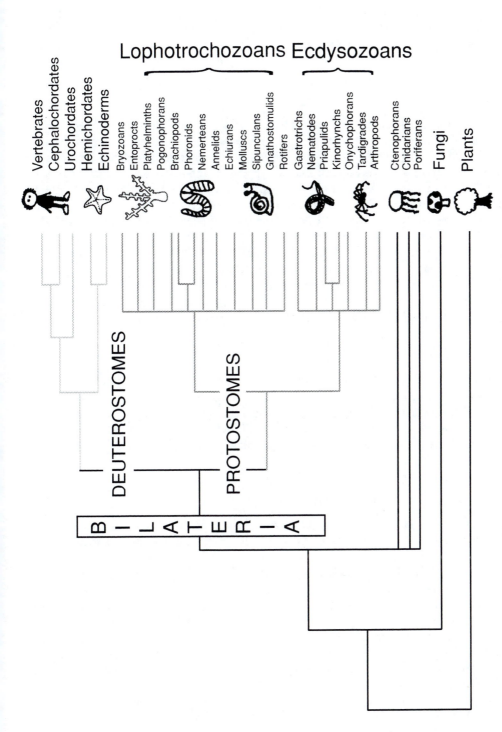

Figure 3.5 Current consensus molecule-based metazoan phylogeny, as described in Adoutte *et al.* (2000).

(Gu 1998) impart a spurious level of confidence. In an attempt to reduce confidence intervals a number of studies have removed so-called 'outliers' before averaging divergence times across the genes (e.g. Kumar and Hedges 1998; Wang et al. 1998). In contrast, the estimates of Bromham et al. (1998) and Cutler (2000) are very large yet more accurately reflect the genetic variability of the sample.

Fossils

Along with the problems of using only one or a few calibration points, by far the majority of metazoan divergence time studies use absolute dates for calibration points. Marshall (1990) presented a method for calculating confidence intervals on divergence times based on the notion that the better the fossil record, the smaller the gap will be between the earliest known fossil and the time of origin of the group. These methods were improved to allow predicted collecting and/or preservation biases to be combined into the calculation of confidence intervals (Marshall 1997; Foote and Sepkoski Jr 1999). More recently work by Tavaré et al. (2002) has combined both molecular and palaeontological approaches to divergence estimation and shows that good agreement between the two can be achieved. Their calculations took into account data on the number of extant species, the mean species lifetime, the age of the bases of the stratigraphic intervals (and the number of fossils found in these), and the relative size of the sampling intensity at each interval.

Standard error of the mean

The standard error of the mean is often used to provide confidence intervals around molecular dating estimates (e.g. Hedges et al. 1996). However the standard error only provides an indication of the certainty of the mean of the population from which the sample was taken. It does not provide a confidence interval for the single true date of divergence (Cooper et al. 2001) and provides error intervals that are unrealistically narrow.

Stochasticity of the molecular clock

Molecular clocks do not tick at regular intervals, but instead at random points in time. In order to characterize these, 'point processes' must be applied – statistical models that describe events occurring at random times (Gillespie 1991). Generally, the 'ticking' of the molecular clock is assumed to follow a simple Poisson distribution of substitution occurrences (Zuckerkandl and Pauling 1965). Most methods assume that the clock adheres strictly to this process. However, confidence intervals calculated for divergence time estimates often ignore the Poisson variance of this distribution, taking instead the standard error of the mean divergence estimates as the total estimate error (Wang et al. 1998). For deeper divergences the 'cumulative' errors are greater and so pose a particular problem when investigating Cambrian divergences. Incorporation of this variance into the confidence intervals greatly widens them but also provides a far more realistic reflection of the true error of the estimate. Likelihood-based estimation methods that incorporate this error have been developed (Rambaut and Bromham 1998; Cutler 2000).

In reality, the molecular clock often appears to violate the assumptions of a Poisson process, suggesting that this may not always be an appropriate model to use. Two cases in which this becomes clear are in the observation of lineage-specific rate variation and in the over-dispersion of the actual calculated substitution variance relative to the Poisson-assumed variance. This latter problem has been analysed by Cutler (2000) who replaced the Poisson process with a more general stationary process (using a central limit theorem approximation). It was found that for mitochondrial genes reconstructing metazoan evolution, the stationary process provides a significantly better fit to the data than the Poisson. It was suggested that this may be a better process to use when both the dataset and the number of substitutions per branch is large.

Non-independence of comparisons

In many studies, the clock is calibrated by averaging all pairwise comparisons among taxa within a group (e.g. Wray *et al.* 1996; Wang *et al.* 1998). However, it is sometimes overlooked that these values are not independent (Felsenstein 1981) because they are based on shared regions of a tree. This lack of independence leads to unrealistic confidence in the correlation between time and divergence and confounds error calculations. Where possible, calibrations should be based on independent lineages. Non-independent comparisons are often a problem when using pairwise distance methods, but processes that incorporate the whole phylogeny, such as maximum parsimony (MP) and ML, can circumvent this problem (Lewis 2001).

Rate variation

Molecular evolutionary rates can differ between taxa and lineages. For example Ayala (1999) reported that the superoxide dismutase gene evolved five times faster within *Drosophila* groups than it did in multicellular organisms in general. Additionally, Rodríguez-Trelles *et al.* (2001) applied divergence time estimation techniques to three variable-rate genes (SOD, GPDH, and XDH) and retrieved a very large range of divergence estimates, illustrating the effects of rate variation on such tests. Friedrich and Tautz (1997) also showed that the substitution rate in the 28S rRNA dipteran stem-lineage underwent marked acceleration in comparison with other insect groups prior to the diversification of the major dipteran subgroups.

Relative rate tests

Relative rate tests compare the amount of change in each of two ingroup sequences with respect to an outgroup. These are commonly used to identify data that do not conform to rate uniformity, which are then removed from the analysis (e.g. Gu 1998). These tests are nearly as old as the molecular clock hypothesis. They first came into use in immunological cross-reaction studies, which utilized the molecular clock hypothesis to date the origin of the hominoids (Sarich and Wilson 1967).

Nowadays the tests are applied to molecular data as demonstrated by Takezaki *et al.* (1995). Multiple nested relative rate tests are performed on different triplets within the phylogenetic tree. This information is then combined in order to remove lineages which have evolved at significantly different rates from the majority consensus of the group, defined by the mean total branch length from root to tip over all lineages.

Trees built in this way are referred to as 'linearized' trees because they make the remaining divergence times linear with distance (Takezaki *et al.* 1995; Hedges *et al.* 1996). However, linearized tree methods do not give reliable divergence estimates when evolutionary rate varies markedly between species groups, for example between orders or classes of vertebrates (e.g. Cao *et al.* 1998; Nei *et al.* 2001). In the Cambrian–Precambrian case the influence of such factors is particularly hard to assess.

Bromham *et al.* (2000) showed that relative rate tests have very limited power when used on short sequences (400 or fewer sites free to vary) and are unlikely to detect moderate levels of rate variation (where the rate of one lineage is 1.5–4 times the other). They found that sequences of at least 1000 sites are required to detect three-fold rate differences and above, and much longer sequences (>2000 sites) are required to detect two-fold rate differences. The inability of relative rate tests to detect moderate to high levels of rate variation on short sequences can provide misleading results in studies where relative rate tests are performed on individual genes.

Datasets will also pass relative rate tests when rates of evolution have accelerated or decelerated and affected all lineages equally (Marshall 1990) and, thus, caution must be taken when using the test. This has particular resonance when studying radiations – where the counter-argument to deep divergences is that some kind of universal rate acceleration may have occurred. This has been claimed in 'Snowball Earth' (Hoffman *et al.* 1998) and 'True Polar Wander' (Kirschvink *et al.* 1997) theories. Both claim that severe stress led to large amounts of genetic change in a short period of time.

Models of sequence substitution

As the number of substitutions between two sequences increases, they become progressively more saturated, as most of the sites changing have already changed previously. Surprisingly, very few studies investigate the saturation levels of the sequences they use. Saturation brings internal tree nodes artefactually closer to one another and can exacerbate long branch attraction (Philippe and Adoutte 1994; Philippe *et al.* 1994).

This problem obviously needs to be addressed when investigating the divergence of lineages as far back as the Cambrian. The saturation of sequences can be tested by plotting the inferred substitutions between all species pairs, against the actual number of differences recorded between the species pairs (Philippe *et al.* 1994). When the number of actual counted differences levels off, while the inferred distance continues to rise, saturation has occurred. In order to minimize the effects of saturation, there are a number of 'distance correction' methods available. These convert observed distances into measures of actual evolutionary distance, thus estimating the amount of evolutionary change that has been 'overprinted'.

Nucleotide models

The Jukes–Cantor (JC69) model was the first nucleotide substitution model proposed and is perhaps the simplest matrix-based model of distance correction (Jukes and Cantor 1969). It assumes that the four bases have equal frequencies and that all substitutions are equally likely. This model was extended to incorporate differing substitution rates between transitions and transversions (K2P) (Kimura 1980) and also to allow varying base frequencies (Felsenstein 1981). These two models were combined in the HKY85 model (Hasegawa *et al.* 1985). The general time-reversible (GTR) model further extended these models to allow all six pairs of substitutions to have differing rates (Rodríguez *et al.* 1990). The LogDet transformation (Lockhart *et al.* 1994) is a para-linear process and was designed to circumvent the problem of variable base composition among sequences. This often provides a better fit to data when sequences have differing base compositions, but is disadvantageous in that LogDet-based models cannot be compared with other models for goodness of fit using likelihood ratio testing (LRT). Furthermore LogDet methods provide branch length estimates which are not interpretable as distances. ML models such as T92+GC (Galtier and Gouy 1998) and N2 (Yang and Roberts 1995) do not assume fixed nucleotide composition values for each taxon but allow them to vary from branch to branch.

Amino acid models

In theory, amino acid-based substitution models should give more satisfactory results when estimating deep divergence times (e.g. Cambrian) than nucleotide substitution models since amino acids evolve more slowly than nucleotides. Additionally, because there are some twenty amino acids in comparison with four nucleotides, saturation is less likely to be encountered. However, amino acid substitution matrices are generally considered to be much less sophisticated than their nucleotide counterparts, as less is known about the factors involved in amino acid substitution (i.e. secondary and tertiary structure and physico-chemical constraints). An amino acid substitution model analogous to the GTR model in its time reversibility is the Dayhoff model (Dayhoff 1978). Amino acid replacement is modelled here with a rate matrix based on the physico-chemical differences of different amino acid groups. The Dayhoff model has been extended many times (Jones *et al.* 1992; Muller and Vingron 2000). Recent work by Lió and Goldman (2002) has provided a model which takes specified secondary structure information into account when creating rate matrices based on mitochondrial amino acid sequence data.

Codon models

Although amino acid substitution models have provided an improvement over nucleotide models when determining deep divergences, they are less realistic in that evolutionary change occurs at the level of DNA sequences, rather than on amino acids. Models of evolution based on codons (61 states) rather than amino acids (20 states) are more representative of reality (Goldman and Yang 1994; Muse and Gaut 1994). These are computationally expensive, though becoming increasingly sophisticated. For example, Pedersen *et al.* (1998) designed a model which takes depressed CpG levels into account.

Maximum likelihood

Using these nucleotide, amino acid or codon substitution models, genetic distances can be estimated and 'corrected' from the data using a ML algorithm (Felsenstein 1981). The ML framework allows the comparison of different models of sequence evolution using likelihood ratio testing (LRT) (Felsenstein 1981; Muse and Gaut 1994). In LRT, these nested hypotheses of sequence evolution can be expressed as a ratio, and if the more specific model provides a significant improvement in likelihood value it is determined to be a better fit to the dataset. In this way the test can ascertain whether incorporating extra parameters into a model provides a better fit to the dataset. Model selectors such as that of Posada and Crandall (2001) utilize LRTs in order to distinguish between the fit of different models to datasets.

Parameters

Additional parameters can be incorporated into substitution matrices in order to relax certain assumptions about rates across sites. Such parameters often provide a significant improvement in fit to data.

Rates of substitution can vary greatly between individual sites (nucleotides, amino acids, or codons) in genes. When this is not allowed for, it causes an underestimation bias in the observed versus true distance, which increases with depth of divergence (Adachi and Hasegawa 1995; Yang 1996). The best models to accommodate the variety of observed patterns are those that incorporate both rate heterogeneity (Yang 1996) and invariant sites data (Steel *et al.* 2000).

Non-stationarity

The stationarity hypothesis proposes constancy of base composition over the whole tree. Non-stationarity is often overlooked because of the computation required to deal with it. Ignoring it can lead to the construction of erroneous phylogenies with high bootstrap support (Phillips *et al.* 2001; Tarrío *et al.* 2001). Galtier and Gouy (1998) reported a method which can accommodate non-stationarity. This has been shown to resolve the correct phylogenetic tree in cases where the LogDet method failed to do so.

Rate heterogeneity

Rate heterogeneity across sites is usually modelled with a discrete gamma distribution, shaped by an estimated alpha value (Yang 1996). When gamma values are included in ML models and estimated from the data they often provide a significant improvement in likelihood. Gamma may not provide an improvement, however, when alpha values are assumed without estimation from the data. Bromham *et al.* (1998) argued this point with respect to work by Ayala *et al.* (1998) who assumed a protein alpha value of two across all genes rather than making an estimation from the dataset. Ayala *et al.* (1998) generated a noticeably shallower divergence estimate (Figure 3.3) compared with studies such as Wray *et al.* (1996) where gamma was estimated from the data. Hidden Markov models (HMM) can also be used to model rate heterogeneity across sites (Felsenstein and Churchill 1996). These allow rate correlations to be made between adjacent sites.

Invariant sites

Incorporation of invariant sites parameters into the substitution model decreases the variance of estimated gamma values (Gu *et al.* 1995). In order to obtain the most unbiased gamma estimate, invariant sites should be included, as this will provide most accurate branch length estimates (Lockhart *et al.* 1994).

Secondary structure

Secondary structure models attempt to account for evolutionary dependence between nucleotide sites opposite one another in RNA stem regions (Muse 1995; Tillier and Collins 1995). The complicated secondary structure of RNA molecules can lead to compensatory changes and thus a single substitution may result in two substitutions (Hickson *et al.* 1996) widely separated in the sequences. The secondary structure of proteins can also influence amino acid replacement rates (Thorne 2000). However, Goldman *et al.* (1998) showed that solvent accessibility correlates more strongly with amino acid replacement than secondary structure. They found that replacement rates at sites on the surface of globular proteins are about twice the rates at sites that are less accessible to solvents. Models have been proposed to account for separate amino acid replacement in differing structural environments (see Thorne *et al.* 1996). Substitution matrices which take these into account provide more realistic branch length estimates.

Methodology for optimizing parameters

Bayesian inference

An increasingly popular approach to determining divergence times involves the use of likelihood models within a Bayesian framework (Huelsenbeck *et al.* 2001; Lewis 2001). Bayesian approaches take prior parameter values as probability distributions, modify them given the data available, and then calculate *a posteriori* distribution, reflecting the level of uncertainty in the parameter after viewing the data. Bayesian approaches can produce probabilities for hypotheses of interest. Likelihoods alone are harder to interpret in this context. These represent the probability of the data given the hypothesis rather than the (Bayesian) probability of the hypothesis given the data. An argument against the Bayesian approach is that the choice of the prior distribution, to which the *a posteriori* distribution is very sensitive, can be too subjective (details in Shoemaker *et al.* 1999). Furthermore, implementing Bayesian methods can be very complicated, especially where the parameter space is complex or the dimensionality high, or both.

Markov Chain Monte Carlo

Markov Chain Monte Carlo (MCMC) methods have made Bayesian methods computationally feasible (see Thorne *et al.* 1998; Huelsenbeck *et al.* 2000). MCMC methods are usually employed to estimate the parameter values determining posterior distributions. In this case, parameter values are continually re-estimated in all combinations, given a particular tree and set of data until the chain 'converges' on a particular set of parameters, at which point the chain can be said to have achieved stationarity. This provides an effective and thorough analysis, although computationally time-consuming. It has been used to model variable evolutionary rates within trees, assuming various distributions including the log-normal (Thorne *et al.* 1998) and compound Poisson (Huelsenbeck *et al.* 2000).

Non-parametric rate smoothing

Along with the parametric methods outlined above for estimating divergence times, non-parametric approaches are also available (e.g. non-parametric rate smoothing, NPRS; Sanderson 1997). This method modifies branch lengths within a given tree, assuming rate correlation between ancestral and descendant branches.

Increased complexity at a price?

As models become more sophisticated and parameter-rich they provide a better fit to the observed patterns of evolution of individual datasets. This is reflected in the general improvement in tree likelihood values with increased model complexity. However, as methods increase in statistical complexity, they also become more computationally expensive and non-user friendly. Also, where assumptions are moot the methods may not be significantly better than more statistically simple alternatives.

As the number of methodological parameters increases, so does the variance of the estimates yielded, as reflected in the ever-widening confidence intervals.

Working example

A small-scale study of metazoan divergence was performed using the above criteria. Taxa chosen were *Homo sapiens*, *Rattus norvegicus*, *Gallus gallus*, *Drosophila melanogaster*, *Caenorhabditis elegans*, *Saccharomyces cerevisiae* and outgroup *Arabidopsis thaliana*. Phosphoglycerate kinase (PGK), replication factor C (RFC), nucleoside diphosphate kinase (NDK) and triose phosphate isomerase (TPI) nuclear gene sequences were obtained from GenBank (accession numbers and alignments available from the authors on request). Sequences were aligned by manual inspection and gapped sections and regions of ambiguous alignment were removed. ML trees were constructed using PAUP (PAUP* 4.0b8, Swofford 1998) from individual genes and concatenated data from first, second, and the two combined codon positions. All analyses were made using the general time-reversible model, and gamma and invariant sites values were estimated from the data. Trees were rooted using plant sequences as outgroups and ML trees were found by an exhaustive search. Topologies shown in Figure 3.2 were compared using the Shimodaira–Hasegawa (SH) test (Shimodaira and Hasegawa 1999) for differences in fit.

Divergence times were estimated in RHINO for both tree topologies and all datasets. RHINO is a modified version of QDATE (Rambaut and Bromham 1998; Cooper *et al.* 2001) which allows for rate heterogeneity in different parts of the ML tree during branch length estimation. In this case rates were allowed to vary between deuterostome taxa and the other metazoans. Trees were constrained to a mammal–bird divergence time of either 288 Ma (Lee 1999) or 310 Ma for comparative purposes, although estimated divergence margins were very similar for both dates and so only the 310 Ma results are shown as ranges in Figure 3.2. Parameter values are shown in Table 3.1. Although these results are based on a very small sample, they

Table 3.1 Parameter values

Genes	Sites[a]	Invariant sites[b]	Alpha value[c]	Tree topology[d]	-LnL[e]	SH test[f]
Individual						
NDK	438	0.073	0.682	A	2670.18	Best
				B	2673.50	0.779
PGK	1182	0.174	1.721	A	7197.49	Best
				B	7204.72	0.276
TPI	738	0.195	0.679	A	4678.84	0.258
				B	4674.50	0.158
RFC	916	0.248	1.369	A	5667.39	0.728
				B	5670.23	0.503
Concatenated						
1st position	1091	0.137	2.694	A	5903.45	Best
				B	5914.17	0.306
2nd position	1091	0.302	∞	A	5109.49	Best
				B	5112.46	0.739
1st + 2nd position	2182	0.290	∞	A	11201.23	Best
				B	11218.86	0.190
All positions	3271	0.148	2.146	A	21946.36	Best
				B	21961.87	0.209

[a] Number of nucleotide sites.
[b] Fraction of those sites that are invariant.
[c] alpha value for the discrete gamma distribution. Both [b] and [c] were estimated from the data.
[d] Tree topologies as shown in Figure 3.2A and B.
[e] Likelihood score for each of the two topologies.
[f] Results of an SH test (Shimodaira and Hasegawa 1999) which compares the likelihood score with that of the ML tree for each dataset. High values denote likelihoods close to the ML value and *vice versa*. 'Best' indicates an ML tree. No tree was found to be significantly worse than the ML tree at 10 per cent.

fall in the middle of the range of previous estimates (Figure 3.3), with an estimated divergence for these genes falling between 687 and 914 Ma, at least 100 myr prior to the Cambrian 'explosion'. It is expected that larger studies incorporating more genes, taxa, and fossil constraints will greatly improve on these estimates.

Methodology appropriate for test cases

From the previous discussions it is apparent that there are still many problems in estimating divergence times. In terms of methodology currently available the following paragraphs summarize the combination of methods that may lead to the most realistic results – which are not necessarily those with the least range of uncertainty. It is inevitable that the levels of uncertainty will increase when divergences as old as the Cambrian 'explosion' are the subject of investigation.

Calibration

Fossil dates should be used to provide lower and upper boundaries for divergence times rather than as single calibration points (Sanderson 1997; Cutler 2000).

Number and type of loci

A range of single copy nuclear genes should be used in preference to 18S or mitochondrial genes when estimating deep divergences (as in Nei *et al.* 2001). Additionally, sequences should be lengthy. While being unsaturated at the level of divergence under scrutiny, an ideal sequence would contain as many informative sites as possible (at least 2 kb worth) to improve the power of relative rates tests (Bromham *et al.* 2000) and the resolution of the tree (Philippe *et al.* 1994).

Branch lengths and phylogeny estimation

ML methods are preferable to pairwise distance methods (Sanderson 1997; Rambaut and Bromham 1998; Cutler 2000).

Model choice

Models that best fit the dataset (using LRT) should be used. These can be selected using model selection programs such as Posada and Crandall (2001). Tests for data skew should be performed to determine whether to apply appropriate corrections. If using a divergence estimation method that does not allow rates to vary within the tree, genes should be scrutinized using saturation plots, relative rates tests, and LRT.

Phylogenetic uncertainty

In case of phylogenetic uncertainty, divergence estimates and their confidence intervals should be calculated for not only the 'best' tree but for all those not significantly worse-fitting in order to get a total interval estimate for all likely phylogenies (as in Cutler 2000).

References

Adachi, J. and Hasegawa, M. (1995) 'Improved dating of the human/chimpanzee separation in the separation in the mitochondrial DNA tree: heterogeneity among amino acid sites', *Journal of Molecular Evolution*, 40: 622–8.

Adoutte, A., Balavoine, G., Lartillot, N., Lespinet, O., Prud'homme, B. and de Rosa, R. (2000) 'The new animal phylogeny: Reliability and implications', *Proceedings of the National Academy of Sciences, USA*, 97: 4453–6.

Aguinaldo, A.M.A., Turbeville, J.M., Linford, L.S., Rivera, M.C., Garey, J.R., Raff, R.A. and Lake, J.A. (1997) 'Evidence for a clade of nematodes, arthropods and other moulting animals', *Nature*, 387: 489–93.

Ahlberg, P.E. (ed.) (2001) *Major events in early vertebrate evolution*, London: Taylor & Francis.

Ahlberg, P.E. and Johansen, Z. (1998) 'Osteolepiforms and the ancestry of tetrapods', *Nature*, 395: 792–4.

Alroy, J. (1999) 'The fossil record of North American mammals: evidence for a Paleocene evolutionary radiation', *Systematic Biology*, 48: 107–18.

Altschul, S.F., Madden, T.L., Schäffer, A.A., Zhang, J., Zhang, Z., Miller, W. and Lipman, D.J. (1997) 'Gapped BLAST and PSI-BLAST: a new generation of protein database search programs', *Nucleic Acids Research*, 26: 3389–402.

Archibald, J.D., Averianov, A.O. and Ekdale, E.G. (2001) 'Late Cretaceous relatives of rabbits, rodents, and other extant eutherian mammals', *Nature*, 414: 62–5.

Ayala, F.J. (1999) 'Molecular clock mirages', *BioEssays*, 21: 71–5.

Ayala, F.J., Rzhetsky, A. and Ayala, F.J. (1998) 'Origin of the metazoan phyla: Molecular clocks confirm paleontological estimates', *Proceedings of the National Academy of Sciences, USA*, 95: 606–12.

Benton, M.J., Wills, M.A. and Hitchin, R. (2000) 'Quality of the fossil record through time', *Nature*, 403: 534–7.

Bergstrom, J. and Hou, X.-G. (1998) 'Chengjiang arthropods and their bearing on early arthropod evolution', in G.E. Edgecombe (ed.) *Arthropod Fossils and Phylogeny*, New York: Columbia University Press, pp. 185–232.

Bromham, L., Rambaut, A., Fortey, R., Cooper, A. and Penny, D. (1998) 'Testing the Cambrian explosion hypothesis by using a molecular dating technique', *Proceedings of the National Academy of Sciences, USA*, 95: 12386–9.

Bromham, L., Penny, D., Rambaut, A. and Hendy, M.D. (2000) 'The power of relative rate tests depends on the data', *Journal of Molecular Evolution*, 50: 296–301.

Brown, R.H., Richardson, M., Boulter, D., Ramshaw, J.A.M. and Jefferies, R.P.S. (1972) 'The amino acid sequence of cytochrome c from *Helix aspera* Müeller (Garden Snail)', *Biochemical Journal*, 128: 971–4.

Budd, G.E. and Jensen, S.A. (2000) 'A critical reappraisal of the fossil record of the bilaterian phyla', *Biological Reviews*, 75: 253–95.

Cao, Y., Adachi, J. and Hasegawa, M. (1998) 'Comment on the Quartet Puzzling Method for finding maximum-likelihood tree topologies', *Molecular Biology and Evolution*, 15: 87–9.

Conway Morris, S. (2000) 'The Cambrian "explosion": Slow-fuse or megatonnage?' *Proceedings of the National Academy of Sciences, USA*, 97: 4426–9.

Cooper, A. and Fortey, R. (1998) 'Evolutionary explosions and the phylogenetic fuse', *Trends in Ecology and Evolution*, 13: 151–6.

Cooper, A. and Penny, D. (1997) 'Mass survival of birds across the Cretaceous–Tertiary boundary: molecular evidence', *Science*, 275: 1109–13.

Cooper, A., Grassly, N. and Rambaut, A. (2001) 'Using molecular data to estimate divergence times', in D.E.G. Briggs and P.R. Crowther (eds) *Palaeobiology II*, Oxford: Blackwell Science, pp. 532–4.

Cutler, D.J. (2000) 'Estimating divergence times in the presence of an overdispersed molecular clock', *Molecular Biology and Evolution*, 17: 1647–60.

Dayhoff, M.O. (1978) *Atlas of Protein Sequence and Structure*, Washington DC: National Biomedical Research Foundation.

Doolittle, R.F., Feng, D.F., Tsang, S., Chao, G. and Little, E. (1996) 'Determining divergence times of the major kingdoms of living organisms with a protein clock', *Science*, 271: 470–7.

Edgecombe, G.E. (ed.) (1998) *Arthropod Fossils and Phylogeny*, New York: Columbia University Press.

Erwin, D.H. (2001) 'Metazoan origins and early evolution', in D.E.G. Briggs and P.R. Crowther (eds) *Palaeobiology II*, Oxford: Blackwell Science, pp. 25–31.

Feduccia, A. (1995) 'Explosive evolution in Tertiary birds and mammals', *Science*, 267: 637–8.

Felsenstein, J. (1981) 'Evolutionary trees from DNA sequences: A maximum likelihood approach', *Journal of Molecular Evolution*, 17: 368–76.

Felsenstein, J. and Churchill, G.A. (1996) 'A hidden Markov model approach to variation among sites in rate of evolution', *Molecular Biology and Evolution*, 13: 93–104.

Feng, D.-F., Cho, G. and Doolittle, R.F. (1997) 'Determining divergence times with a protein clock: update and reevaluation', *Proceedings of the National Academy of Sciences, USA*, 94: 13028–33.

Field, K.G., Olsen, G.J., Lane, D.J., Giovannoni, S.J., Ghiselin, M.T., Raff, E.C., Pace, N.R. and Raff, R.A. (1988) 'Molecular phylogeny of the animal kingdom', *Science*, 239: 748–53.

Foote, M. and Sepkoski Jr, J.J. (1999) 'Absolute measures of the completeness of the fossil record', *Nature*, 398: 415–17.

Fortey, R.A., Briggs, D.E.G. and Wills, M.A. (1996) 'The Cambrian evolutionary 'explosion': Decoupling cladogenesis from morphological disparity', *Biological Journal of the Linnean Society*, 57: 13–33.

—— (1997) 'The Cambrian evolutionary 'explosion' recalibrated', *BioEssays*, 19: 429–34.

Foster, P.G. and Hickey, D.A. (1999) 'Compositional bias may affect both DNA-based and protein-based phylogenetic reconstructions', *Journal of Molecular Evolution*, 48: 284–90.

Friedrich, M. and Tautz, D. (1997) 'An episodic change of rDNA nucleotide substitution rate has occurred during the emergence of the insect order Diptera', *Molecular Biology and Evolution*, 14: 644–53.

Galtier, N. and Gouy, M. (1998) 'Inferring pattern and process: Maximum-likelihood implementation of a nonhomogeneous model of DNA sequence evolution for phylogenetic analysis', *Molecular Biology and Evolution*, 15: 871–9.

Gillespie, J.H. (1991) *The Causes of Molecular Evolution*, New York: Oxford University Press.

Giribet, G., Edgecombe, G.E. and Wheeler, W.C. (2001) 'Arthropod phylogeny based on eight molecular loci and morphology', *Nature*, 413: 157–61.

Goldman, N. and Yang, Z. (1994) 'A codon-based model of nucleotide substitution for protein coding sequences', *Molecular Biology and Evolution*, 11: 725–36.

Goldman, N., Thorne, J.L. and Jones, D.T. (1998) 'Assessing the impact of secondary structure and solvent accessibility on protein evolution', *Genetics*, 149: 445–58.

Gu, X. (1998) 'Early metazoan divergence was about 830 million years ago', *Journal of Molecular Evolution*, 47: 369–71.

Gu, X., Fu, Y.-X. and Li, W.-H. (1995) 'Maximum likelihood estimation of the heterogeneity of substitution rate among nucleotide sites', *Molecular Biology and Evolution*, 12: 546–57.

Hasegawa, M., Kishino, H. and Yano, T.-A. (1985) 'Dating of the human–ape splitting by a molecular clock of mitochondrial DNA', *Journal of Molecular Evolution*, 22: 160–74.

Hausdorf, B. (2000) 'Early evolution of the Bilateria', *Systematic Biology*, 49: 130–42.

Hedges, S.B., Parker, P.H., Sibley, C.G. and Kumar, S. (1996) 'Continental breakup and the ordinal diversification of birds and mammals', *Nature*, 381: 226–9.

Hickson, R.E., Simon, C., Cooper, A., Spicer, G.S., Sullivan, J. and Penny, D. (1996) 'Conserved sequence motifs, alignment, and secondary structure for the third domain of Animal 12S rRNA', *Molecular Biology and Evolution*, 13: 150–69.

Hillis, D.M., Moritz, C. and Mable, B. (eds) (1996) *Molecular Systematics*, Sunderland, Massachusetts: Sinauer Associates.

Hoffman, P.F., Kaufman, A.J., Halverson, G.P. and Schrag, D.P. (1998) 'A Neoproterozoic snowball Earth', *Science*, 281: 1342–6.

Hooker, J.J. (2001) 'Tarsals of the extinct insectivoran family Nyctitheriidae (Mammalia): evidence for archontan relationships', *Zoological Journal of the Linnean Society*, 132: 501–29.

Huelsenbeck, J.P., Larget, B. and Swofford, D.L. (2000) 'A compound Poisson process for relaxing the molecular clock', *Genetics*, 154: 1879–92.

Huelsenbeck, J.P., Ronquist, F., Nielsen, R. and Bollback, J.P. (2001) 'Bayesian inference of phylogeny and its impact on evolutionary biology', *Science*, 294: 2310–14.

Hwang, U.I., Friedrich, M., Tautz, D., Park, K.J. and Kim, W. (2001) 'Mitochondrial protein phylogeny joins myriapods and chelicerates', *Nature*, 413: 155–7.

Jablonski, D. and Bottjer, D.J. (1990) 'The origin and diversification of major groups: environmental patterns and macroevolutionary lags', in P.D. Taylor and G.P. Larwood (eds) *Major Evolutionary Radiations*, Oxford: Clarendon Press, pp. 17–58.

Jones, D.T., Taylor, W.R. and Thornton, J.M. (1992) 'The rapid generation of mutation data matrices from protein sequences', *Computer Applications in the Biosciences*, 8: 275–82.

Jukes, T.H. and Cantor, C.R. (1969) 'Evolution of protein molecules', in H.N. Munro (ed.) *Mammalian Protein Metabolism III*, New York: Academic Press, pp. 21–132.

Kimura, M. (1980) 'A simple method for estimating evolutionary rates of base substitutions through comparative studies of nucleotide sequences', *Journal of Molecular Evolution*, 16: 111–20.

Kirschvink, J.L., Ripperdan, R.L. and Evans, D.A. (1997) 'Evidence for a large-scale reorganisation of Early Cambrian continental masses by Inertial Interchange', *Science*, 277: 541–5.

Kluessendorf, J. and Doyle, P. (2000) '*Pohlsepia mazonensis*, an early 'octopus' from the Carboniferous of Illinois, USA', *Palaeontology*, 43: 919–26.

Kumar, S. (1996) 'Patterns of nucleotide substitution in mitochondrial protein coding genes of vertebrates', *Genetics*, 143: 537–48.

Kumar, S. and Hedges, S.B. (1998) 'A molecular timescale for vertebrate evolution', *Nature*, 392: 917–19.

Landing, E., Bowring, S.A., Davidek, K., Westrop, S.R., Geyer, G. and Heldmeir, W. (1998) 'Duration of the Early Cambrian: U-Pb ages of volcanic ashes from Avalon and Gondwana', *Canadian Journal of Earth Sciences*, 35: 329–38.

Lee, M.S.Y. (1999) 'Molecular clock calibrations and metazoan divergence dates', *Journal of Molecular Evolution*, 49: 385–91.

Lewis, P.O. (2001) 'Phylogenetic systematics turns over a new leaf', *Trends in Ecology and Evolution*, 16: 30–7.

Lió, P. and Goldman, N. (2002) 'Modeling mitochondrial protein evolution using structural information', *Journal of Molecular Evolution*, 54: 519–29.

Lockhart, P.J., Steel, M.A., Hendy, M.D. and Penny, D. (1994) 'Recovering evolutionary trees under a more realistic model of sequence evolution', *Molecular Biology and Evolution*, 11: 605–12.

Lynch, M. (1999) 'The age and relationships of the major animal phyla', *Evolution*, 53: 319–25.

Marshall, C.R. (1990) 'The fossil record and estimating divergence times between lineages: Maximum divergence times and the importance of reliable phylogenies', *Journal of Molecular Evolution*, 30: 400–8.

—— (1997) 'Confidence intervals on stratigraphic ranges with nonrandom distributions of fossil horizons', *Paleobiology*, 23: 165–73.

Martin, A.P. and Palumbi, S.R. (1993) 'Body size, metabolic rate, generation time, and the molecular clock', *Proceedings of the National Academy of Sciences, USA*, 90: 4087–91.

Muller, T. and Vingron, M. (2000) 'Modeling amino acid replacement', *Journal of Computational Biology*, 7: 761–76.

Muse, S.V. (1995) 'Evolutionary analyses of DNA sequences subject to constraints on secondary structure', *Genetics*, 139: 1429–39.

Muse, S.V. and Gaut, B.S. (1994) 'A likelihood approach for comparing synonymous and nonsynonymous nucleotide substitution rates, with applications to the chloroplast genome', *Molecular Biology and Evolution*, 11: 715–24.

Nei, M., Xu, P. and Glazko, G. (2001) 'Estimation of divergence times from multiprotein sequences for a few mammalian species and several distantly related organisms', *Proceedings of the National Academy of Sciences, USA*, 98: 2497–502.

Nichols, R. (2001) 'Gene trees and species trees are not the same', *Trends in Ecology and Evolution*, 16: 358–64.

Page, R.D.M. (2000) 'Extracting species trees from complex gene trees: Reconciled trees and vertebrate phylogeny', *Molecular Phylogenetics and Evolution*, 14: 89–106.

Pedersen, A.M.K., Wiuf, C. and Christiansen, F.B. (1998) 'A codon-based model designed to describe lentiviral evolution', *Molecular Biology and Evolution*, 15: 1069–81.

Penny, D., Hasegawa, M., Waddell, P.J. and Hendy, M.D. (1999) 'Mammalian evolution: Timing and implication from using the LogDeterminant transform for protein of differing amino acid composition', *Systematic Biology*, 48: 76–93.

Philippe, H. and Adoutte, A. (1994) 'What can phylogenetic patterns tell us about the evolutionary processes generating biodiversity?', in M. Hochberg, J. Clobert and R. Barbault (eds) *Aspects of the Genesis and Maintenance of Biological Diversity*, Oxford: Oxford University Press, pp. 41–59.

—— (1998) 'The molecular phylogeny of Eukaryota: solid facts and uncertainties', in G. Coombs, K. Vickerman, M. Sleigh and A. Warren (eds) *Evolutionary Relationships among Protozoa*, London: Chapman & Hall, pp. 25–56.

Philippe, H. and Germot, A. (2000) 'Phylogeny of eukaryotes based on ribosomal RNA: Long-branch attraction and models of sequence evolution', *Molecular Biology and Evolution*, 17: 830–4.

Philippe, H., Chenuil, A. and Adoutte, A. (1994) 'Can the Cambrian explosion be inferred through molecular phylogeny?' *Development*, 120: 15–25.

Phillips, M.J., Lin, Y.-H., Harrison, G.L. and Penny, D. (2001) 'Mitochondrial genomes of a bandicoot and a brushtail possum confirm the monophyly of australidelphian marsupials', *Proceedings of the Royal Society, London*, B268: 1533–8.

Posada, D. and Crandall, K.A. (2001) 'Selecting the best fit model of nucleotide substitution', *Systematic Biology*, 50: 580–601.

Rambaut, A. and Bromham, L. (1998) 'Estimating divergence dates from molecular sequences', *Molecular Biology and Evolution*, 15: 442–8.

Rickards, R.B. (2000) 'The age of the earliest clubmosses: the Silurian *Baragwanathia* flora in Victoria, Australia', *Geological Magazine*, 157: 207–9.

Rodríguez, F., Oliver, J.L., Marin, A. and Medina, J.R. (1990) 'The general stochastic model of nucleotide substitution', *Journal of Theoretical Biology*, 142: 485–501.

Rodríguez-Trelles, F., Tarrío, R. and Ayala, F.J. (2001) 'Erratic overdispersion of three molecular clocks: GPDH, SOD and XDH', *Proceedings of the National Academy of Sciences, USA*, 98: 11405–10.

Runnegar, B. (1982) 'A molecular-clock date for the origin of the animal phyla', *Lethaia*, 15: 199–205.

—— (1986) 'Molecular palaeontology', *Palaeontology*, 29: 1–24.

Sanderson, M.J. (1997) 'A nonparametric approach to estimating divergence times in the absence of rate constancy', *Molecular Biology and Evolution*, 14: 1218–31.

Sarich, V.M. and Wilson, A.C. (1967) 'Rates of albumin evolution in primates', *Proceedings of the National Academy of Sciences, USA*, 58: 142–8.

Shimodaira, H. and Hasegawa, M. (1999) 'Multiple comparisons of log-likelihoods with applications to phylogenetic inference', *Molecular Biology and Evolution*, 16: 1114–16.

Shoemaker, J.S., Painter, I.S. and Weir, B.S. (1999) 'Bayesian statistics in genetics. A guide for the uninitiated', *Trends in Genetics*, 15: 354–8.

Shu, D.-G., Luo, H.-L., Conway Morris, S., Zhang, X.-L., Hu, S.-X., Chen, L., Han, J., Zhu, M., Li, Y. and Chen, L.-Z. (1999) 'Lower Cambrian vertebrates from South China', *Nature*, 402: 42–6.

Siveter, D.J., Williams, M. and Walossek, D. (2001) 'A phosphatocopid crustacean with appendages from the Lower Cambrian', *Science*, 293: 479–80.

Smith, A.B. and Peterson, K.J. (2002) 'Dating the time of origin of major clades: molecular clocks and the fossil record', *Annual Review of Earth and Planetary Science*, 30: 65–88.

Smith, A.B., Gale, A.S. and Monks, N.E.A. (2001) 'Sea level change and rock record bias in the Cretaceous: a problem for extinction and biodiversity studies', *Paleobiology*, 27: 241–53.

Sole, R.V., Manrubia, S.C. and Benton, M. (1999) 'Criticality and scaling in evolutionary ecology', *Trends in Ecology and Evolution*, 14: 156–60.

Steel, M., Huson, D. and Lockhart, P.J. (2000) 'Invariable sites model and their use in phylogeny reconstruction', *Systematic Biology*, 49: 225–32.

Swofford, D.L. (1998) *PAUP*4.0 – Phylogenetic Analysis Using Parsimony (*and other methods)*, Sunderland, Massachusetts: Sinauer Associates.

Takezaki, N., Rzhetsky, A. and Nei, M. (1995) 'Phylogenetic test of the molecular clock and linearized trees', *Molecular Biology and Evolution*, 12: 823–33.

Tarrío, R., Rodríguez-Trelles, F. and Ayala, F.J. (2001) 'Shared nucleotide composition biases among species and their impact on phylogenetic reconstructions of the Drosophilidae', *Molecular Biology and Evolution*, 18: 1464–73.

Tavaré, S., Marshall, C.R., Will, O., Soligo, C. and Martin, R.D. (2002) 'Using the fossil record to estimate the age of the last common ancestor of extant primates', *Nature*, 416: 726–9.

Thorne, J.L. (2000) 'Models of protein sequence evolution and their applications', *Current Opinion in Genetics and Development*, 10: 602–5.

Thorne, J.L., Goldman, N. and Jones, D.T. (1996) 'Combining protein evolution and secondary structure', *Molecular Biology and Evolution*, 13: 666–73.

Thorne, J.L., Kishino, H. and Painter, I.S. (1998) 'Estimating the rate of evolution of the rate of molecular evolution', *Molecular Biology and Evolution*, 15: 1647–57.

Tillier, E.R.M. and Collins, R.A. (1995) 'Neighbouring joining and maximum likelihood with RNA sequences: addressing the interdependence of sites', *Molecular Biology and Evolution*, 12: 7–15.

Wang, D.Y.-C., Kumar, S. and Hedges, S.B. (1998) 'Divergence time estimates for the early history of animal phyla and the origin of plants, animals and fungi', *Proceedings of the Royal Society, London*, B266: 163–71.

Wheeler, W.C. (1998) 'Sampling, groundplans, total evidence and the systematics of arthropods', in R.A. Fortey and R.H. Thomas (eds) *Arthropod Relationships*, Amsterdam: Kluwer, pp. 87–96.

Wills, M.A. and Fortey, R.A. (2000) 'The shape of life: how much written in stone?' *BioEssays*, 22: 1142–52.

Wray, G.A., Levinton, J.S. and Shapiro, L.H. (1996) 'Molecular evidence for deep Precambrian divergences among metazoan phyla', *Science*, 274: 568–72.

Yang, Z. (1996) 'Among-site variation and its impact on phylogenetic analyses', *Trends in Ecology and Evolution*, 11: 367–72.

Yang, Z. and Roberts, D. (1995) 'On the use of nucleic acid sequences to infer early branchings in the tree of life', *Molecular Biology and Evolution*, 12: 451–8.

Yoder, A.D. and Yang, Z. (2000) 'Estimation of primate speciation dates using local molecular clocks', *Molecular Biology and Evolution*, 17: 1081–90.

Zhuravlev, A.Y. and Riding, R. (eds) (2001) *The Ecology of the Cambrian Radiation*, New York: Columbia University Press.

Zuckerkandl, E. and Pauling, L. (1965) 'Evolutionary divergence and convergence in proteins', in V. Bryson and H. J. Vogel (eds) *Evolving Genes and Proteins*, New York: Academic Press, pp. 97–166.

Chapter 4

The quality of the fossil record

Michael J. Benton

ABSTRACT

Ever since the days of Charles Darwin, palaeontologists have been concerned about the quality of the fossil record. New concerns have arisen from two themes: (1) the finding that molecular dates of origin of certain major clades are often twice as old as the oldest fossils, and (2) the discovery that much of the variation in diversity, origination, and extinction signals from the fossil record can be explained by sampling. The molecular age-doubling phenomenon may be real, or it could be explained by either major gaps in the fossil record or by the inability of molecular techniques to discount unequivocally the possibility of rapid clock rates during times of divergence. The rock record certainly controls much of the fine detail of diversity and extinction plots, but mass extinctions, and the overall rise in diversity through time, may be real. Comparison of molecular and morphological phylogenies with the order and spacing of events in the rock record shows congruence, and hence suggests that much of the biotic signal in the fossil record is not misleading.

Introduction

The quality of the fossil record is a focal issue in current debates about the timing of origins of major groups. Some molecular estimates place the origins of Metazoa (animal phyla), green plants, angiosperms, and modern orders of birds and mammals at points up to twice as old as the oldest representative fossils (e.g. Hedges *et al.* 1996; Wray *et al.* 1996; Cooper and Penny 1997; Kumar and Hedges 1998; Heckman *et al.* 2001; Nei *et al.* 2001; van Tuinen and Hedges 2001; Wray 2001). The range of molecular estimates for the origin of metazoans is 600–1200 Ma (million years ago), with most estimates closer to 1000 Ma than 600 Ma. The range of molecular estimates for the origin and basal splitting of placental mammals, and of modern birds, is 130–70 Ma, again with more estimates nearer 120 Ma than 70 Ma. The first fossils date, respectively, from around 600 and 70 Ma.

It is unclear how widespread this substantial mismatch of age estimates is. Bleiweiss (1999) has already alluded to this when he linked the molecular age-doubling phenomenon in these broad-scale examples to an identical finding for species of birds in the Quaternary (Klicka and Zink 1997). However, most estimates of a ages from molecules match the fossil ages closely (Hedges and Kumar 2003). Does the mismatch happen only for origins of major clades or is it more widespread?

Palaeontologists have long debated the quality of the fossil record. Since the time of Darwin, and before, a serious theme has been 'the incompleteness of the fossil record', and authors have repeatedly emphasized the obvious fact that only a tiny fraction of those organisms that have ever lived are preserved as fossils, and only a tiny fraction of those fossils will ever become objects of scientific scrutiny (Raup 1972; reviewed in Donovan and Paul 1998). Currently, palaeontologists fall into two camps, those who are content that the fossil record is adequate to show the broad outlines of the history of life (e.g. Sepkoski *et al.* 1981; Benton 1995, 1999a,b; Foote 1997; Miller 1998; Benton *et al.* 2000), and those who believe that sampling problems overwhelm the signal in rocks older than perhaps 20 or 30 Ma (e.g. Alroy *et al.* 2001; Smith 2001; Peters and Foote 2001, 2002). For example, Smith (2001; Smith and Peterson 2002) and Peters and Foote (2002) have demonstrated that diversity signals from the fossil record vary with the amount of exposed rock and with sea level change, and hence may not contain much of an original biological signal.

The purpose of this chapter is to consider approaches to estimating sampling in the fossil record, and whether they can assess quality. The apparent mismatch of fossil and molecular dates for the radiation of major clades is explored and, for some groups at least, a rapprochement appears to have taken place. The proposition that the fossil record contains more of a sampling than a biotic signal will be considered. Finally, it will be suggested that the fossil record is adequate, and robust in the face of sampling problems, at certain scales, based on semi-independent phylogenetic investigations.

Molecular age doubling and error

If the molecular age-doubling phenomenon is real, there are two possible explanations for the apparent mismatch between molecular and fossil dates for the origin of major clades. One is based on the assumption that the fossil record is adequate, the other that it is not. If it is assumed that the fossil record is good enough, then the first half of the history of many (most?) major clades has evidently been cryptic. The organisms remained small, soft-bodied, or restricted geographically for a long span of time, before they finally flourished and became detectable as fossils. This is the 'phylogenetic fuse' idea of Cooper and Fortey (1998). A simpler assumption is that the fossil record is inadequate (early fossils too small or delicate to be preserved, appropriate rocks absent or in parts of the world that have not been sampled). This has been the general view of many molecular analysts (e.g. Hedges *et al.* 1996; Wray *et al.* 1996; Cooper and Penny 1997; Kumar and Hedges 1998; Easteal 1999; Heckman *et al.* 2001; Nei *et al.* 2001; van Tuinen and Hedges 2001; Wray 2001; Hedges, Chapter 2).

If the molecular age-doubling phenomenon is questionable, then one has to consider the methods of assigning dates in molecular phylogenetic analysis. Fossil and molecular evidence are both subject to error. Where fossils are correctly identified, palaeontological evidence will always underestimate the maximum age of a clade, whereas molecular evidence can both over- and underestimate ages. Several authors (e.g. Vermeij 1996; Benton 1999b; Foote *et al.* 1999; Lee 1999; Conway Morris 2000) have stressed the propensity of molecular methods to overestimate the timing of origin of major clades since they do not fully take account of the possibility that molecular rates speed up enormously during times of major diversification ('adaptive radiation'). Extrapolating with a constant-rates model over a time of enhanced rates means that the point of origin is projected too far back in time.

Current molecular clock techniques take account of rate variation across the tree (e.g. maximum likelihood techniques, the quartet method, and the use of multiple calibration points), but it is not clear that they can yet assess the validity of molecular age doubling at the radiation of major clades.

(1) Maximum likelihood techniques (Cavalli-Sforza and Edwards 1967; Huelsenbeck and Rannala 1997; Whelan *et al.* 2001) may be used to calculate differing rates of evolution in extant lineages, but there are many available models, and arbitrary choices among possible models have to be made (Siddall and Whiting 1999). Differing rates between lineages within a clade may be detected, but a decisive test between a model that posits explosive diversification of all lineages in a clade at one time, and a model that does not include such a dramatic change of rate, cannot be made (Huelsenbeck and Rannala 1997).

(2) The quartet method (Rambaut and Bromham 1998) compares subsets of four taxa extracted from a tree. Those quartets that show significant rate variation between the two pairs of taxa are rejected. Surviving quartets then provide numerous estimates of the date of a common basal node. The method can then allow analysts to calculate the amount of error associated with such a basal date, but it cannot take account of a situation where molecular rates were *all* uniformly faster during a time of explosive diversification (Smith and Peterson 2002).

(3) Multiple calibration points allowed Springer (1997) to detect changes in rates of molecular evolution across the tree of placental mammals. If the confirmed dates, based on fossil data, are scattered densely enough across a molecular phylogeny, they can provide constraints on other, undated branching points. However, extrapolating from multiple dates high in a tree to determine dates of branching low in a tree still does not address the possibility of a uniform explosive rate of evolution early in a diversification event. The difficulty in establishing calibration points that tightly bracket a clade radiation is illustrated by Paton *et al.* (2002). To date the origin of modern birds, they use one distant low date, the split of the bird and crocodile line in the Triassic (245 Ma), one distant high date, the split of the emu and cassowary (35 Ma, based on the oldest emu fossil of 25 Ma), and one 'close' date, the Galloanserae divergence (85 Ma). These three dates give a range of estimates, from 108–155 Ma, for the radiation of modern birds. Each of the three reference dates can be criticized, the first for being too distant, the second for being based on an arbitrary addition of 10 Ma to a known fossil date, and the third for being itself a molecular estimate that might involve similar error to the date being assessed.

Hence, it is frustratingly clear that none of these approaches can test unequivocally whether or not certain past diversification events were marked by rates of molecular change that speeded up dramatically for a short time across a whole clade. There are other reasons that molecular estimates tend to overestimate branching dates, just as palaeontological estimates underestimate dates (Benton and Ayala 2003).

Sampling methods

Two main approaches have been used by palaeontologists in assessing the completeness of their fossil records: confidence intervals and group sampling. Both methods may suffer from circularity in reasoning – if the input sampling distributions are incomplete, the estimates too will be incomplete.

Confidence intervals

Estimation of confidence intervals is an intuitive approach. It is based on the assumption that, if fossils are known from many geological horizons within a known stratigraphic range, then it is likely that very much older (or younger) fossils will not be found. Potential range extensions, at a particular probability level, will be small. If, on the other hand, fossils are scattered sparsely through a known range, it is likely that unknown fossils may occur far below the known oldest fossil (or far above the known youngest fossil, for an extinct group). The method of gap analysis is a statistical expression of this intuitive assumption, where the probability, P, is the confidence level (say 0.95) that a hypothetical interval added to a known range will include the true stratigraphic range,

$$P = 1 - (1 + a)^{-(n-1)}$$

where a is the confidence interval expressed as a fraction of the observed stratigraphic range, and n is the number of known fossiliferous horizons (Strauss and Sadler 1989). The method was developed to deal with local rock sections, but it may be applied to global examples of this kind (Marshall 1990), providing that the distribution of known fossiliferous horizons within the overall range is random and independent. If the distribution of potentially fossiliferous rocks is not random and independent, then the appropriate statistical tests, generalized confidence intervals (Marshall 1997), must be applied, but these are statistically much less powerful.

In an example of this approach, Bleiweiss (1998) looked at the fossil records of three bird groups, the Strigiformes (owls), Caprimulgiformes (goatsuckers), and Apodiformes (swifts, hummingbirds), and documented all known fossils in each order. His purpose was to compare fossil and molecular evidence for the origin of those orders. The oldest fossils are dated at 58–54.5 Ma for each of the groups, definitively within the Tertiary, whereas molecular estimates (Hedges et al. 1996; Cooper and Penny 1997) placed modern bird ordinal origins at 80–100 Ma, well down in the Cretaceous. Bleiweiss (1998) found that fossils in each of the orders have been reported from some 20–30 separate horizons from the date of these oldest fossils to the present day (Figure 4.1). He tested for randomness, and calculated that the maximum possible range extensions, based on the known fossil records of the owls, goatsuckers, and swifts, would hardly even take these orders into the latest Cretaceous, let alone the mid-Cretaceous. The 95 per cent confidence intervals estimated for the base of each of the three ranges were 62 Ma for swifts, 67 Ma for goatsuckers, and 63 Ma for owls. When all three groups were combined, producing a more densely sampled composite record, the range extension, at 95 per cent confidence, was back to only 61 Ma.

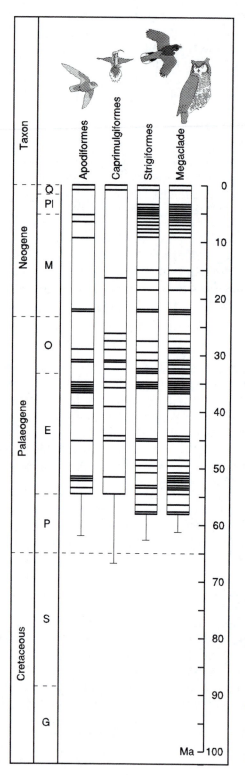

Figure 4.1 The known fossil record of modern bird groups gives no indication that fossil finds can be expected in the Cretaceous. Based on the distribution in time of known fossils of three modern bird orders, the apodiforms (swifts and hummingbirds), caprimulgiforms (goatsuckers), and strigiforms (owls), confidence intervals (95 per cent) are confined mainly to the Palaeogene (early Tertiary), with just one barely creeping into the latest Cretaceous. The 'megaclade' consists of all three orders summed together. The levels of known finds are indicated by horizontal bars (which may represent one find or many dozens of the same age). The more closely spaced the known finds, the shorter the confidence interval. Abbreviations: G, Gallic; S, Senonian; P, Palaeocene; E, Eocene; O, Oligocene; M, Miocene; Pl, Pliocene; Q, Quaternary. Based on data in Bleiweiss (1998).

Corroborating evidence (Bleiweiss 1999) comes from morphological studies that suggest rapid divergence of modern bird orders. Earliest Tertiary forms may be assigned to modern orders, but they generally have only one or two synapomorphies of those orders (Dyke 2001), and the full suite of distinguishing characters of the modern orders was acquired piecemeal through the early Tertiary. Indeed some of the earliest modern bird fossils have traditionally been very hard to assign to orders (Feduccia 1996). Had the orders been divergent for 40 myr or so before the first fossils occur, they might be expected to have accumulated much more character difference by the beginning of the Tertiary.

In a response to the study by Bleiweiss (1998), Marshall (1999) suggested that the confidence intervals method falls down if the global abundance and diversity of a group varies through time. Hence, he suggests, if the three bird orders in question had a long early history when they were rare, then the confidence intervals would expand. Using his generalized confidence interval approach, Marshall (1999) found that the 95 per cent confidence intervals estimated by Bleiweiss as 4–10 myr grew to 40 myr. This insight, that early parts of clades are less likely to produce fossils than later parts of clades, provided the basis for a model used by Tavaré et al. (2002) to estimate true points of origin for the primates from the fossil record. Using a logistic diversification model, and taking information from the modern diversity of a group, mean species duration, and the distribution of known fossil finds, the calculated point of origin of a group then falls well before the first fossil. How far before depends on the input relative sampling intensity (the lower the value, the longer the range extension) which depends on the shape of the logistic curve and on assumptions about early members of a clade being small, rare, and geographically restricted.

Corroborating evidence in favour of this view is that most molecular evidence suggests long internodes between at least some basal bird groups (e.g. Cooper and Penny 1997; Harlid and Arnason 1999), although Bleiweiss et al. (1994, 1995) found the opposite, based on DNA hybridization studies. Also, the dataset used by Bleiweiss (1998) was geographically biased, consisting of 71 sites in western Europe, and only 23 from elsewhere. The bias was inevitable since this represents the current knowledge of fossils of the bird groups in question. But the critic can claim that Bleiweiss (1998) has produced predictions only for future finds in Europe, where there is a demonstrable bias in the rock record (Smith and Peterson 2002), even though he used the best currently available data.

If Bleiweiss (1998) is right, then the gap analysis approach has predicted rather short 95 per cent confidence intervals, and that the fossil record is adequate. On the other hand, if Marshall (1999) is right, then the potential range extension for the bird orders is very large, and completely in keeping with the molecular estimates, and the fossil record of birds is evidently not adequate to show major features of their early history. The method of Tavaré et al. (2002) would presumably produce a figure somewhere between the two extremes. Both viewpoints contain assumptions that are hard to test. The best evidence in favour of Marshall's (1999) view would be finds of definitively mid- and Late Cretaceous birds of modern type. There is no equivalent test of the view expressed by Bleiweiss (1998), since the absence of such finds supports his view, but of course 'absence of evidence is not evidence of absence'. This imbalance in the possibility of confirming evidence was noted by Benton (1999b): one fossil find could confirm the 'early origins' view, but there is no such simple confirmation of the 'late origins' view.

Group sampling

Foote and Raup (1996) developed a simple method to derive an empirical estimate of sampling at group level, which they termed FreqRat. This depends on a knowledge of the distribution of frequencies of species or genera of particular durations within a larger clade, and follows the formula:

$$R = \frac{[f(2)^2]}{[f(1)f(3)]}$$

where R is the probability that a taxon will be preserved at least once in a time unit, and f(1), f(2), and f(3) are the recorded frequencies of taxa spanning one, two, and three equal-length intervals, respectively. This is a simplification of a much more complex set of equations that take account of relative extinction probabilities of different taxa, distributions of occurrences within ranges, and other factors, but empirically the relationship works for exponential ('hollow curve') distributions, where there are relatively large numbers of taxa with short ranges, and rapidly falling numbers of taxa with longer durations. Foote (1997) developed the method further for continuous (rather than discrete) ranges, and for situations where there might be a sample-size bias, but the FreqRat formula is a good approximation for most typical cases.

Foote and Raup (1996) found values in the range of 60–90 per cent for the completeness of different groups – the proportions of species of trilobites, bivalves, and mammals, and the proportions of genera of crinoids preserved. They confirmed that incompleteness of these readily fossilizable groups was a result of the loss of fossiliferous rock rather than the failure of species to enter the fossil record in the first place. Foote and Sepkoski (1999) presented a wider array of estimates of the probability of preservation of genera of different animal groups, ranging from 5 per cent for polychaete worms, to 40–50 per cent for sponges, corals, crinoids, gastropods, bivalves, and ostracods, to essentially 100 per cent for brachiopods.

Foote et al. (1999) applied their technique to the fossil record of mammals in North America, to assess whether molecular estimates for the origin of the orders (130–70 Ma) were more or less likely than fossil estimates (oldest fossils, 70–50 Ma). They modelled typical patterns of branching evolution, and then applied imaginary filters to cut out species. In other words, they decreased the value of R, the preservation probability, until all fossils disappeared over a set span of time, the situation implied by the molecular age-doubling hypothesis for the initial radiation of modern mammals. The preservation probability of North American Cenozoic mammal species is 0.25 per 0.7 Ma interval, corresponding to a completeness of 58 per cent (Foote and Raup 1996), whereas values predicted for the complete or virtual absence of modern mammals in the mid- to Late Cretaceous are two orders of magnitude lower, a level that Foote et al. (1999) find to be lower than any other calculated preservation probabilities for any taxa, and hence most unlikely.

Is this a valid test? Smith and Peterson (2002) argue that there is a major flaw, that Foote et al. (1999) were mistaken to calculate preservation probabilities from a sampling of the fossil record that was overwhelmingly dominated by the Campanian and Maastrichtian record of North America. Indeed, Foote et al. (1999) included only limited evidence about mammalian faunas from other parts of the world, and it would be worthwhile to repeat their experiment, but with fuller documentation. It is wrong to assume that all we know about continental vertebrates in the Late Cretaceous comes from North America. Sea levels were high worldwide, but continental units with fossil vertebrates are known (Weishampel 1990) from around the Mediterranean (Portugal, Spain, France, Romania, Middle East), from Asia (Uzbekistan, Tadzhikistan, Kazakhstan, Russia, India, Mongolia, China, Japan, Laos), from South America (Colombia, Bolivia, Peru, Brazil, Uruguay, Chile, Argentina), and from Africa (Morocco, Algeria, Egypt, Niger, Kenya, South Africa, Madagascar). Mammal fossils are known from all these areas, with spectacular examples from numerous horizons in the Late Cretaceous of Uzbekistan, Kazakhstan, Mongolia, China, Argentina, and Madagascar. Expanding the sampling worldwide might not in fact invalidate the findings by Foote *et al.* (1999).

Oldest fossils

The current literature about origins of major clades often includes discussion about the oldest relevant fossils. Does close scrutiny of the fossil evidence indicate rapprochement between molecular early dates and palaeontological late dates of origin? This does not seem to be the case for the earliest metazoans, green plants, angiosperms, or birds, but there is now good agreement for the basal radiation of mammals.

Mismatch

The first metazoan fossils are generally accepted to date from the earliest Cambrian, the great burst of expansion of skeletonized animal groups from about 545 Ma, long after the molecular date of around 1000 Ma. Fossil evidence for metazoans is known from the Precambrian, but nothing before about 600 Ma. This evidence includes possible sponges and cnidarians from the Ediacara faunas, exquisitely preserved fossil embryos, and simple creeping trails (Valentine *et al.* 1999; Conway Morris 2000).

The first vascular land plants are found as fossils in the Silurian, and earlier evidence from spores extends the range back to the Ordovician (475 Ma; Kenrick and Crane 1997), considerably younger than a molecular estimate of 700 Ma (Heckman *et al.* 2001). A similar gap exists for angiosperms, with the oldest generally accepted fossils being from the Early Cretaceous, pollen records dated at about 130 Ma, and abundant pollen and macroplant fossils from 120 Ma onwards (Crane *et al.* 1995). Older putative fossil angiosperms, from the Jurassic, and even from the Triassic, have not been generally accepted. DNA sequence evidence places the divergence of angiosperms in the Mid-Jurassic (175 Ma; Wikström *et al.* 2001). However, genealogical evidence actually suggests a much more ancient date of origin, back in the Carboniferous at 290 Ma (Kenrick 1999), if it turns out that the sister group of angiosperms is the gymnosperms.

In the case of the origin of modern birds, many supposed Cretaceous representatives of modern bird orders have been cited (e.g. Cooper and Penny 1997; van Tuinen *et al.* 2000; Paton *et al.* 2002), but most have been disputed, because the fossils are generally isolated elements (Dyke 2001; Chiappe and Dyke 2002). Hence, the oldest uncontroversial fossils of modern bird orders date from the latest Cretaceous or Palaeocene (70–60 Ma), much younger than most molecular estimates of origins, at 70–120 Ma.

Rapprochement

The dating of the radiation of modern placental (= eutherian) mammals seemed to be similarly fraught until a year or two ago. The traditional, palaeontological view (e.g. Carroll 1988; Benton 1990) was that placentals split from marsupials some time in the Early Cretaceous (144–99 Ma), and modern orders split at the end of the Cretaceous and in the early Tertiary (70–55 Ma). The first molecular dates (Hedges *et al.* 1996; Kumar and Hedges 1998; Easteal 1999) seemed much older: origin of eutherians in the Late Jurassic (c. 150 Ma), split of major placental groups in the Early Cretaceous (c. 110–120 Ma), and split of modern placental orders in the mid- to Late Cretaceous (c. 80–100 Ma). It seemed there was a major problem.

Since 1996, however, there has been a rapprochement, and palaeontological and molecular evidence now seem to agree. The change has happened because of a better understanding of what the fossils show, and because molecular age estimates have been revised upwards.

First, the oldest fossils of modern placental mammals are not entirely basal Tertiary (Archibald 1996; Archibald *et al.* 2001), but it is important to distinguish group membership. Modern placental mammals are divided into 18 orders, and these fall into four larger superorders, the Xenarthra (edentates), Afrotheria (elephants, hyraxes, sirenians, tenrecs, golden moles, and the aardvark), Euarchontoglires (rodents, rabbits, flying lemurs, tree shrews, primates), and Laurasiatheria (insectivores, bats, pangolins, carnivores, perissodactyls, artiodactyls, whales). An extraordinary Late Cretaceous locality in Uzbekistan, dated at 85–90 Ma, has yielded no representatives of modern placental *orders*, but has produced specimens that are assigned to basal parts of placental *superorders*, the Glires and Ungulatomorpha (part of Laurasiatheria) (Archibald *et al.* 2001). The recent discovery of a beautifully preserved basal placental mammal from the Early Cretaceous (c. 125 Ma) of China (Ji *et al.* 2002) could point either way. It could be a late survivor of a split that happened in the Jurassic, but it is just as likely that it confirms the split of placentals and marsupials in the Early Cretaceous.

Since 1997, the molecular estimates have been revised upwards (Hedges and Kumar 1999; Eizirik *et al.* 2001; Murphy *et al.* 2001; Springer *et al.* 2003). Superordinal diversification is dated at 64–104 Ma (mean 84 Ma), and ordinal diversification at 50–83 Ma, entirely in line with the fossils. Note that different probability modelling approaches either support the new consensus (Foote *et al.* 1999; Archibald and Deutschmann 2001), or dispute it by suggesting the possibility of much older ordinal divergence dates (Tavaré *et al.* 2002). Is the mammal example an indicator of possible rapprochement in the future over other disputed dates of origination?

Evidence for a poor-quality fossil record

Heterogeneity

It is clear that the fossil record contains a biotic and an abiotic signal. The distribution of fossils in the rocks consists of a combination of the record of the history of life and the vicissitudes of the history of the rocks. Most palaeontologists would like to believe that the history of life is a robust enough signal to stand out from the background smearing. Critics, however, believe that sampling overwhelms the biotic signal; the heterogeneity of the temporal and geographical distribution of rocks masks the real story.

Smith (2001) and Peters and Foote (2001, 2002) have argued that the distribution of sedimentary rocks controls the preservation of fossils and that much (?most) of the standard plots of diversifications and extinctions from the fossil record (e.g. Sepkoski 1984, 1996; Benton 1995) are artefacts. For example, in a study of the marine fossil record of the post-Palaeozoic, Smith (2001) found that outcrop area and sea level changes correlated with some aspects of diversity change, and Peters and Foote (2001, 2002) made the same observation for the whole of the Phanerozoic. Small-scale changes in diversity and in origination rate were related to the surface area of outcrop, and these authors stress that it would be foolhardy to interpret every rise and fall in the global diversity, extinction, and origination signals as biologically meaningful.

Mass extinctions represent a particular issue. Smith (2001) found that most peaks in extinction did not correspond to changes in outcrop area, but two occurred towards the culmination of stacked transgressive system tracts and close to system bases. One of these, falling at the Cenomanian–Turonian boundary, and representing a well-known postulated mass extinction in the sea (e.g. Raup and Sepkoski 1984; Hallam and Wignall 1997), may then be truly an artefact of sampling and sea level change (Smith *et al.* 2001). Peters and Foote (2001, 2002) found that all such 'lesser' global extinction events disappeared when the effect of sampling was taken into account. Most of the 'big five' mass extinctions also seem to be equivocal, or at least to be exaggerated by sampling. These are rather startling findings.

A criticism of the work by Smith (2001) and Peters and Foote (2001) could be that they use limited datasets on sampling. Smith (2001) used map areas of rocks of different age from Britain and France only for comparison with the global biodiversity signal, while Peters and Foote (2001) used a lexicon of numbers of named stratigraphic formations in North America. However, both studies, using such different samples of sampling, came to the same conclusion, and Peters and Foote (2002) have taken a broader sample of named marine stratigraphic units from around the world, and the results are the same.

Both Smith (2001; Smith *et al.* 2001) and Peters and Foote (2001, 2002) tested for the relative roles of abiotic and biotic factors, and they found that changes in rock surface area explained most of the variance. So, the results indicate that rock outcrop area drives the record of the diversification of life and that extinction events are largely artefacts of the appearance and disappearance of rock rather than of organisms. Alternatively, as both teams stressed, an additional factor, perhaps sea level change, could drive both signals, that marine rock area rises and falls as sea level rises and falls, and marine biodiversity expands and contracts as does the volume of the sea. Either way of course the fossil record is much weakened as an accurate document of the history of life.

The finding that short-term rises and falls in marine biodiversity mirror the rock record is convincing. Most palaeontologists already knew that most of the postulated Mesozoic 'mass extinctions' required by the hypothesis of periodicity in mass extinction (Raup and Sepkoski 1984), such as those in the Early and Mid-Jurassic, the Jurassic–Cretaceous extinction, and the Early Cretaceous event were artefactual to a greater or lesser extent (Hallam 1986; Benton 1995; Little and Benton 1995). Does any biotic signal survive?

The dramatic diversification of life in the sea, and the even more dramatic diversification of life on land over the past 250 Ma has been noted by many authors (Sepkoski 1984, 1996; Benton 1995, 1997, 2001). Smith (2001) found that the overall rise in marine generic and familial diversity through the Mesozoic and Cenozoic could not be explained by sea level change, and hence was probably real. Peters and Foote (2001) question this, however, suggesting that even the massive diversification of marine life in the past 250 Ma could be an artefact of low turnover and the pull of the Recent, confirming Raup's (1972) earlier suggestion that marine life diversified dramatically early in the Palaeozoic, and has remained at a constant level ever since (see below, 'Bias').

The studies so far have focused on life in shallow seas. An interesting further test would be to compare the results with the deep sea, where habitats are less heterogeneous and sedimentation is more continuous, and with continental settings, where habitats are diverse and sedimentation is often supposed to be even more sporadic than in shallow seas. Smith (2001) argued for a strong bias in the case of terrestrial tetrapods, and Smith and Peterson (2002) indeed show a strong correlation between the rock record in western Europe and the number of bird families recorded through time. On a broader scale, Peters and Foote (2001) noted a correlation of numbers of terrestrial animal families and terrestrial formations in North America. However, the correlation was weaker than for marine animal families and genera, contrary to expectations. In a more detailed study, Fara (2002) actually found no evidence for a correlation for continental tetrapods: as sea levels fell, and continental areas expanded, there is no evidence for a matching expansion in the diversity of land animals.

What about the 'big five' mass extinctions? Smith (2001) does not question the reality of the end-Permian and K–T events, although he notes an interaction of a biotic (rapid extinction) and abiotic (major sea level change and reduction in surface area of preserved onshore facies) signal that must be disentangled. In this case, correlation of the biotic and abiotic signals need not indicate that the first is an artefact of the second, but that both are part of the global cataclysms associated with times of mass extinction. Peters and Foote (2002) leave the issue of mass extinctions much more open. By their modelling approach, they argue that all of the big five mass extinctions *could* be explained as artefacts of sampling. The first two, in the Late Ordovician and Late Devonian, might be real, but only if global generic extinction rates are modelled as constant, rather than declining. The end-Permian mass extinction, the largest ever, does appear above the noise if extinction rates are modelled as declining, but it is swamped by sampling in a constant-extinction-rate model. The same is true of the end-Triassic mass extinction. Only the K–T mass extinction shines through in all models where sampling is accounted for. Peters and Foote (2002) say that this result could indicate one of two things, either that mass extinctions are merely artefacts of variations in available rock volume, or that mass extinctions and reductions in rock volume are associated with an additional common factor, such as major sea level change, as Smith (2001) suggested.

If the method can reject mass extinctions, perhaps there is a problem with the method. Peters and Foote (2002) meant to challenge conventional assumptions, and they are careful to outline potential pitfalls in their data and their models. The explanation of minor variations in the biotic signal as abiotic artefacts (Peters and Foote 2001, 2002; Smith 2001; Smith *et al.* 2001) makes a great deal of sense. However, there is so much geological, geochemical, and palaeontological evidence, in addition to the broad-brush diversity plots that are under scrutiny, for the end-Permian, end-Triassic, and K–T crises that these events can probably be accepted as real. Hence if a statistical method is capable of rejecting their reality, one has to look closely at the statistical method since it may be too crude.

This new work on heterogeneity gives mixed messages about scaling of time and taxa. Smith (2001) stressed that the key geological driver is at the level of major sequence stratigraphic cycles of 20–50 myr, not 1–10 myr. Peters and Foote (2001) showed a close linkage of the biotic and geological signals at the level of epochs (2–42 myr, mean 19 myr), but in their later paper (Peters and Foote 2002), the scaling was at stage level (2–20 myr, mean 7.1 myr). Is the proposal that scaling is fractal, and every biotic signal can be shown to follow a geological signal slavishly? Or can palaeontologists expect that observations on certain timescales may be free of geological control? Care is required in seeking fossil versus rock correlations: if genera are sampled, and the time bins are too broad, then each occurrence is effectively a single point, and it is then most likely that the number of fossils will be controlled by the rock area or volume, but the linkage would be largely an artefact of the method. Finer time divisions will allow true ranges of genera to be assessed. We have found (Fara and Benton 2000) that known gaps in the continental rock record of the Cretaceous are bridged by new discoveries on either side, if the taxonomic scale is appropriate to allow spanning (we chose to look at families and stratigraphic stages). So, at family level, the biotic signal is robust to global strati-graphic gaps. Had we chosen genera, then they could never have spanned the known gaps (stages, 2–13 myr, mean 6.4 myr), and then rock area could have been said to drive the biotic signal.

Sea level changes can clearly produce artificial extinctions, but could they hide a diversification for tens of millions of years, as postulated by the molecular age-doubling observation? This seems most unlikely because of timescale considerations. Known habitat shifts and hiatuses in the global rock record account for a few million years at most. Such heterogeneity could not delete 30 or 40 Ma of the his-tory of a group. So sea level changes can create false extinction events, but it is hard to see how they alone could hide real diversifications.

Bias

A related, but more extreme, view has been that a combination of factors render the fossil record poorer and poorer the further back in time one goes. Raup (1972) argued that the fossil record suffers from a number of biases, such as the evident loss in volume of rocks, and the exposed area of rocks, as a result of the cumulative effects of burial, metamorphism, subduction, erosion, and covering through geological time. In addition, fossils from ever older rocks belong to groups that are less and less like modern forms, and are hence harder and harder to identify to species level. Practically speaking also, relatively fewer palaeontologists work on more ancient rocks and fossils than more recent ones.

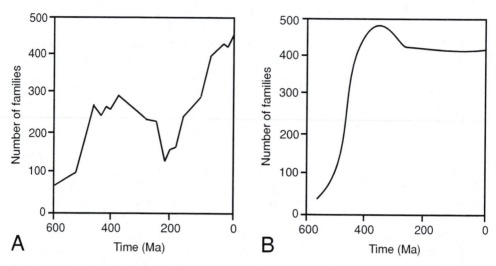

Figure 4.2 Comparison of empirical (A) and bias-simulation models (B) for diversification of well-skeletonized marine invertebrates through the Phanerozoic. The empirical pattern (A) is a literal reading of changes in diversity of families, and the bias-simulation model (B) is a theoretical construct that purports to show the true pattern of diversification after corrections for the poorer Palaeozoic fossil record and lower levels of study of such materials. Based on data in Valentine 1969 (A) and Raup 1972 (B).

Raup (1972) presented his view of the likely true shape of the diversification of life in the sea (Figure 4.2A), his so-called 'bias-simulation model', and contrasted it with the 'empirical' model of Valentine (1969). Based on theoretical considerations, deriving from competition studies in ecology and the Macarthur and Wilson's (1967) *Theory of Island Biogeography*, Raup (1972) argued that, following the Cambrian explosion, the sea filled up with families and species, and reached its carrying capacity within a geologically short span of time. After an overshoot, and some adjustment, a dynamic equilibrium level was achieved, which is the present diversity of life in the sea, and this level has been sustained for some 500 Ma.

Further consideration of these polarized views in the 1970s led to a reconciliation in which the empirical model was considered to be nearer the truth than Raup's (1972) bias-simulation model (Sepkoski *et al.* 1981) based on a comparison of a number of independently compiled datasets. Since then, palaeontologists have felt that they could legitimately study diversification and extinction on the basis of global-scale compilations of data on the fossil record, and that the broad patterns were correct (e.g. Raup and Sepkoski 1982, 1984; Niklas *et al.* 1983; Sepkoski 1984, 1996; Benton 1985, 1995, 1997, 2001; Miller 1998).

Alroy *et al.* (2001) have now suggested that Raup (1972) might well have been correct. In the first publication from the Paleobiology Database (PD) project, in which a sample-based approach is adopted, global levels of diversity extrapolated from fossil samples from the Palaeozoic appear to be comparable with those from the Cenozoic. The empirical finding by all authors that diversity increased dramatically at the levels of families and genera through the past 250 myr (e.g. Valentine 1969; Sepkoski *et al.* 1981; Sepkoski 1984, 1996; Benton 1995; Miller 1998) must then be explained as an artefact of dramatically improved sampling during that interval, and sampling that improves steadily from the Triassic to the present. As Alroy *et al.* (2001) make clear, their preliminary results are based on the sample of fossil collection data that has been accrued in the database so far, and it cannot yet be assessed whether that sample might include Palaeozoic collections that exaggerate apparent generic diversity (taxa oversplit, samples based on large 'localities', high levels of time-averaging, localities with sparse faunas omitted) when compared with the Cenozoic collections. Broadly put, the rarefaction approach adopted by the PD team requires unbiased environmental sampling through time, an objective that will be hard to achieve. For these reasons, Jackson and Johnson (2001) urge caution in the use of such a database based on random samples instead of comprehensive databases.

Failure of statistical approaches?

On the face of it, current standpoints on the quality of the fossil record could not be more extreme, and resolving these differences might seem an insurmountable problem. Available statistical approaches such as confidence intervals and group sampling are based on internalized assessments of the data which are being assessed, so they are not true statistical tests, where the data would be compared with an external standard. Smith and Peterson (2002) argue that these approaches cannot test whether temporal and geographical heterogeneity in the distribution of sedimentary rocks are not controlling patterns in the fossil record, but does this mean that the techniques should be abandoned?

Probably not. The critique of the confidence intervals approach is clearly correct: Marshall (1997, 1999) has argued that case already, and his generalized confidence intervals method can deal with heterogeneous preservation probability. However, the group sampling approach should not be rejected simply because it assumes homogeneous preservation probability. Foote (1997) showed that his methods can be misled by heterogeneity in the distribution of rocks, and in extreme cases, when suitable rocks are largely absent, the preservation probability is overestimated. However, this failure applies only in extreme cases, and Foote *et al.* (1999, note 31) claim that fluctuating preservation rate, associated with changes in sea level and other factors, is not likely to distort substantially either the overall probability of species preservation or estimates of preservation rate by the FreqRat and associated methods. Foote and Sepkoski (1999) and Foote *et al.* (1999) make a strong case that their methods are valid for estimating general broad-scale fossil record quality.

None the less, rock-record heterogeneity clearly causes problems for all statistical approaches to fossil sampling. Is there an alternative approach that might allow palaeontologists to escape from the risk of circularity in using internal measures of the fossil record to assess the quality of that same fossil record?

Clade versus age techniques

Independence

A sideways leap provides a partial answer, and that is to use a source of data on the history of life that is independent of the rocks, namely phylogenetics. It has been argued

(Platnick 1979; Patterson 1982; Smith and Littlewood 1994; Benton and Hitchin 1996, 1997) that phylogenetic data, whether from the cladistic analysis of morphological characters or from molecular phylogenetic reconstruction, are essentially independent of stratigraphic (geological age, rock distribution) evidence.

In cladistics, characters are determined and polarized (primitive → derived) according to their distribution among a group of organisms, living and extinct, and without reference to geological age. Many analysts these days do not even polarize the characters, so they include no directional information prior to the analysis. Molecular phylogeny reconstruction is even more obviously divorced from stratigraphy in that all the organisms under investigation are extant, and characters are generally not polarized. The implied history of the group is then subtended from the present-day with no reference to fossil taxa.

Trees, whether cladistic or molecular, cannot entirely escape from time-related aspects. A small input of stratigraphic bias may be associated with the choice of outgroups (comparator standards), but outgroups can readily be changed and analyses re-run. In addition, there are certain unavoidable temporal biases in phylogenies that relate to their geometry and the completeness of taxon sampling (Wagner 2000a) and to the relative timings of acquisition of apomorphies and homoplasies (Wagner 2000b). These issues do not substantially modify the geometry of a tree and the relative order of branching points, and hence independence between stratigraphy and tree shape is sustained (Benton 2001; Wills 2002). More serious though is that the age versus clade measures cannot detect major hiatuses in the rock record. So, if all early members of a clade are unknown as fossils, the relative order of appearance of lineages within that clade then becomes meaningless.

The hypothesis behind all age versus clade comparisons is that congruence indicates the true pattern. It is accepted of course that the fossil record is subject to bias, as are the techniques of cladistics and molecular phylogenetics. But the biases that might affect these three approaches are clearly different, and unlikely to reinforce each other. So, if a phylogenetic tree is congruent with the order of fossils in the rocks, it is most likely that both are correct. If incongruence is found, then it cannot be said whether the tree or the fossils, or both, are incorrect. The test for congruence does not mean that the entire tree is congruent with stratigraphy, but that it is more congruent than random, and certainly not significantly incongruent.

Metrics

Trees and fossil sequences may be compared in various ways, and several metrics have been proposed (reviewed in Benton et al. 1999). First attempts to compare clade and age data concentrated simply on the rank order of first fossils and branching points. The age and clade rank orders could then be compared by the use of simple correlation measures, such as Spearman rank correlation. Rank-order approaches are crude, however, since they can cope only with single runs of digits, and they cannot code the more complex patterns of typical cladograms with their multiple branches except by dividing them. Furthermore, in cases where numerous fossils occur close together in time, or where many branching events happened in a short time, the rank order is hard to sort out. For these reasons, we have abandoned use of this approach (Benton and Hitchin 1996; Hitchin and Benton 1997; Wagner 1998).

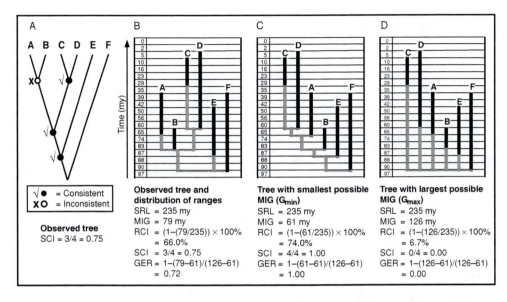

Figure 4.3 Calculation of the three congruence metrics for age versus clade comparisons, the stratigraphic consistency index (SCI), the relative completeness index (RCI), and the gap excess
ratio (GER). SCI is the ratio of consistent to inconsistent nodes in a cladogram. RCI is,

$$RCI = 1 - \left[\frac{\sum MIG}{\sum SRL}\right] \times 100\%$$

where MIG is the minimum implied gap, or ghost range, and SRL is the standard range
length, the known fossil record. GER is,

$$GER = 1 - \frac{(MIG - G_{min})}{(G_{max} - G_{min})}$$

where G_{min} is the minimum possible sum of ghost ranges and G_{max} the maximum, for any
given distribution of origination dates. A, The observed tree with SCI calculated according to the distribution of ranges in B. B, The observed tree and observed distribution of
stratigraphic range data, yielding an RCI of 54.6 per cent. GER is derived from G_{min} and
G_{max} values calculated in C and D. C, The stratigraphic ranges from B rearranged on a
pectinate tree to yield the smallest possible MIG or G_{min}. D, The stratigraphic ranges from
B rearranged on a pectinate tree to yield the smallest possible MIG or G_{max}.

We have used three measures to assess age versus clade congruence (Figure 4.3):
the stratigraphic consistency index (SCI; Huelsenbeck 1994), the relative completeness
index (RCI; Benton and Storrs 1994), and the gap excess ratio (GER; Wills 1999). The
first measure looks at the branching points (nodes) in a cladogram and their relation
to each other. A minimum date is assigned to each branching point by assessment of
the oldest known fossils of each of the subtended sister taxa. The consistency of each
node is then assessed by determining whether it is younger than, or the same age as,
the node immediately below. The SCI is the ratio of consistent to inconsistent nodes,
and it can range from 0 to 1.0 in a fully pectinate (unbalanced) tree, but the minimum value lies between 0 and 0.5 in balanced trees (Siddall 1996; Wills 1999).

The RCI and GER depend on numerical age estimates of the branching points on a cladogram, and the calculation of 'ghost ranges'. The ghost lineage (Norell 1992) is the implied missing evolutionary line indicated by the difference in age between the oldest known fossils of two sister taxa, and that missing span of time is termed the ghost range, or the minimum (cladistically) implied gap (MIG; Benton and Storrs 1994). It is based on the observation that a node in a cladogram represents a single point in time, but that the oldest fossil representatives of the two lineages branching from that node are most often not of the same age (Smith and Patterson 1988). The RCI is assessed as the ratio of the sum of ghost ranges to the sum of recorded fossil ranges in any cladogram. The GER focuses solely on the estimated dates of origin of groups, and compares the sum of actual ghost ranges in a cladogram with the theoretical minimum and maximum ghost ranges if the various branches in the cladogram are rearranged. Values for the GER range from 0.0 (no congruence) to 1.0 (perfect congruence), while values for the RCI range from 0–100 per cent where MIG < SRL. However, the RCI can range to $-\infty$ when the known ranges (SRL) are point occurrences, and the sum of ghost ranges (MIG) is large.

Many additional clade versus age congruence metrics have been proposed, and all are related to one or other of the metrics we use. Norell (1992) proposed the Z statistic, one minus the ratio of the sum of ghost ranges divided by the number of taxa to the sum of known ranges. Smith and Littlewood (1994) proposed the implied gap (IG) metric, the ratio of summed ghost ranges to summed ghost ranges plus summed observed ranges in a cladogram. Weishampel (1996) used sums of ghost lineage durations (GLDs). Siddall (1998) presented his Manhattan stratigraphic measure (MSM), which uses Manhattan distances between stratigraphic ages. Brochu and Norell (2000) proposed SMIG, the sum of minimum implied gaps (i.e. ghost ranges).

Results

Based on various samples of published trees, age versus clade comparisons have yielded a number of results. Norell and Novacek (1992) found that 75 per cent of their sample of trees of mammals showed congruence, confirmation of the validity of the tree-making methods and of the fossil record. Later studies, based on larger samples of trees, and for wider groups of organisms, found lower proportions of congruent trees based on Spearman rank correlation (36–50 per cent; Benton and Hitchin 1997), but the rather more realistic metrics outlined above confirmed that roughly three-quarters of published trees *are* congruent with the fossil record (Benton *et al.* 1999, 2000).

Comparisons of different sectors of the data (the current dataset stands at 1000 trees; http://palaeo.gly.bris.ac.uk/cladestrat/cladestrat.html) showed no clear bias. For example, marine and continental organisms show equivalent levels of congruence between trees and the fossil record (Benton and Simms 1995; Benton and Hitchin 1996; Benton 2001). Different taxonomic groups on the whole also show equivalence, with no clear evidence that plants or animals, invertebrates or vertebrates, or whatever, are uniformly better preserved, or have uniformly better analysed cladograms than any other (Benton and Simms 1995; Benton and Hitchin 1996; Benton 2001). Wills (2001) showed that arthropods on the whole show poorer age–clade congruence than vertebrates, but he could not readily explain this observation.

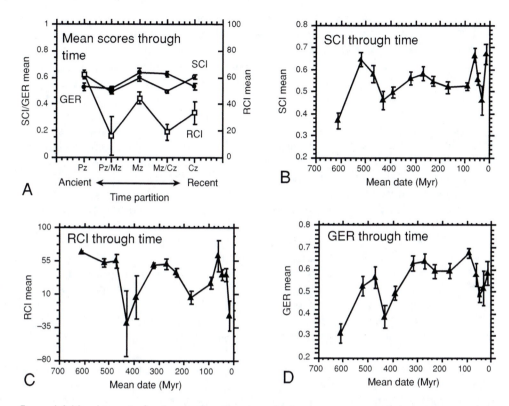

Figure 4.4 No change in fossil record quality through time, mean scores of the age versus clade metrics for finer-scale divisions of geological time. A, Stratigraphic consistency index (SCI), the relative completeness index (RCI), and the gap excess ratio (GER) for five time partitions of the dataset of 1000 cladograms, namely cladograms with origins solely in the Palaeozoic (Pz), cladograms with origins spanning the Palaeozoic and Mesozoic (Pz/Mz), cladograms with origins solely in the Mesozoic (Mz), cladograms with origins spanning the Mesozoic and Cenozoic (Mz/Cz), and cladograms with origins solely in the Cenozoic (Cz). B–D, Age versus clade metrics for cladograms partitioned into geological periods and epochs showing temporal variations in the SCI (B), RCI (C), and GER (D). The age versus clade metrics are explained in the caption to Figure 4.3. There is no statistically significant secular trend for the broad-scale time divisions (A), nor for the period-by-period assessments by the SCI (B) or RCI (C). The GER values (D) do improve through time (0.5 > P > 0.025), but the regression becomes non-significant if the low Vendian value (based on 34 trees) is omitted.

Comparisons of the change in palaeontological knowledge through time show a statistically significant reduction in ghost ranges over a 25 year sample period (Benton and Storrs 1994). This confirms that new fossil discoveries do not extend fossil ranges in unexpected ways, but they tend to fill predicted gaps.

Finally, there is no evidence for a decline in the quality of the fossil record back through geological time (Benton *et al.* 2000). The sample of 1000 cladograms was divided into various time bins, and the age versus clade metrics assessed. The expectation was that Palaeozoic trees would show poorer congruence values than, say, Cenozoic trees. Actually, the two time-independent metrics (SCI, GER) showed essentially no change through time (Figure 4.4), while the RCI worsened through time (but that was expected, since the RCI is a ratio of ghost range to known range, and known stratigraphic ranges for groups originating in the Palaeozoic may be much longer than those originating in the Cenozoic).

Are the age versus clade methods subject to the problem of temporal and geographical heterogeneity of the rock record, as are the confidence interval and group sampling methods? The answer is no, where heterogeneity is at 'normal' scales:

(1)   Heterogeneity in the rock record is purely a geological issue, and it is related to all the other biases outlined by Raup (1972). Therefore, in assessing congruence of the order of fossils in the rocks with the patterns of cladograms, there is no linkage.

(2)   The available set of published cladograms includes trees for soft-bodied organisms and those with hard parts alike. There is no evidence for a substantial difference in expectations of congruence between groups with readily fossilizable parts (such as vertebrates with their bones, or echinoderms with their calcite skeletons) and those with less robust skeletons (e.g. plants, arthropods, with thinner skeletons or organic cuticles). Of course, the comparison cannot be extended to entirely soft-bodied organisms since, in the absence of any fossils, it is impossible to make any age versus clade comparisons. However, there is no reason to assume that entirely soft-bodied organisms were any more or less abundant in the past than they are now, and hence that the pattern of evolution of groups with hard parts can be accepted as a proxy for the evolution of groups lacking such hard parts (Valentine 1969).

In the case of large-scale heterogeneity, however, the methods cannot function (Andrew Smith, Natural History Museum, London, pers. obs.). For example, if there were truly a gap of 30–40 myr in the Late Cretaceous where no bird fossils are found, then any calculations of age versus clade metrics based on a cladogram of major bird lineages plotted against time would be meaningless. I question, however, how often such vast gaps in the record actually occur.

Peters and Foote (2001, pp. 597–8) have stressed that the age versus clade metrics cannot be used to assess absolute or overall completeness of the fossil record, a point made also by Benton *et al.* (2000). The metrics can only compare known and postulated parts of the record that exist in the rocks, termed the intrinsic completeness of the fossil record. Peters and Foote (2001) contrast these two aspects as 'global' and 'local', but their terms have common geographical meanings, and perhaps 'absolute' and 'intrinsic' are preferable.

Conclusion

The fossil record is, as ever, under close scrutiny. Current viewpoints range from gung-ho to abject despair. The gung-ho view is that things have never been better: the fossil record may be read like a book that documents every nuance of the history of life. The despairing position is that the fossil record can never say much about the history of life since it is so riddled with bias and error, much of which can never be estimated and corrected.

Our age versus clade studies may be read to confirm something midway between these two stances, but definitely tending to the gung-ho end of the scale. The two key results have been that most trees are congruent with the fossil record, and that there is no evident large-scale time bias through the past 500–600 myr or so, at the scale of eras (65–300 myr) and periods (40–80 myr). The latter result is counter-intuitive, since it is evident that many factors must act as time-related biases: rock volume, rock area, metamorphism, erosion, study levels. However, these biases evidently affect things only at the lower level of focus, when one considers individual specimens or species in particular localities at fine-scale stratigraphic divisions. At the taxonomic level of families and above, and the stratigraphic level of geological periods, the patterns may well be sound.

In support of the molecular age doubling found for modern orders of birds and mammals, Hedges *et al.* (1996), Cooper and Penny (1997), Kumar and Hedges (1998), Easteal (1999) and others have suggested three reasons why the fossils have not been found, but I doubt these:

(1)   Ancestral forms were cryptic, or did not display all synapomorphies. This idea is that somehow molecular and morphological evolution are uncoupled, and that molecular divergence between major clades could happen tens of millions of years before full morphological differentiation. There is no evidence for such substantial uncoupling of molecular and morphological evolution, indeed rather the opposite (Omland 1997), and it is hard to see how the suggestion could ever be tested.

(2)   Ancestors were unpreservable (too small, soft-bodied). This might be true for basal metazoans in the late Precambrian, but the ancestors of modern bird and mammal groups were most unlikely to have been unpreservable: dozens of localities through the Early and Late Cretaceous have yielded tiny, delicate skeletons of birds and mammals, but none of them pertains to extant orders (Benton 1999b; Fara and Benton 2000).

(3)   Ancestors lived in hitherto unexplored parts of the world, such as the southern continents. This might be true, but current work is opening up richly fossiliferous sites in the Cretaceous of Gondwana, in South America, southern Africa, Madagascar, and India, and not a hint of a modern bird or mammal has been found. Similarly, for the early origins of Metazoa case, palaeontologists are working actively in hitherto palaeontologically unexplored parts of the world (China, Australia, Africa), so the chances that the fossils required by the age-doubling molecular argument will be found are diminishing. I termed this the 'living mastodon' argument (Benton 1999b), after expectations in the 18th century that mastodons, represented by abundant fossils from North America, might yet be found living in the Wild West.

Acknowledgements

Supported by Leverhulme Grant F/182/AK. Many thanks to Andrew Smith for his critical comments and advice on the manuscript.

References

Alroy, J., Marshall, C.R., Bambach, R.K., Bezusko, K., Foote, M., Fürsich, F.T., Hansen, T.A., Holland, S.M., Ivany, L.C., Jablonski, D., Jacobs, D.K., Jones, D.C., Kosnik, M.A., Lidgard, S., Low, S., Miller, A.I., Novack-Gottshall, P.M., Olszewski, T.D., Patzkowsky, M.E., Raup, D.M., Roy, K., Sepkoski, J.J., Sommers, M.G., Wagner, P.J. and Webber, A. (2001)

'Effects of sampling standardization on estimates of Phanerozoic marine diversification', *Proceedings of the National Academy of Sciences, USA*, 98: 6261–6.

Archibald, J.D. (1996) 'Fossil evidence for a Late Cretaceous origin of "hoofed" mammals', *Science*, 272: 1150–3.

Archibald, J.D. and Deutschmann, D.H. (2001) 'Quantitative analysis of the timing of the origin and diversification of extant placental orders', *Journal of Mammalian Evolution*, 8: 107–24.

Archibald, J.D., Averianov, A.O. and Ekdale, E.G. (2001) 'Late Cretaceous relatives of rabbits, rodents, and other extant eutherian mammals', *Nature*, 414: 62–5.

Benton, M.J. (1985) 'Mass extinction among non-marine tetrapods', *Nature*, 316: 811–14.

—— (1990) *Vertebrate Palaeontology*, London: Unwin Hyman.

—— (1995) 'Diversification and extinction in the history of life', *Science*, 268: 52–8.

—— (1997) 'Models for the diversification of life', *Trends in Ecology and Evolution*, 12: 490–5.

—— (1999a) 'The history of life, large data bases in palaeontology', in D.A.T. Harper (ed.) *Statistical Methods in Palaeobiology*, London: Wiley, pp. 249–83.

—— (1999b) 'Early origins of modern birds and mammals: molecules vs. morphology', *BioEssays*, 21: 1043–51.

—— (2001) 'Biodiversity on land and in the sea. *Geological Journal*, 36: 211–30.

Benton, M.J. and Ayala, F.J. (2003) 'Dating the Tree of Life', *Science*, 300: 1698–700.

Benton, M.J. and Hitchin, R. (1996) 'Testing the quality of the fossil record by groups and by major habitats', *Historical Biology*, 12: 111–57.

—— (1997) 'Congruence between phylogenetic and stratigraphic data on the history of life', *Proceedings of the Royal Society, London*, B264: 885–90.

Benton, M.J. and Simms, M.J. (1995) 'Testing the marine and continental fossil records', *Geology*, 23: 601–4.

Benton, M.J. and Storrs, G.W. (1994) 'Testing the quality of the fossil record, paleontological knowledge is improving', *Geology*, 22: 111–14.

Benton, M.J., Hitchin, R. and Wills, M.A. (1999) 'Assessing congruence between cladistic and stratigraphic data', *Systematic Biology*, 48: 581–96.

Benton, M.J., Wills, M.A. and Hitchin, R. (2000) 'Quality of the fossil record through time', *Nature*, 403: 534–7.

Bleiweiss, R. (1998) 'Fossil gap analysis supports early Tertiary origin of trophically diverse avian orders', *Geology*, 26: 323–6.

—— (1999) 'Fossil gap analysis supports early Tertiary origin of trophically diverse avian orders, Comment and Reply', *Geology*, 27: 95–6.

Bleiweiss, R., Kirsch, J.A.W. and Lapointe, F.-J. (1994) 'DNA–DNA hybridization-based phylogeny for higher nonpasserines, reevaluating a key portion of the avian family tree', *Molecular Phylogenetics and Evolution*, 3: 248–55.

Bleiweiss, R., Kirsch, J.A.W. and Shafi, N. (1995) 'Confirmation of a portion of the Sibley and Ahlquist "tapestry" ', *Auk*, 112: 87–97.

Brochu, C.A. and Norell, M.A. (2000) 'Temporal congruence and the origin of birds', *Journal of Vertebrate Paleontology*, 20: 197–200.

Carroll, R.L. (1988) *Vertebrate Paleontology and Evolution*, San Francisco: W. H. Freeman.

Cavalli-Sforza, L.L. and Edwards, A.W.F. (1967) 'Phylogenetic analysis, models and estimation procedures', *Evolution*, 21: 550–70.

Chiappe, L.M. and Dyke, G.J. (2002) 'The Mesozoic radiation of birds', Annual Reviews of Ecology and Systematics, 33: 91–124.

Conway Morris, S. (2000) 'The Cambrian "explosion", slow-fuse or megatonnage?', *Proceedings of the National Academy of Sciences, USA*, 97: 4426–9.

Cooper, A. and Fortey, R. (1998) 'Evolutionary explosions and the phylogenetic fuse', *Trends in Ecology and Evolution*, 13: 151–6.

Cooper, A. and Penny, D. (1997) 'Mass survival of birds across the Cretaceous–Tertiary boundary, molecular evidence', *Science*, 275: 1109–13.

Crane, P.R., Friis, E.M. and Pedersen, K.R. (1995) The origin and early diversification of angiosperms', *Nature*, 374: 27–33.

Donovan, S.K. and Paul, C.R.C. (eds) (1998) *The Adequacy of the Fossil Record*, New York: Wiley.

Dyke, G.J. (2001) 'The evolutionary radiation of modern birds, systematics and patterns of diversification', *Geological Journal*, 36: 305–15.

Easteal, S. (1999) 'Molecular evidence for the early divergence of placental mammals', *BioEssays*, 21: 1052–8.

Eizirik, E., Murphy, W.J. and O'Brien, S.J. (2001) 'Molecular dating and biogeography of the early placental mammal radiation', *Journal of Heredity*, 92: 212–19.

Fara, E. (2002) 'Sea-level variations and the quality of the continental fossil record', *Journal of the Geological Society, London*, 159: 489–91.

Fara, E. and Benton, M.J. (2000) 'The fossil record of Cretaceous tetrapods', *Palaios*, 15: 161–5.

Feduccia, A. (1996) *The Origin and Evolution of Birds*, New Haven: Yale University Press.

Foote, M. (1997) 'Estimating taxonomic durations and preservation probability', *Paleobiology*, 23: 278–300.

Foote, M. and Raup, D.M. (1996) 'Fossil preservation and the stratigraphic ranges of taxa', *Paleobiology*, 22: 121–40.

Foote, M. and Sepkoski, J.J. Jr. (1999) 'Absolute measures of the completeness of the fossil record', *Nature*, 398: 415–17.

Foote, M., Hunter, J.P., Janis, C. and Sepkoski, J.J. Jr. (1999) 'Evolutionary and preservational constraints on origins of biologic groups, divergence times of eutherian mammals', *Science*, 283: 1310–14.

Hallam, A. (1986) 'The Pliensbachian and Tithonian extinction events', *Nature*, 319: 765–8.

Hallam, A. and Wignall, P.B. (1997) *Mass Extinctions and their Aftermath*, Oxford: Oxford University Press.

Harlid, A. and Arnason, U. (1999) 'Analyses of mitochondrial DNA nest ratite birds within the Neognathae, supporting a neotenous origin of ratite morphological characters', *Proceedings of the Royal Society, London*, B266: 305–9.

Heckman, D.S., Geiser, D.M., Eidell, B.R., Stauffer, R.L., Kardos, N.L. and Hedges, S.B. (2001) 'Molecular evidence for the early colonization of land by fungi and plants', *Science*, 293: 1129–33.

Hedges, S.B. and Kumar, S. (1999) 'Technical comments: divergence times of eutherian mammals', *Science*, 285: 2031a.

—— (2003) 'Genomic clocks and evolutionary timescales', *Trends in Genetics*, 19: 200–6.

Hedges, S.B., Parker, P.H., Sibley, C.G. and Kumar, S. (1996) 'Continental breakup and the ordinal diversification of birds and mammals', *Nature*, 381: 226–9.

Hitchin, R. and Benton, M.J. (1997) 'Congruence between parsimony and stratigraphy, comparisons of three indices', *Paleobiology*, 23: 20–32.

Huelsenbeck, J.P. (1994) 'Comparing the stratigraphic record to estimates of phylogeny', *Paleobiology*, 20: 470–83.

Huelsenbeck, J.P. and Rannala, B. (1997) 'Phylogenetic methods come of age, testing hypotheses in an evolutionary context', *Science*, 276: 227–32.

Jackson, J.B.C. and Johnson, K.G. (2001) 'Measuring past biodiversity', *Science*, 293: 2401–4.

Ji, Q., Luo, Z., Yuan, C.-X., Wible, J.R., Zhang, J.-P. and Georgi, J.A. (2002) 'The earliest known eutherian mammal', *Nature*, 416: 816–22.

Kenrick, P. (1999) 'The family tree flowers', *Nature*, 402: 358–9.

Kenrick, P. and Crane, P.R. (1997) 'The origin and early evolution of plants on land', *Nature*, 389: 33–9.

Klicka, J. and Zink, R.M. (1997) 'The importance of Recent ice ages in speciation, a failed paradigm', *Science*, 277: 1666–9.

Kumar, S. and Hedges, S.B. (1998) 'A molecular timescale for vertebrate evolution', *Nature*, 392: 917–20.

Lee, M.S.Y. (1999) 'Molecular clock calibrations and metazoan divergence dates', *Journal of Molecular Evolution*, 49: 385–91.

Little, C.T.S. and Benton, M.J. (1995) 'Early Jurassic mass extinction: a global long-term event', *Geology*, 23: 495–8.

Macarthur, R.H. and Wilson, E.O. (1967) *The Theory of Island Biogeography*, Princeton: Princeton University Press.

Marshall, C.R. (1990) 'Confidence intervals on stratigraphic ranges', *Paleobiology*, 16: 1–10.

—— (1997) 'Confidence intervals on stratigraphic ranges with nonrandom distribution of fossil horizons', *Paleobiology*, 23: 165–73.

—— (1999) 'Fossil gap analysis supports early Tertiary origin of trophically diverse avian orders, Comment and Reply', *Geology*, 27: 95.

Miller, A.I. (1998) 'Biotic transitions in global marine diversity', *Science*, 281: 1157–60.

Murphy, W.J., Eizirik, E., O'Brien, S.J., Madsen, O., Scally, M., Douady, C.J., Teeling, E., Ryder, O.A., Stanhope, M.J., de Jong, W.W. and Springer, M.S. (2001) 'Resolution of the early placental mammal radiation using Bayesian phylogenetics', *Science*, 294: 2348–51.

Nei, M., Xu, P. and Glazko, G. (2001) 'Estimation of divergence times from multiprotein sequences for a few mammalian species and several distantly related organisms', *Proceedings of the National Academy of Sciences, USA*, 98: 2497–502.

Niklas, K.J., Tiffney, B.H. and Knoll, A.H. (1983) 'Patterns in vascular land plant diversification', *Nature*, 303: 614–16.

Norell, M.A. (1992) 'Taxic origin and temporal diversity, the effect of phylogeny', in M.J. Novacek and Q.D. Wheeler (eds) *Extinction and Phylogeny*, New York: Columbia University Press, pp. 89–118.

Norell, M.A. and Novacek, M.J. (1992) 'The fossil record and evolution, comparing cladistic and paleontologic evidence for vertebrate history', *Science*, 255: 1690–3.

Omland, K.E. (1997) 'Correlated rates of molecular and morphological evolution', *Evolution*, 51: 1381–93.

Paton, T., Haddrath, O. and Baker, A.J. (2002) 'Complete mitochondrial DNA genome sequences show that modern birds are not descended from transitional shorebirds', *Proceedings of the Royal Society, London*, B269: 839–46.

Patterson, C. (1982) 'Morphological characters and homology', in C. Patterson (ed.) *Problems of Phylogenetic Reconstruction*, Systematics Association Special Volume 21, London: Academic Press, pp. 21–74.

Peters, S.E. and Foote, M. (2001) 'Biodiversity in the Phanerozoic, a reinterpretation', *Paleobiology*, 27: 583–601.

—— (2002) 'Determinants of extinction in the fossil record', *Nature*, 416: 420–4.

Platnick, N.L. (1979) 'Philosophy and the transformation of cladistics', *Systematic Zoology*, 28: 537–46.

Rambaut, A. and Bromham, L. (1998) 'Estimating divergence dates from molecular sequences', *Molecular Biology and Evolution*, 15: 442–8.

Raup, D.M. (1972) 'Taxonomic diversity during the Phanerozoic', *Science*, 177: 1065–71.

Raup, D.M. and Sepkoski, J.J. Jr. (1982) 'Mass extinctions in the marine fossil record', *Science*, 215: 1501–3.

—— (1984) 'Periodicity of extinctions in the geologic past', *Proceedings of the National Academy of Sciences, USA*, 81: 801–5.

Sepkoski, J.J. Jr. (1984) 'A kinetic model of Phanerozoic taxonomic diversity. III. Post-Paleozoic families and mass extinctions', *Paleobiology*, 10: 246–67.

—— (1996) 'Patterns of Phanerozoic extinction, a perspective from global data bases', in O.H. Walliser (ed.) *Global Events and Event Stratigraphy*, Berlin: Springer, pp. 35–51.

Sepkoski, J.J. Jr., Bambach, R.K., Raup, D.M. and Valentine, J.W. (1981) 'Phanerozoic marine diversity and the fossil record', *Nature*, 293: 435–7.

Siddall, M.E. (1996) 'Stratigraphic consistency and the shape of things', *Systematic Biology*, 45: 111–15.

—— (1998) 'Stratigraphic fit to phylogenies: A proposed solution', *Cladistics*, 14: 201–8.

Siddall, M.E. and Whiting, M.F. (1999) 'Long-branch abstractions', *Cladistics*, 15: 9–24.

Smith, A.B. (2001) 'Large-scale heterogeneity of the fossil record, implications for Phanerozoic biodiversity studies', *Philosophical Transactions of the Royal Society, London*, B356: 1–17.

Smith, A.B. and Littlewood, D.T.J. (1994) 'Paleontological data and molecular phylogenetic analysis', *Paleobiology*, 20: 259–73.

Smith, A.B. and Patterson, C. (1988) 'The influence of taxonomic method on the perception of patterns of evolution', *Evolutionary Biology*, 23: 127–216.

Smith, A.B. and Peterson, K.J. (2002) 'Dating the time of origin of major clades, molecular clocks and the fossil record', *Annual Reviews in Ecology and Systematics*, 30: 65–88.

Smith, A.B., Gale, A.S. and Monks, N. (2001) 'Sea-level change and rock record bias in the Cretaceous: a problem for extinction and biodiversity studies'. *Paleobiology*, 27: 241–53.

Springer, M.S. (1997) 'Molecular clocks and the timing of the placental and marsupial radiations in relation to the Cretaceous–Tertiary boundary', *Journal of Mammalian Evolution*, 4: 285–302.

Springer, M.S., Murphy, M.J., Eizirik, E. and O'Brien, S.J. (2003) 'Placental mammal diversification and the Cretaceous-Tertiary boundary', *Proceedings of the National Academy of Sciences, USA*, 100: 1056–61.

Strauss, D. and Sadler, P.M. (1989) 'Classical confidence intervals and Bayesian probability estimates for ends of local taxon ranges', *Mathematical Geology*, 21: 411–27.

Tavaré, S., Marshall, C.R., Will, O., Soligo, C. and Martin, R.D. (2002) 'Using the fossil record to estimate the age of the last common ancestor of extant primates', *Nature*, 416: 726–9.

Valentine, J.W. (1969) 'Patterns of taxonomic and ecological structure of the shelf benthos during Phanerozoic time', *Palaeontology*, 12: 684–709.

Valentine, J.W., Jablonski, D. and Erwin, D.H. (1999) 'Fossils, molecules and embryos, new perspectives on the Cambrian explosion', *Development*, 126: 851–9.

van Tuinen, M. and Hedges, S.B. (2001) 'Calibration of avian molecular clocks', *Molecular Biology and Evolution*, 18: 206–13.

van Tuinen, M., Sibley, C.G. and Hedges, S.B. (2000) 'The early history of modern birds inferred from DNA sequences of nuclear and mitochondrial ribosomal genes', *Molecular Biology and Evolution*, 17: 451–7.

Vermeij, G.J. (1996) 'Animal origins', *Science*, 274: 525–6.

Wagner, P.J. (1998) 'Phylogenetic analyses and the quality of the fossil record', in S.K. Donovan and C.R.C. Paul (eds) *The Adequacy of the Fossil Record*, New York: Wiley, pp. 165–87.

—— (2000a) 'The quality of the fossil record and the accuracy of phylogenetic inferences about sampling and diversity', *Systematic Biology*, 49: 65–86.

—— (2000b) 'Exhaustion of morphologic character states among fossil taxa', *Evolution*, 54: 365–86.

Weishampel, D.B. (1990) 'Dinosaurian distribution', in D.B. Weishampel, P. Dodson and H. Osmólska (eds) *The Dinosauria*, Berkeley: University of California Press, pp. 63–139.

—— (1996) 'Fossils, phylogeny, and discovery, a cladistic study of the history of tree topologies and ghost lineage durations', *Journal of Vertebrate Paleontology*, 16: 191–7.

Whelan, S., Liò, P. and Goldman, N. (2001) 'Molecular phylogenetics: state-of-the-art methods for looking at the past', *Trends in Genetics*, 17: 262–72.

Wikström, N., Savolainen, V. and Chase, M.W. (2001) 'Evolution of the angiosperms, calibrating the family tree', *Proceedings of the Royal Society, London*, B268: 2211–20.

Wills, M.A. (1999) 'The gap excess ratio, randomization tests, and the goodness of fit of trees to stratigraphy', *Systematic Biology*, 48: 559–80.

—— (2001) 'How good is the fossil record of arthropods? An assessment using the stratigraphic congruence of cladograms', *Geological Journal*, 36: 187–210.

—— (2002) 'The tree of life and the rock of ages, are we getting better at estimating phylogeny?', *BioEssays*, 24: 203–7.

Wray, G.A. (2001) 'Dating branches on the Tree of Life using DNA', *Genome Biology*, 3: 1–7.

Wray, G.A., Levinton, G.S. and Shapiro, L.H. (1996) 'Molecular evidence for deep Precambrian divergences among metazoan phyla', *Science*, 274: 568–73.

Chapter 5

Ghost ranges

Christopher R.C. Paul

ABSTRACT

Ghost ranges occur where phylogenetic reconstructions predict fossil ranges beyond the known stratigraphic occurrence of the fossils. Cladistic methodology refers all closest relationships to sister groups and assumes they arose at the same time. Where the real relationship is ancestor–descendant, the ancestral taxon must have existed for some time before the descendant taxon evolved and cladistic analysis will produce an artificial ghost range. Alternatively, real ghost ranges are caused by gaps in the fossil record. First records of fossil species can only occur in the wrong order with respect to the true evolutionary order if the species coexisted. Estimates indicate that at least 95 per cent of Phanerozoic fossil species did not coexist at any time. This defines the minimum level of reliability of the stratigraphic sequence of fossils, a result which is independent of the completeness of the record. Discrepancies between stratigraphic occurrences of fossils and predictions of cladograms are a valid test of the cladograms, not of the incompleteness of the fossil record. If real ghost ranges were common, then first occurrences of fossils would frequently move back in time as new, earlier occurrences were detected. This provides a preliminary method for investigating the frequency of real ghost ranges in the fossil record.

Introduction

Opinions differ as to the usefulness of stratigraphic data in reconstructing phylogenies. On the one hand there are those who argue that stratigraphic data are irrelevant (e.g. Patterson 1981). In contrast, Huelsenbeck and Ranala (1997) have published a maximum likelihood method for deriving cladograms from stratigraphic data alone. Others (e.g. Clyde and Fisher 1997; Smith 1994; Smith and Littlewood 1994; Wagner 1995, 1998) argue that the best phylogenies derive from both stratigraphic data and character analysis. Tests for congruence have been developed (e.g. Huelsenbeck 1994) and often suggest close comparison between branching patterns in cladograms derived from character analysis and the order of stratigraphic occurrence of fossils (e.g. Benton and Storrs 1994; Benton and Simms 1995; Benton and Hitchin 1997; Hitchin and Benton 1997; Benton 1998). Norrell (1992) introduced the term 'ghost taxa' for taxa predicted by cladistics, but not actually known. The term 'ghost lineage' has also been used in a similar sense (e.g. Wagner 1998). Here I use the term 'ghost range' (Benton and Storrs 1996), because often what is really being implied is

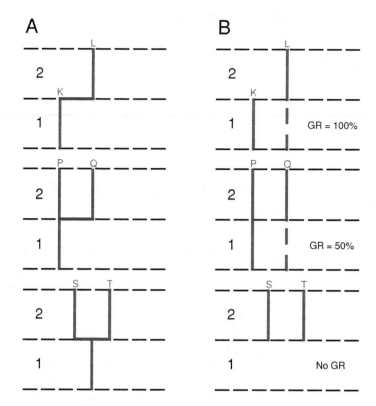

Figure 5.1 Diagrams illustrating the origin of artificial ghost ranges. A. Real phylogenies of three pairs of species (solid lines). B. Corresponding stratigraphic ranges (solid lines) and artificial ghost ranges (broken lines, GR), which can be any length from zero to 100 per cent of the stratigraphic range of the ancestral species.

that the stratigraphic range of a fossil taxon should be extended back below the stratigraphically earliest occurrence of the taxon. Techniques have been developed to put confidence intervals on known stratigraphic ranges (e.g. Paul 1982; Strauss and Sadler 1989; Marshall 1990, 1994, 1997, 1998) and therefore the reality of ghost ranges should be amenable to study.

Ghost ranges

Ghost ranges occur where phylogenetic reconstructions predict fossil ranges beyond the known stratigraphic occurrence of the fossils. In particular, cladistic methodology refers all closest relationships to sister groups and assumes that the sister groups arose at the same time. For example, Figure 5.1A shows the true phylogenies of three pairs of species. Figure 5.1B shows the cladograms for these species, which produce ghost ranges from zero to fully 100 per cent of the stratigraphic range of the ancestral species. For the purposes of this argument I am assuming that we know both the true phylogeny and total stratigraphic range of all the species. Clearly ghost ranges can be an artefact of cladistic methods. I propose to call these artificial ghost ranges. On

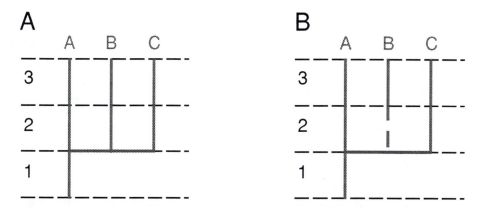

Figure 5.2 Diagrams illustrating one possible origin of real ghost ranges. A. Real phylogeny (solid lines) with known stratigraphic occurrences (thickened). B. Corresponding real ghost range for taxon B (broken line).

the other hand, ghost ranges can arise from the imperfections of the fossil record. These I shall refer to as genuine ghost ranges. Theoretically it is possible to detect a genuine ghost range, that is, a gap in the fossil record. Figure 5.2A shows the true phylogeny of three fossil species with their known stratigraphic ranges (thickened lines). Figure 5.2B identifies a minimum ghost range that must be due to a gap in the record rather than an artefact of cladistic methodology. Almost certainly, genuine ghost ranges occur in the fossil record. The problem lies in distinguishing between artificial and genuine ghost ranges because we have no way of knowing the true phylogenies of fossil species and hence whether or not a ghost range has been produced purely by cladistic methods.

However, when discrepancies occur between known stratigraphic ranges of fossil species and the inferred order of their occurrence deduced from cladistics, it seems to me unwise to use the cladogram as a measure of the incompleteness of the fossil record, because we know that cladistic methodology can produce artificial ghost ranges. We can never know for certain whether we have discovered an artificial ghost range or a genuine gap in the fossil record, but because cladistics *can* sometimes produce artificial ghost ranges, cladograms cannot be used as a test of the completeness of the fossil record. On the other hand, there are sound reasons for believing that the order in which species occur in the fossil record is accurate (irrespective of the completeness or otherwise of the fossil record; Paul 1982, 1985). Hence it is legitimate to use the order of occurrence of fossil species as a test of inferred cladograms.

Reliability of the fossil record

So, how can we be sure that the order in which species occur in the fossil record is the correct order in which they actually evolved? In Figure 5.3A the vertical lines represent the entire period of existence of species E and F (i.e. their durations in the sense of Foote and Raup, 1996), not just their stratigraphic ranges. Assume that species E is a Cambrian trilobite and species F is a Pleistocene snail, so there are literally

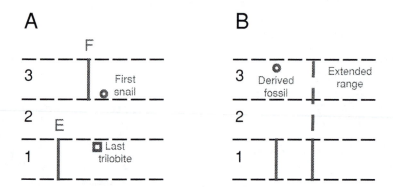

Figure 5.3 A. Total durations of two fossil taxa that do not overlap. Note that their first occurrences in the fossil record cannot be in the wrong order with respect to their real order of origin. B. Reworked fossils merely extend apparent ranges (dashed line); they do not alter the order of first occurrences.

hundreds of millions of years between the extinction of the trilobite and the evolution of the snail (= interval 2). The trilobite cannot possibly supply skeletal parts to be included in the fossil record after it became extinct. Equally, there cannot be any snail shells available for fossilization until the base of the Pleistocene only two million years ago. It follows that there is no possibility whatsoever that the first preserved specimen of the snail can occur in the fossil record below the first (or for that matter the last) preserved specimen of the trilobite. It is true that reworked fossils may extend apparent stratigraphic ranges (Figure 5.3B). Nevertheless, the only way reworked fossils can affect the order of first occurrences of fossils is if we fail to recognize them as being reworked *and* if we know of no examples from their original stratigraphic range. While by no means impossible, the combination of these two circumstances will be relatively rare.

If species coexisted then there is a possibility that the stratigraphic order of first (or last) occurrences may be incorrect with respect to their real times of origin (or extinction). Figure 5.4 illustrates an example where the stratigraphically earliest known record of the earlier species (V, solid circle) occurs after that of the later species (W, solid square). However, as soon as we discover an example of species V from level one (open circle), it is no longer possible for the stratigraphic order to be wrong. We can never know when this happens, but nevertheless sometimes it must occur and, of course, the more specimens we collect the more likely we are to discover an example of species V in level 1.

So when species did not coexist they cannot be preserved in the fossil record in the wrong order with respect to the order in which they evolved. However, when two species did coexist it is possible that they might be preserved in the fossil record in the wrong order. Take the conservative view that when species did coexist the fossil record is unreliable and so we should reject it. We will only accept the fossil record of species which did not coexist. Then the probability that the fossil record is reliable depends on the proportion of all fossil species that did not coexist. Assuming five million years for the average lifetime of a species and about 500 million years

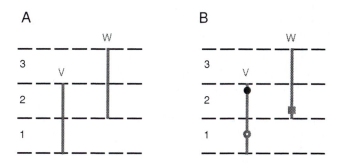

Figure 5.4 A. Total durations of two fossil taxa that do overlap. B. An example where the first occurrence of taxon V (solid circle) could occur above (later than) the first occurrence of taxon W (solid square) (i.e. in the wrong order with respect to their real order of origin). Note, however, that as soon as an example of V is found in level one (open circle), it is no longer possible for the first occurrences of these taxa to be in the wrong order with respect to their real order of origin.

since the base of the Cambrian, and making due allowance for changes in biodiversity through the Phanerozoic (e.g. from Sepkoski 1978; Sepkoski *et al.* 1981), this suggests that 3–5 per cent of fossil species coexisted at any given time in the Phanerozoic. Put another way, in 95–97 per cent of random comparisons between pairs of fossil species there is no chance whatsoever that they could be preserved in the wrong stratigraphic order. Recently, Peters and Foote (2001) have questioned the reality of the diversity increase apparent in Sepkoski (1978) and Sepkoski *et al.*'s (1981) analyses. If their re-interpretation is correct, this would reduce the proportion of taxa that coexisted in the Phanerozoic even further and strengthen the above arguments. Furthermore, this is a conservative estimate of the reliability of the fossil record. Even when species did coexist, the probabilities are that their first known occurrences will still be in the correct stratigraphic order. More detailed arguments to support these conclusions are presented in Paul (1982).

This argument becomes weaker when applied to higher taxa, largely because the periods of coexistence of higher taxa tend to be large relative to the differences between their times of origin. For example, most phyla have existed at least from the Cambrian to the present day. Detecting their order of origin in the Cambrian reliably is significantly more difficult because sampling this relatively brief interval adequately for all phyla is less likely than sampling the 500+ million years since they evolved. However, first stratigraphic records of all phyla are based on species. Equally, even cladograms depicting relationships between higher taxa use one or a few 'representative' species to characterize the higher taxa. To my mind a similar problem exists in both cases. Does the stratigraphically earliest species in a phylum closely approximate to the real time of origin, that is, is it truly representative of the time of origin? Similarly, are the species chosen to represent phyla or classes in cladograms truly representative of the higher taxa analysed? I would argue that if the species chosen are extant, they cannot possibly be representative of the character set that existed within the higher taxon 500 million years ago in the Cambrian. In both cases this is an argument for dealing with species rather than higher taxa, because we can apply more rigorous analytical tests to data on species.

The conclusion that the succession of species in the fossil record is reliable does not depend on its completeness. If the fossil record consisted of just two different fossils, the probability that they would occur in the correct stratigraphic order would still be overwhelmingly large. If you think about it, this conclusion should not be so surprising. If the order of species in the fossil record were not reliable, we could not correlate using fossils. So there really are sound reasons for believing that the order in which species occur in the fossil record is accurate (irrespective of its completeness). Hence it is legitimate to use the order of occurrence of fossil species as a test of inferred cladograms.

Detecting genuine ghost ranges

However, the problem still remains as to how we distinguish between an artificial ghost range and a genuine gap in the fossil record. Irrespective of cladistics, some ghost ranges must be genuine. In these cases as more fossil specimens are collected the first known occurrences of the species should become earlier in time, that is the ghost range should decrease. This is another example of investigating growth of knowledge or the 'test of time'. Paul (1980, 1982) used the idea to estimate the relative completeness of the fossil records of species, genera, and families of cystoids, and to show that an example from the neontological record (black smoker faunas) was probably less completely known than the fossil records of molluscs and echinoderms (Paul 1998). Foote (1997) used this approach to investigate growth of knowledge of morphological diversity. It has long been used in the form of species/area curves in plant ecology (e.g. Cain 1938) and is often called the 'collector curve'. Benton (1998) provided an example for dinosaur species. A curve of the general shape shown in Figure 5.5 should result. The curve is initially steep, but flattens out as more specimens are collected. In the case of ghost ranges, early discoveries should add to fossil ranges, but extending stratigraphic ranges will become progressively more difficult as more and more specimens are discovered.

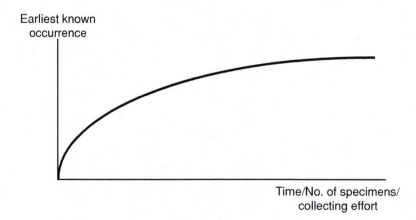

Figure 5.5 Hypothetical collector curve. As more effort is expended it becomes progressively more difficult to discover stratigraphically earlier occurrences.

Method

Data for first records of fossil families are adequate for the present purposes and readily available in the literature. Initially, I planned to compare the earliest records of families of fossils using the two versions of 'The Fossil Record' (Harland *et al.* 1967; Benton 1993) to see what proportion of fossil families had had their stratigraphic ranges extended back in time over the interval between 1967 and 1993. However, in many major groups there were significantly more families in the later volume, more than twice as many for teleosts and mammals, for example. Even with a group such as the graptolites, where the number of families was almost the same in both publications, there were still differences in the precise composition of many families due to taxonomic refinements and changes in classification between the two volumes. Thus the original approach proved impractical without a thorough knowledge of changes in classification for all major taxonomic groups over the last forty years or so.

An alternative method is to accept the classification in the later volume (Benton 1993) and record when historically the earliest record for a species within a family of fossils was first published. If stratigraphic ranges are known very well and genuine ghost ranges are rare, then it is probable that the earliest known stratigraphic record of a given family was discovered in the 19th century. Alternatively, if the fossil record is full of ghost ranges we should still be regularly discovering ever older stratigraphic occurrences within families. Thus the historical distribution of publication dates for the earliest stratigraphic occurrence of any species within a family should give us a measure of how fossil ranges of families have been extended over time. Fortunately Benton (1993) is a good source for the relevant information because most chapters include dates of publication of the earliest species or of the earliest stratigraphic record of species within families. It soon became apparent, however, that actual curves tend to be concave up, rather than convex up as the theory suggests. They have shallow initial portions and steep later portions (see Figure 5.7). The most likely reason for this is that the theoretical curve assumes the same search effort throughout, whereas there are far more palaeontologists actively publishing stratigraphic data now than there were, for example, in the 19th century.

Standard curve

A measure of the level of palaeontological activity is needed against which to compare real discovery curves. In this case the rate of description of new taxa is not an adequate measure, unlike the arguments used to estimate the completeness of the cystoid fossil record (Paul 1980, 1982). To extend a stratigraphic range requires at least two publications; the first to establish the stratigraphic occurrence and the second to extend the known range. Furthermore, this is an open-ended experiment. The fact that a second publication did not extend a stratigraphic range does not mean that subsequent publications will not. What is needed is a measure of total stratigraphic palaeontological effort. The measure I have used is the size of annual volumes of *Zoological Record* (Figure 5.6). This is not an ideal measure, if only because *Zoological Record* contains information on living animals as well as fossils, but it is a starting point. The curve in Figure 5.6 represents a cumulative measure of the total page size of annual volumes of *Zoological Record* taken from 1864 and then at decadal

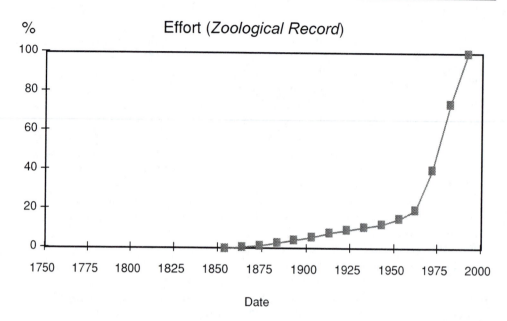

Figure 5.6 Cumulative palaeontological effort as measured by the size (total printed page area) of annual volumes of *Zoological Record*, at decadal intervals from 1864 to 1993.

intervals from 1873 to 1993. Total page size was used rather than page number because zoological record has changed page size from octavo to quarto to A4, the last being more than twice the print area per page of the first. Thus page number alone would seriously underestimate total zoological effort in more recent volumes. However, I have not attempted to allow for the change from single to double column page format, nor for any changes in font or font size. The curve is a relatively conservative estimate of the increase in zoological effort. It shows a sharp increase in publication rate from the 1960s onwards (Figure 5.6). I have not attempted a breakdown of publication rate into subdisciplines or taxonomic groupings.

Data

The data are taken from Benton (1993). Some comments are necessary since the data were not originally compiled with this particular aim in mind. I can make no comment on the mammals because the data refer to genera within families, not species, and it is possible that the precise record on which the first occurrence of a genus is based was described many years after the original description of the genus. Thus, for these purposes, the date of description of a genus is an unreliable source. Similarly, a number of records are based on specifically indeterminate material and commonly do not have an accompanying date. In addition, dates of description are occasionally omitted. For major taxa where such indeterminate records exceed five per cent of the total, both the total number of records and the number of 'unknowns' are given in the relevant figure (e.g. echinoderms in Figure 5.7c). Where two dates are given, the first is the date of description of the species and I have assumed that the second (later) date is the date of discovery of an earlier stratigraphic occurrence of the species. Occasionally, however, the later date is for a compendium of stratigraphic occurrences and may not reflect an extension of range. To be on the conservative side, I have consistently taken the later date as the time of discovery of the earliest occurrence of the fossil species. Where only one date is given, I have assumed that this is the date of discovery of the earliest occurrence of the species with one exception. A few records are of species described by Linné (1758). I have assumed here that these are Recent species discovered later as Pleistocene or Pliocene fossils. Subject to these caveats, Benton (1993) provides an invaluable source of data for most major taxonomic groups, with only a small proportion of 'doubtful' records.

Results

Results are shown in Figure 5.7A–D and are based on the relevant chapters in Benton (1993). The key points to look for are where the curves start and how steeply they rise towards the present day. In general, curves to the left of the 'effort' curve represent a 'good' record, those to the right a 'poor' record (e.g. conodonts, Figure 5.7D). Equally, the more concave a curve is, or the steeper the slope of the most recent portion, the 'worse' the fossil record. The sponges and to a lesser extent the cnidarians, show interesting subdivisions. Sponges plus stromatoporoids have an apparently good record, whereas archaeocyathids have a significantly worse record. It is perhaps significant that both 'problematica' and 'miscellanea' have curves closest to the effort curve, suggesting that these enigmatic fossils are discovered in exact proportion to the effort expended in looking for fossils in general.

The bivalves and teleost fish have curves that begin before 1775, the non-ostracod Crustacea, Zoantharia, and 'other cnidarians' have curves that begin before 1800. However, most groups (24 out of 35 total) have curves that start between 1800 and 1850. The archaeocyathid curve starts between 1900 and 1925, whereas that of conodonts not until between 1925 and 1950. To some extent this may reflect the relatively recent change from element-based conodont taxa to apparatus-based taxa. However, this cannot be the entire explanation, since newly recognized apparatus-based taxa will still take the oldest available name for any element in the apparatus, in the same way that the conodont animal, first recognized in 1983 (Briggs et al. 1983), was tentatively referred to a species described in 1969.

The conodonts are an example of a somewhat unexpected result that stratigraphically important groups appear to have relatively poor records. Only the foraminifera and nautiloids have curves that start before 1825, ammonites and trilobites start before 1850, ostracods and graptolites before 1875 and conodonts not until after 1925. Of the stratigraphically important groups, the graptolites seem to have the best record, with the steepest part of the curve between 1925 and 1950. The foraminifera and nautiloids have the next best records with the steepest part of the curves between 1950 and 1975. All the other groups have the steepest parts of their curves between 1975 and 2000, suggesting that stratigraphic ranges of their constituent families are still being extended back in time on a regular basis. Perhaps range extension is not so unexpected in groups whose stratigraphic occurrence is considered significant and therefore recorded accurately and repeatedly.

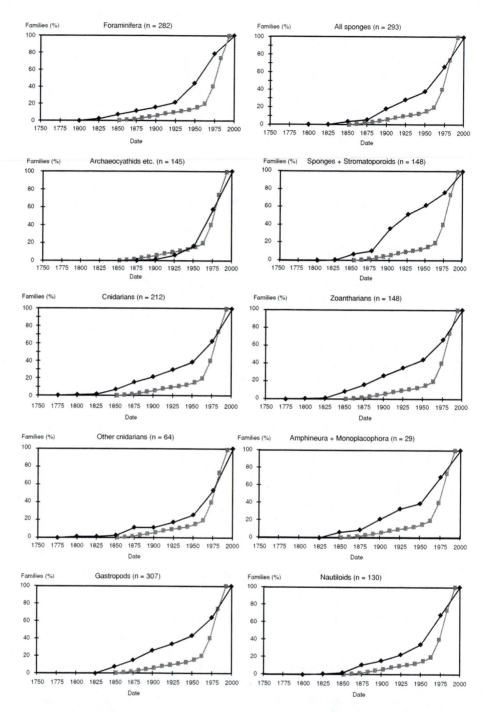

Figures 5.7A–D Historical distributions of publication of the earliest stratigraphic records for fossil families (diamonds). Data from Benton (1993) plotted at 25-year intervals and compared with the palaeontological effort curve (squares). Almost invariably fossil groups have curves to the left of the effort curve, suggesting that they are better known than would be expected from random discovery. See text for further explanation.

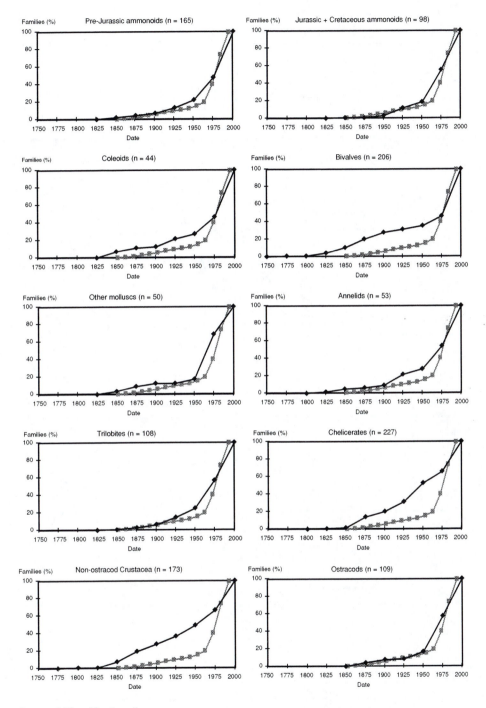

Figures 5.7B   (Continued)

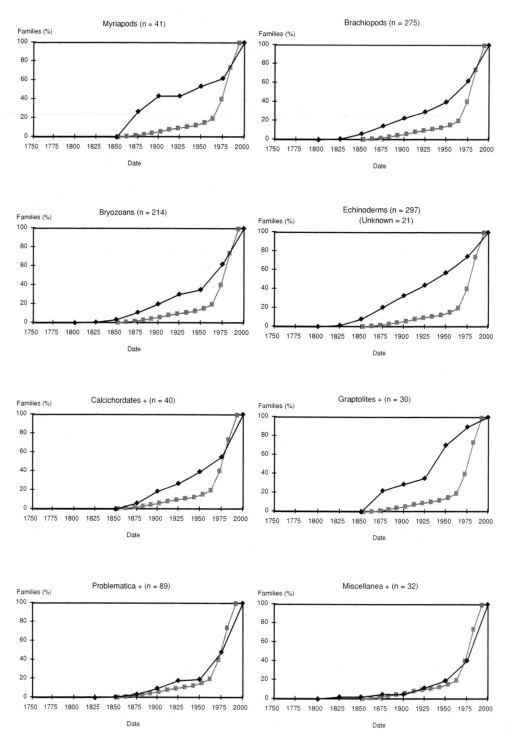

Figures 5.7C   (Continued)

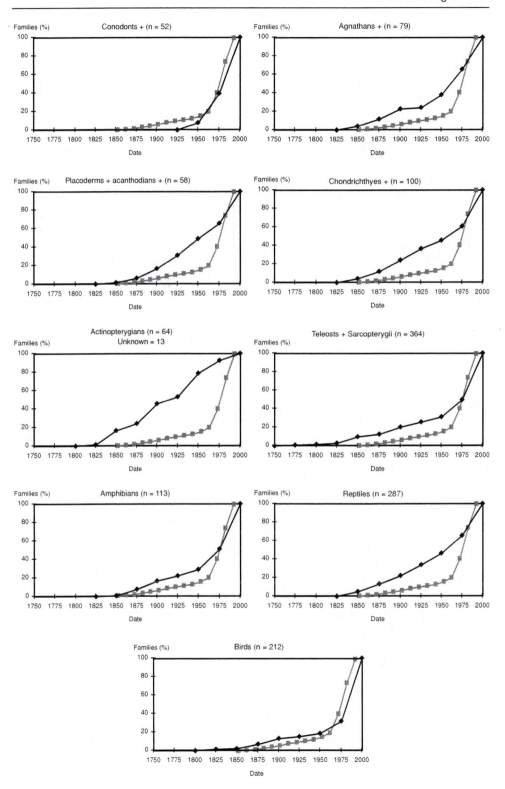

Figures 5.7D   (Continued)

Discussion

The prime problem with techniques such as this is to derive a standard against which a specific curve can be compared. The *Zoological Record* curve reflects total zoological effort, not stratigraphic effort on fossil data alone. This may explain why most curves for fossil groups lie to the left of the *Zoological Record* curve. Furthermore, the *Zoological Record* curve only starts in 1864 which is later than about half the curves for fossil taxa. However, this does seem to represent a promising approach. If a curve could be derived that reflected stratigraphic effort reasonably accurately, one could use it to predict the probability that a taxon of a given size would have a record beginning at any specified time.

An alternative approach is to consider the shape of the curve for all fossil groups. At least then one could say which taxa were better (or worse) than the average for the fossil record as a whole. However, it would still be impossible to say whether or not the data for the fossil record as a whole are better or worse than predicted on an hypothesis of random discovery. Note, too, that the fossil data used here were not collected for this particular purpose and are somewhat suspect.

Conclusions

(1)   Ghost ranges may be artefacts of cladistic methodology or the result of genuine gaps in the fossil record.

(2)   The stratigraphic order in which fossils occur accurately reflects the order in which they evolved, irrespective of the completeness (or otherwise) of the fossil record.

(3)   It follows from the above that it is unwise to use cladistic methodology to test the completeness of the fossil record, but congruence with known stratigraphic occurrences is a powerful test of cladograms.

(4)   If the fossil record contains many ghost ranges, earliest stratigraphic occurrences of fossil families would continually be extended back in time as new discoveries are made.

(5)   Such collector curves offer a potential test of the reliability of the fossil record, but deriving a standard for comparison is not easy. Here I have used the size of annual volumes of the *Zoological Record* at decadal intervals from 1864 to 1993.

(6)   Most fossil groups have curves that lie to the left of the *Zoological Record* curve, but a few start very late and/or parallel the *Zoological Record* curve closely.

(7)   A few fossil groups show signs of flattening out towards the present day despite the increased effort in recent years shown by the *Zoological Record* curve.

(8)   Both of these points suggest that the stratigraphic ranges of at least some fossil groups are better known than would be expected from random discovery (= collecting effort).

References

Benton, M.J. (1993) *The Fossil Record 2*, London: Chapman & Hall.
—— (1998) 'The quality of the fossil record of the vertebrates', in S.K. Donovan and C.R.C. Paul (eds) *The Adequacy of the Fossil Record*, Chichester: John Wiley & Sons, pp. 269–303.

Benton, M.J. and Hitchin, R. (1997) 'Congruence between phylogenetic and stratigraphic data on the history of life', *Proceedings of the Royal Society, London*, B264: 885–90.

Benton, M.J. and Simms, M.J. (1995) 'Testing the marine and continental fossil records', *Geology*, 23: 601–4.

Benton, M.J. and Storrs, G.W. (1994) 'Testing the quality of the fossil record: paleontological knowledge in improving', *Geology*, 22: 111–14.

—— (1996) 'Diversity in the past: comparing cladistic phylogenies and stratigraphy', in M.E. Hochberg, J. Clobert and R. Barbault (eds) *Aspects of the Genesis and Maintenance of Biological Diversity*, Oxford: Oxford University Press, pp. 19–40.

Briggs, D.E.G., Clarkson, E.N.K. and Aldridge, R.J. (1983) 'The conodont animal', *Lethaia*, 16: 1–14.

Cain, S.A. (1938) 'The species–area curve', *American Midland Naturalist*, 19: 573–81.

Clyde, W.C. and Fisher, D.C. (1997) 'Comparing the fit of stratigraphic and morphologic data in phylogeny analysis', *Paleobiology*, 23: 1–19.

Foote, M. (1997) 'Sampling, taxonomic description, and our evolving knowledge of morphological diversity', *Paleobiology*, 23: 181–206.

Foote, M. and Raup, D.M. (1996) 'Fossil preservation and the stratigraphic ranges of taxa', *Paleobiology*, 22: 121–40.

Harland, W.B., Holland, C.H., House, M.R., Hughes, N.F., Reynolds, A.B., Rudwick, M.J.S., Satterthwaite, G.E., Tarlo, L.B.H. and Willey, E.C. (eds) (1967) *The Fossil Record*, London: The Geological Society.

Hitchin, R. and Benton, M.J. (1997) 'Congruence between parsimony and stratigraphy: comparisons of three indices', *Paleobiology*, 23: 20–32.

Huelsenbeck, J.P. (1994) 'Comparing the stratigraphic record to estimates of phylogeny', *Paleobiology*, 20: 470–83.

Huelsenbeck, J.P. and Rannala, B. (1997) 'Maximum likelihood estimation of phylogeny using stratigraphic data', *Paleobiology*, 23: 174–80.

Linné, C. von (1758) *Systema naturae, sive regna tria naturae systematice proposita per classes, ordines, genera et species* (10th edn), Holmiae.

Marshall, C.R. (1990) 'Confidence intervals on stratigraphic ranges', *Paleobiology*, 16: 1–10.

—— (1994) 'Confidence intervals on stratigraphic ranges: partial relaxation of the assumption of randomly distributed fossil horizons', *Paleobiology*, 20: 459–69.

—— (1997) 'Confidence intervals on stratigraphic ranges with non-random distributions of fossil horizons', *Paleobiology*, 23: 165–73.

—— (1998) 'Determining stratigraphic ranges', in S.K. Donovan and C.R.C. Paul (eds) *The Adequacy of the Fossil Record*, Chichester: John Wiley & Sons, pp. 23–53.

Norrell, M. (1992) 'Taxic origin and temporal diversity: the effect of phylogeny', in M.J. Novacek and Q.D. Wheeler (eds), *Extinction and Phylogeny*, New York: Columbia University Press, pp. 89–118.

Patterson, C. (1981) 'Significance of fossils in determining evolutionary relationships', *Annual Review of Ecology and Systematics*, 12: 195–223.

Paul, C.R.C. (1980) *The Natural History of Fossils*, London: Weidenfeld & Nicholson.

—— (1982) 'The adequacy of the fossil record', in K.A. Joysey and A.E. Friday (eds) *Problems of Phylogenetic Reconstruction*, Systematics Association Special Volume 21, London: Academic Press, pp. 75–117.

—— (1985) 'The adequacy of the fossil record reconsidered', in J.C.W Cope and P.W. Skelton (eds) Evolutionary case histories from the fossil record, *Special Papers in Palaeontology*, 33: 7–16.

—— (1998) 'Adequacy, completeness and the fossil record', in S.K. Donovan and C.R.C. Paul (eds) *The Adequacy of the Fossil Record*, Chichester: John Wiley & Sons, pp. 1–22.

Peters, S.E. and Foote, M. (2001) 'Biodiversity in the Phanerozoic: a reinterpretation', *Paleobiology*, 27: 583–601.

Sepkoski, J.J. Jr (1978) 'A kinematic model of Phanerozoic taxonomic diversity I. Analysis of marine orders', *Paleobiology*, 4: 223–51.

Sepkoski, J.J. Jr, Bambach, R.K., Raup, D.M. and Valentine, J.W. (1981) 'Phanerozoic marine diversity patterns and the fossil record', *Nature*, 293: 435–7.

Smith, A.B. (1994) *Systematics and the fossil record – documenting evolutionary patterns*, Oxford: Blackwell Scientific.

Smith, A.B. and Littlewood, D.T.J. (1994) 'Paleontological data and molecular phylogenetic analysis', *Paleobiology*, 20: 259–73.

Strauss, D. and Sadler, P.M. (1989) 'Classical confidence intervals and Bayesian probability estimates for ends of local taxon ranges', *Mathematical Geology*, 21: 411–27.

Wagner, P.J. (1995) 'Stratigraphic tests of cladistic hypotheses', *Paleobiology*, 21: 153–78.

Wagner, P.J. (1998) 'A likelihood approach for evaluating estimates of phylogenetic relationships among fossil taxa', *Paleobiology*, 24: 430–49.

Chapter 6

Episodic evolution of nuclear small subunit ribosomal RNA gene in the stem-lineage of Foraminifera

Jan Pawlowski and Cédric Berney

ABSTRACT

Several studies provide evidence that the rates of nucleotide substitution vary across lineages within evolutionary radiations. Little is known, however, about the episodic variations of rates in the stem-lineages leading to these radiations. Phylogenetic analyses of nuclear small subunit ribosomal RNA gene of Foraminifera, calibrated with fossil data, demonstrate extreme rates variations between different lineages within the foraminiferan radiation. Our present data also show a significant increase of substitution rates in the stem-lineage of Foraminifera. The duration of this stem-lineage was estimated at a maximum of 150–250 million years. Based on this conservative calibration, we calculated that the rate of substitution of the foraminiferan stem-lineage averaged 1.0–1.65 substitutions/1000 sites per 10^6 years. This is more than 30 times faster than the typical rates observed within the radiation of benthic Foraminifera. Such high rates can be related to a G/C bias in base composition, which apparently led to a reinforcement of the secondary structure of the foraminiferan SSU rRNA. Our study shows that an acceleration of substitution rates in a stem-lineage may lead to an erroneous inference of the phylogenetic position of a group and to an overestimate of its divergence time.

Introduction

Accurate prediction of evolutionary time from molecular data depends largely on the homogeneity of nucleotide substitution rates. However, with increasing DNA sequence data, there is growing evidence that the substitution rates vary across lineages and over time. Numerous studies demonstrate the important variations of evolutionary rates between and within different taxonomic groups (Britten 1986; Li et al. 1987; Bousquet et al. 1992; Philippe et al. 1994; Sorhannus 1996; Ayala 1997; Friedrich and Tautz 1997; Hwang et al. 1998; Van de Peer et al. 2000). These variations have been proposed to be due to evolutionary changes in DNA replication or repair mechanisms (Britten 1986), to the generation-time effect (Li et al. 1996), or to changes of metabolic rates (Martin and Palumbi 1993), but none of these hypotheses seems to be sufficiently confirmed by available data (Li 1993).

Several studies show an episodic acceleration of evolutionary rates in protein coding genes. Acceleration–deceleration patterns were reported in the evolution of globin and cytochrome c in vertebrates (Goodman 1981), the vesicular stomatitis virus in cattle (Nichol *et al.* 1993), lysozymes in primates (Messier and Stewart 1997), and the GPDH gene in the Diptera (Ayala *et al.* 1996). Bursts of rapid changes were also observed in the growth hormone in primates (Liu *et al.* 2001; Wallis *et al.* 2001) and other protein hormones in mammals (Wallis 2000, 2001). Episodic changes of substitution rates in proteins are usually explained by positive adaptive selection, and often viewed as an argument against the neutral allele theory and the molecular clock hypothesis (Gillespie 1984, 1993). However, the evidence for a functional impact of rapid changes is not always clear, suggesting that positive selection may not be the only cause of episodic acceleration of substitution rates. For example, the episodic evolution of xanthine dehydrogenase was suggested to be related to the particularities of the genomes in which the locus is embedded (Rodríguez-Trelles *et al.* 2001).

Episodic variations of substitution rates have also been reported in large- and small-subunit ribosomal RNA (LSU and SSU rRNA) genes, the universally conserved function of which excludes positive selection. A 10-fold acceleration of substitution rates, probably caused by a change of DNA base composition, was reported in ribosomal genes of the dipteran stem-lineage (Friedrich and Tautz 1997). Here we present the example of an even faster (> 30-fold) episodic change in evolutionary rates of the SSU rRNA gene in Foraminifera.

Foraminifera as a tool to study evolutionary rates

Foraminifera are the most important group of microfossils widely used in micropalaeontology for the stratigraphic analysis of ancient sediments and for palaeoecological and palaeoclimatic reconstructions (Culver 1993; Wilson and Norris 2001). Foraminiferan tests are abundant and widespread in sediment samples. Species identification is sometimes difficult, but the recognition of genera and higher taxonomic units is usually relatively easy. This makes Foraminifera an excellent tool for stratigraphic purposes. The quality of the foraminiferan fossil record is particularly good for planktonic species, in which occurrences in marine strata have been precisely calibrated by radiometric dating and tested with independent biochronological datasets (Berggren *et al.* 1995).

Despite the abundance and diversity of Foraminifera, their use in molecular evolutionary studies is obstructed by technical problems. Obtaining large amounts of foraminiferan DNA is hindered by their slow growth and the difficulty of getting them to reproduce in laboratory cultures. Moreover, because many shallow water Foraminifera live in association with symbiotic algae and epiphytic micro-organisms, their DNA extractions are often contaminated with foreign eukaryotic DNA. At present, the foraminiferan DNA database is composed mainly of ribosomal genes, for which specific foraminiferan PCR primers exist (Pawlowski 2000) and whose large number of copies allows amplification even from single cells. The first sequences of foraminiferan protein-coding genes, including actin, tubulin, and RNA polymerase, have been obtained only recently (Pawlowski *et al.* 1999; unpublished data). For this reason, the present study is limited to analyses of ribosomal data.

Heterogeneity of the substitution rates in foraminiferan rRNA genes

Analyses of partial SSU rDNA sequences show important differences in the evolutionary rates between major taxonomic groups of Foraminifera (Pawlowski *et al.* 1997). Results of a re-analysis of previously published data (Pawlowski *et al.* 1997; de Vargas and Pawlowski 1998), with additional pairs of benthic and planktonic foraminiferan genera and species, and using an enhanced alignment comprising 620 unambiguously aligned positions, are summarized in Table 6.1.

Pairwise-distance rates were calculated using distances corrected according to Kimura's two-parameter model of substitutions; however, as all pairs of species are closely related, very similar results were obtained using the more complex GTR model of substitutions, taking into account a proportion of invariant sites and a gamma distribution of the rates of substitution across sites, with an alpha distribution shape parameter calculated from the dataset (data not shown). The values obtained are slightly different from those previously published. Observed discrepancies may be explained by the fact that calculation of rates of substitution depends directly on the choice of sites used in the analyses. In this study, we used a slightly larger number of sites in order to avoid very small numbers of differences between the slowest evolving lineages. However, the differences between the values presented here (Table 6.1) and previously published data are smaller than 50 per cent and the ratios between substitution rates of different lineages remain similar.

In the Foraminifera, the slowest rates of substitution (0.02–0.1 substitutions/1000 sites per 10^6 years) are observed in the clade of textulariids and rotaliids, two groups that are characterized by multi-chambered tests, having agglutinated (Textulariida) or calcareous, perforate (Rotaliida) walls. Miliolids, a group of multi-chambered Foraminifera with calcareous, porcellaneous tests, exhibit rates that are five to 10 times higher than in the textulariids and rotaliids. But the most rapid substitution rates are observed in planktonic species; these rates are more than 100 times faster than in the most slowly evolving benthic lineages (Table 6.1).

Such variations contrast sharply with values calculated in other eukaryotes. Friedrich and Tautz (1997) showed the fast-evolving Diptera to have only two- to four-fold higher rates of substitution than other holometabolous insects for LSU and SSU rRNA genes. Similarly, Smith *et al.* (1992) reported statistically significant differences in evolutionary rates between different lineages of echinoids, based on LSU rRNA data. But again, the rate difference between lineages did not exceed a three-fold magnitude. More important differences can be observed in protein-coding genes. Li *et al.* (1987) estimated substitution rates in rodents to be four to 10 times higher than those in higher primates, and two to four times higher than those in artiodactyls. Exceptionally high variations (up to 138 per cent for non-synonymous substitutions) were observed by Bousquet *et al.* (1992) in rbcL sequences of seed plants, which nevertheless remain small compared with those in the Foraminifera.

The rates seem to be much more uniform within some foraminiferan lineages, suggesting the existence of local molecular clocks. In the large porcellaneous Foraminifera of the superfamily Soritacea (Miliolida), the rates of SSU rRNA evolution vary from 0.25–0.5 substitutions/1000 sites per 10^6 years, with a mean value of 0.34 substitutions/1000 sites per 10^6 years (Table 6.1). The rRNA rates are also

Table 6.1 Heterogeneity of rates of substitution in pairs of phylogenetically related genera and species

	Time of divergence (T)	K2-dist. (K)[a]	PDR[b]	ILR 1[c]	LR 2[c]
(1) Rotaliids					
Nummulites – Cyclochypeus	37–66 Ma	0.004688	0.036–0.063	0.03	0.081
Operculina – Heterostegina	58–66 Ma	0.007849	0.06–0.068	0.038	0.095
Islandiella – Globocassidulina	40–66 Ma	0.007839	0.059–0.098	0.091	0.05
Bolivina – Rosalina	> 70 Ma	0.040160	< 0.287	< 0.286	< 0.314
(2) Globigerinids					
Globigerinoides conglobatus – G. ruber	5–10 Ma	0.11080	5.54–11.08	4.1–8.2	7.1–14.2
Globigerinoides sacculifer – Orbulina universa	17–22 Ma	0.21178	4.81–6.23	6.27–8.12	3.98–5.15
(3) Globorotaliids					
Globorotalia inflata – G. scitula	17–18.5 Ma	0.043624	1.18–1.28	0.649–0.706	2.43–2.65
Neogloboquadrina duterrei – Pulleniatina obliquiloculata	6 Ma	0.010997	0.916	1	0.833
Globorotalia hirsuta – G. truncatulinoides	5.1–6 Ma	0.14298	11.91–14.02	2.33–2.75	21–24.71
(4) Textulariids					
Textularia – Bigenerina	58–66 Ma	0.003125	0.024–0.027	0.03	0.026
Ammobaculites – Ammotium	> 130 Ma	0.025351	< 0.098	< 0.115	0.077
Trochammina – Spiroplectammina	300–350 Ma	0.012587	0.018–0.021	0.013	0.023
(5) Miliolids					
Borelis – Alveolinella	17–40 Ma	0.019006	0.238–0.559	0.325	0.353
Archaias – Cyclorbiculina	24–37 Ma	0.018197	0.246–0.379	0.338	0.25
Sorites – Marginopora	16–24 Ma	0.015793	0.329–0.494	0.361	0.5
Sorites – Amphisorus	24 Ma	0.013418	0.28	0.25	0.312

[a] Number of substitutions per site between the two sequences out of 620 sites, after correction for multiple hits according to Kimura's two-parameter model.

[b] Pairwise-distance rate (in substitutions/1000 sites per 10^6 years) calculated as PDR = K/2T (Li and Graur 1991).

[c] Individual lineage rates (in substitutions/1000 sites per 10^6 years) of the first, respectively the second taxon of the pair, calculated directly on maximum likelihood tree reconstructions by attributing divergence times to either terminal or terminal and internal branch lengths (as proposed by Hillis et al. 1996).

relatively stable in a family of planktonic Foraminifera, the Globigerinidae (de Vargas *et al.* 1997). The mean evolutionary rate in the Globigerinidae was shown to average 4.3 substitutions/1000 sites per 10^6 years, with values ranging from 4.0–4.6 substitutions/1000 sites per 10^6 years in pairwise comparisons, and from 3.2–6.1 substitutions/1000 sites per 10^6 years, when they were calculated on individual lineages (de Vargas and Pawlowski 1998). The existence of local molecular clocks was also observed by Smith *et al.* (1992) for LSU rRNA in individual echinoid lineages, and by O'h Uigin and Li (1992) for numerous protein-coding genes in rodents.

On the other hand, much higher SSU rRNA rate variations are observed in another family of planktonic Foraminifera, the Globorotaliidae, where pairwise divergence rates were shown to vary from 0.3–5.3 substitutions/1000 sites per 10^6 years (de Vargas and Pawlowski 1998). Re-examination of the data in one of the rapidly evolving species, *Globorotalia truncatulinoides*, by including additional species of Globorotaliidae and with a revision of their divergence times, suggests even higher values of 19.7–24.6 substitutions/1000 sites per 10^6 years, about 20 times faster than the rates of the most slowly evolving representatives of this family (de Vargas *et al.* 2001; see also Table 6.1). Interestingly, the exceptional acceleration of the substitution rate in *G. truncatulinoides* seems to result from a progressive increase of the rates in the phylogenetic lineage leading to this species, from 2.25 substitutions/1000 sites per 10^6 years in *Globorotalia scitula* to 4.1–5.2 substitutions/1000 sites per 10^6 years in *Globorotalia hirsuta*, the two species the most closely related to *G. truncatulinoides* (de Vargas *et al.* 2001).

Acceleration of the foraminiferan stem-lineage

The most striking example of rate variation in the Foraminifera is the acceleration of their stem-lineage, as evidenced by analyses of 1175 unambiguously aligned positions of complete SSU rRNA sequences. When sequences of Foraminifera are compared with those of other eukaryotes, their stem-lineage appears as a very long branch leading to a radiation of all foraminiferan lineages (Figure 6.1).

Fossil calibration

It is impossible to calibrate precisely the foraminiferan stem-lineage acceleration because no fossil data exist either for the divergence of the Foraminifera from their eukaryotic ancestor or for the beginning of the radiation of extant foraminiferan lineages. We can only propose a very conservative estimate of the time of the foraminiferan stem-lineage evolution. The divergence between Foraminifera and their eukaryotic ancestor probably occurred at less than 1000 Ma. This estimation corresponds to the radiation of 'crown' Eukaryotes, based on the palaeontological record (Knoll 1992). As the Foraminifera apparently evolved from some filose sarcodinids within this radiation (Keeling 2001; unpublished data), it seems reasonable to consider that their divergence occurred at around this time.

On the other hand, the radiation of Foraminifera probably occurred at 750–850 Ma. Although the first fossil Foraminifera appear only in the Early Cambrian, about 550 Ma (Culver 1991; McIlroy *et al.* 2001), there is molecular evidence for a much earlier radiation of non-fossilized lineages of naked or organic-walled,

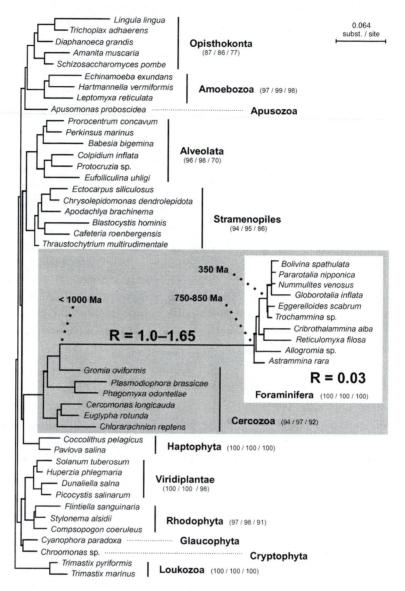

Figure 6.1 Phylogenetic position of the Foraminifera among 'crown' Eukaryotes, inferred from 50 complete SSU rRNA sequences with the maximum likelihood method, using fastDNAml (Olsen *et al.* 1994). Numbers next to the names of the eukaryotic clades are their bootstrap support values after 200, 800, and 400 replicates for maximum likelihood, neighbour-joining, and maximum parsimony analyses of the data, respectively. Bootstrap support percentages between and within these clades are omitted for clarity. The Foraminifera are most probably derived from within the Cercozoa (grey box), as indicated by the high support values. Probable divergence times are given according to fossil data or as deduced from partial SSU rRNA tree calibration; calculated rates of substitution (in changes/1000 sites per 10^6 years) are indicated for the stem-lineage leading to the Foraminifera and for the radiation of the group (see text for more details). The tree was arbitrarily rooted with two sequences from the genus *Trimastix*. All branches are drawn to scale.

single-chambered allogromiid Foraminifera that existed far before their skeletonized remnants were found and identified in the fossil record (Pawlowski *et al.* 1999, 2002; unpublished data). Indeed, the high genetic and morphological diversity of allogromiid Foraminifera suggests that their diversification probably occurred in the Neoproterozoic.

In order to obtain a more precise estimate for the timing of this radiation, we used a global phylogeny of Foraminifera based on partial SSU rRNA sequences (data not shown). We first performed a relative rate test (Robinson-Rechavi and Huchon, 2000) in order to exclude all sequences evolving significantly slower or faster than the others at a one per cent level, so that we could approximate a global molecular clock on the dataset. Then we calibrated the tree using the time of divergence of the first multi-chambered textulariid Foraminifera, at 350 Ma (Ross and Ross, 1991). We calculated the mean rate of substitution in the clade of textulariid and rotaliid Foraminifera and applied it to estimate the time of radiation of all foraminiferan lineages. However, the position of the root in foraminiferan phylogenies is quite arbitrary, because complete SSU rRNA sequences from many single-chambered lineages are still missing. Furthermore, because of the lack of fossil data for the allogromiid Foraminifera, we cannot exclude that the rates of substitution in these lineages are globally higher than the rates of substitution in the textulariid + rotaliid clade. So after testing different root positions we retain a proposed interval of 750–850 Ma for the initial foraminiferan radiation.

Estimation of substitution rates

According to our calibration, duration of evolution of the foraminiferan stem-lineage can be estimated to a maximum of 150–250 million years. Using the topology of Figure 6.1 and divergence times inferred from fossil and molecular data as discussed above, we calculated that the rate of substitution in the most conserved regions of SSU rRNA genes averaged at least 1.0 to 1.65 substitutions/1000 sites per 10^6 years during the stem-lineage evolution of the Foraminifera. This rapid burst of the evolutionary rate was followed by a return to a 'normal' value of about 0.03 substitutions/1000 sites per 10^6 years during the subsequent radiation of the different foraminiferan lineages. This value is comparable with evolutionary rates calculated in other groups of Eukaryotes (see for instance Sorhannus 1996).

A substantial slowdown of rates might even be observed in some lineages of Foraminifera, for example in the clade of textulariids and rotaliids, the two groups where the slowest rates of substitution were observed (Table 6.1). Conversely, secondary accelerations occurred in some groups, especially planktonic taxa, as shown by their high rates of substitution for ribosomal genes (Table 6.1). Most spectacular is the progressive increase of the rates in the phylogenetic lineage leading to *G. truncatulinoides* (de Vargas *et al.* 2001), as this species is less than six million years old.

Characterization of the stem-branch leading to the Foraminifera

A more detailed study of the complete SSU rRNA sequences used in Figure 6.1, taking into account the secondary structure model proposed by Neefs *et al.* (1993), reveals that 170 particular nucleotides in specific positions define the Foraminifera, when compared with 'crown' Eukaryotes. These particular nucleotides are not randomly scattered throughout the molecule: 57 per cent of them are the result of substitutions in more variable, external regions of the SSU rRNA secondary structure (49 per cent of the alignment), whereas 24 per cent of them are the result of substitutions in internal single-strand regions (21 per cent of the alignment), and only 19 per cent of them are the result of substitutions in the most conserved, internal stems (30 per cent of the alignment). Besides, several of the substitutions that occurred in stem regions, especially at the periphery of the molecule, apparently led to local modifications of the secondary structure of the SSU rRNA, which can concern only a few base pairings or extend to a whole helix. Furthermore, expansions specific to the Foraminifera can be observed in several terminal and internal loops (unpublished data).

The 170 substitutions that occurred in the stem-lineage leading to Foraminifera affected the base composition of the molecule. The mean G/C content of the 40 non-foraminiferan sequences that we used in our analyses is 44.7 per cent, ranging from 43 per cent in some ciliates to 47 per cent in the haptophytes, whereas it is 48.2 per cent for the foraminiferan sequences, ranging from 47.2 per cent in *Globorotalia inflata* to 48.9 per cent in *Reticulomyxa filosa*. However, a spatial zonation can be observed in the G/C content of foraminiferan SSU rRNA. In the most conserved internal helices and loops, it ranges from 48 to 50 per cent, whereas in the external hairpins, it ranges from 45.5 to 47.8 per cent. Finally, in the universal variable regions V1 to V9, which were excluded from our phylogenetic analyses, the mean G/C content is very low (24.5 per cent, ranging from 20 per cent in some allogromiids to 39 per cent in some rotaliids), and in the regions of expansion specific to Foraminifera, it can even get close to zero in some groups.

These observations suggest that in the stem-lineage leading to Foraminifera, a base composition bias towards G/C led to a reinforcement of the internal secondary structure of the SSU rRNA. Then, an inverse, secondary base composition bias towards A/T occurred in the external, more variable regions of the molecule. However, this secondary bias did not, apparently, affect all lineages in equal proportion, and did not affect the more conserved regions of the SSU rRNA, where the specific substitutions that accumulated during the stem-lineage acceleration became fixed in most species. This would mean that different bias and constraints affect different parts of the molecule.

Implications for phylogenetic inference and estimation of divergence times

The episodic evolution of ribosomal genes in the foraminiferan stem-lineage has a strong impact on the inference of the phylogenetic position of the group. In previous phylogenetic analyses of SSU and LSU rRNA genes, Foraminifera are placed in the lower part of the eukaryotic tree, among the earliest mitochondrial protists (Pawlowski *et al.* 1994, 1996). This position was interpreted as the result of rapid rates in LSU and SSU rRNA (Sogin 1997). Indeed, the analysis of foraminiferan actin-coding gene sequences suggests a more recent origin for the group (Pawlowski *et al.* 1999), which probably evolved from filose protists belonging to the clade Cercozoa (Keeling 2001). The close relationship between Foraminifera and cercozoans is also confirmed by analysis of SSU sequences in the absence of fast evolving lineages (unpublished data; see Figure 6.1).

The Foraminifera are not the only example of misplacement in rRNA-based phylogenies of protists, the best known being the case of Microsporidia, first considered to be very primitive Eukaryotes (Vossbrinck et al. 1987), and now regarded as highly derived, secondarily amitochondriate members of the Fungi (Hirt et al. 1999; Keeling et al. 2000). However, the lack of a fossil record for this group makes it very difficult to determine whether their rapid evolution is due to an episodic change of substitution rates or to other factors. The same uncertainty exists for most other protists exhibiting high rates of substitution, as groups with a satisfactory fossil record are unfortunately very scarce.

The rate acceleration in the foraminiferan stem-lineage also provides an example of the kind of bias that can be introduced in the estimation of divergence times. If we calculate the time of divergence between the Foraminifera and other Eukaryotes using the rates estimated for different foraminiferan lineages, we will obtain the absurd value of more than 9 Ga. In general, any episodic acceleration in a stem-lineage will lead to overestimating the divergence time by the same factor for all groups belonging to this lineage, each time the calibration is done using the sister group. Curiously, the variation of rates in stem-lineages is usually not taken into account in the methods that have been developed to estimate divergence times (see for instance Rambaut and Bromham 1998).

The Foraminifera and the Diptera are certainly not the only lineage of Eukaryotes where such acceleration–deceleration patterns in the substitution rates of rRNA genes might be observed, although in most cases it is probably not as striking – and not as easy to reveal – as it is with Foraminifera. Several groups of organisms can be found at all taxonomic levels, whose stem-lineage appears as a very long branch in rRNA-based trees, and in which episodic rate changes are very likely to have happened. These include for example the bilaterian metazoans (Giribet and Wheeler 1999), the florideophycean red algae (Ragan et al. 1994), Plasmodium (Van de Peer and De Wachter 1997), and of course even the Eukaryotes as a whole (Winker and Woese 1991; Morin 2000). Nevertheless, the putative episodic nature of the observed differences in substitution rates of SSU rRNA genes is generally not discussed in these studies.

Today, it is a common practice to discard the fast evolving species or lineages from phylogenetic analyses and estimations of divergence times. Such a radical process merely avoids the problem, rather than solving it. We think that a better understanding of the patterns of molecular rate variations, including their episodic character, may be more useful for accurate prediction of evolutionary origins and divergences.

Acknowledgements

The authors thank Maria Holzmann, Colomban de Vargas, and José Fahrni for co-operation in molecular data collecting. This study was supported by Swiss National Science Foundation grant 31–59145.99.

References

Ayala, F.J. (1997) 'Vagaries of the molecular clock', Proceedings of the National Academy of Sciences, USA, 94: 7776–83.

Ayala, F.J., Barrio, E. and Kwiatowski, J. (1996) 'Molecular clock or erratic evolution? A tale of two genes', *Proceedings of the National Academy of Sciences, USA*, 93: 11729–34.

Berggren, W.A., Kent, D.V., Aubry, M.P. and Hardenbol, J. (eds) (1995) *Geochronology, Time Scales and Global Stratigraphic Correlations*, Tulsa, USA: SEPM, Special Publication no. 54, Society for Sedimentary Geology.

Bousquet, J., Strauss, S.H., Doersken, A.H. and Price, R.A. (1992) 'Extensive variation in evolutionary rate of rbcL gene sequences among seed plants', *Proceedings of the National Academy of Sciences, USA*, 89: 7844–8.

Britten, R.J. (1986) 'Rates of DNA sequence evolution differ between taxonomic groups', *Science*, 231: 1393–8.

Culver, S.J. (1991) 'Early Cambrian Foraminifera from West Africa', *Science*, 254: 689–91.

—— (1993) 'Foraminifera', in J.H. Lipps (ed.) *Fossil Prokaryotes and Protists*, Boston, MA: Blackwell, pp. 203–47.

de Vargas, C. and Pawlowski, J. (1998) 'Molecular versus taxonomic rates of evolution in planktonic Foraminifera', *Molecular Phylogenetics and Evolution*, 9: 463–9.

de Vargas, C., Zaninetti, L., Hilbrecht, H. and Pawlowski, J. (1997) 'Phylogeny and rates of molecular evolution of planktonic Foraminifera: SSU rDNA sequences compared to the fossil record', *Journal of Molecular Evolution*, 45: 285–94.

de Vargas, C., Renaud, S., Hilbecht, H. and Pawlowski, J. (2001) 'Pleistocene adaptive radiation in *Globorotalia truncatulinoides*: genetic, morphologic, and environmental evidence', *Paleobiology*, 27: 104–25.

Friedrich, M. and Tautz, D. (1997) 'An episodic change of rDNA nucleotide substitution rate has occurred during the emergence of the insect order Diptera', *Molecular Biology and Evolution*, 14: 644–53.

Gillespie, J.H. (1984) 'The molecular clock may be an episodic clock', *Proceedings of the National Academy of Sciences, USA*, 81: 8009–13.

—— (1993) 'Episodic evolution of RNA viruses', *Proceedings of the National Academy of Sciences, USA*, 90: 10411–12.

Giribet, G. and Wheeler, W.C. (1999) 'The position of arthropods in the animal kingdom: Ecdysozoa, islands, trees, and the "parsimony ratchet"', *Molecular Phylogenetics and Evolution*, 13: 619–23.

Goodman, M. (1981) 'Decoding the pattern of protein evolution', *Progress in Biophysics and Molecular Biology*, 37: 105–64.

Hillis, D.M., Mable, B.K. and Moritz, C. (1996) 'Applications of molecular systematics: the state of the field and a look to the future', in D.M. Hillis, C. Moritz and B.K. Mable (eds) *Molecular Systematics*, Sunderland, Massachusetts: Sinauer Associates, pp. 515–43.

Hirt, R.P., Logsdon Jr., J.M., Healy, B., Dorey, M.W., Doolittle, W.F. and Embley, T.M. (1999) 'Microsporidia are related to Fungi: evidence from the largest subunit of RNA polymerase II and other proteins', *Proceedings of the National Academy of Sciences, USA*, 96: 580–5.

Hwang, U.W., Kim, W., Tautz, D. and Friedrich, M. (1998) 'Molecular phylogenetics at the Felsenstein zone: approaching the Strepsiptera problem using 5.8S and 28S rDNA sequences', *Molecular Phylogenetics and Evolution*, 9: 470–80.

Keeling, P.J. (2001) 'Foraminifera and Cercozoa are related in actin phylogeny: two orphans find a home?', *Molecular Biology and Evolution*, 18: 1551–7.

Keeling, P.J., Luker, M.A. and Palmer, J.D. (2000) 'Evidence from beta-tubulin phylogeny that Microsporidia evolved from within the Fungi', *Molecular Biology and Evolution*, 17: 23–31.

Knoll, A.H. (1992) 'The early evolution of Eukaryotes: a geological perspective', *Science*, 256: 622–7.

Li, W.-H. (1993) 'So, what about the molecular clock hypothesis?', *Current Opinions in Genetics and Development*, 3: 896–901.

Li, W.-H. and Graur, D. (1991) *Fundamentals of Molecular Evolution*, Sunderland, Massachusetts: Sinauer Associates.

Li, W.-H., Tanimura, M. and Sharp, P.M. (1987) 'An evaluation of the molecular clock hypothesis using mammalian DNA sequences', *Journal of Molecular Evolution*, 25: 330–42.

Li, W.-H., Ellsworth, D.L., Krushka, J., Chang, B.H.-J. and Hewett-Hemmett, D. (1996) 'Rates of nucleotide substitution in primates and rodents and the generation-time effect hypothesis', *Molecular Phylogenetics and Evolution*, 5: 182–7.

Liu, J.-C., Makova, K.D., Adkins, R.M., Gibson, S. and Li, W.-H. (2001) 'Episodic evolution of growth hormone in primates and emergence of the species specificity of human growth hormone receptor', *Molecular Biology and Evolution*, 18: 945–53.

Martin, A.P. and Palumbi, S.R. (1993) 'Body size, metabolic rate, generation time, and the molecular clock', *Proceedings of the National Academy of Sciences, USA*, 90: 4087–91.

McIlroy, D., Green, O.R. and Brasier, M.D. (2001) 'Palaeobiology and evolution of the earliest agglutinated Foraminifera: *Platysolenites*, *Spirosolenites* and related forms', *Lethaia*, 34: 13–29.

Messier, W. and Stewart, C.-B. (1997) 'Episodic adaptive evolution of primate lysozymes', *Nature*, 385: 151–4.

Morin, L. (2000) 'Long branch attraction effects and the status of "basal Eukaryotes": phylogeny and structural analysis of the ribosomal RNA gene cluster of the free-living diplomonad *Trepomonas agilis*', *Journal of Eukaryotic Microbiology*, 47: 167–77.

Neefs, J.-M., Van de Peer, Y., De Rijk, P., Chapelle, S. and De Wachter, R. (1993) 'Compilation of small subunit RNA structures', *Nucleic Acids Research*, 21: 3025–49.

Nichol, S.T., Rowe, J.E. and Fitch, W.M. (1993) 'Punctuated equilibrium and positive Darwinian evolution in vesicular stomatitis virus', *Proceedings of the National Academy of Sciences, USA*, 90: 10424–8.

O'h Uigin, C. and Li, W.-H. (1992) 'The molecular clock ticks regularly in muroid rodents and hamsters', *Journal of Molecular Evolution*, 35: 377–84.

Olsen, G.J., Matsuda, H., Hagstrom, R. and Overbeek, R. (1994) 'FastDNAml: a tool for construction of phylogenetic trees of DNA sequences using maximum likelihood', *Computer Applications in the Biosciences*, 10: 41–8.

Pawlowski, J. (2000) 'Introduction to the molecular systematics of Foraminifera', *Micropaleontology*, 46 (supplement 1): 1–12.

Pawlowski, J., Bolivar, I., Guiard-Maffia, J. and Gouy, M. (1994) 'Phylogenetic position of Foraminifera inferred from LSU rDNA gene sequences', *Molecular Biology and Evolution*, 11: 929–38.

Pawlowski, J., Bolivar, I., Fahrni, J., Cavalier-Smith, T. and Gouy, M. (1996) 'Early origin of Foraminifera suggested by SSU rDNA gene sequences', *Molecular Biology and Evolution*, 13: 445–50.

Pawlowski, J., Bolivar, I., Fahrni, J., de Vargas, C., Gouy, M. and Zaninetti, L. (1997) 'Extreme differences in rates of molecular evolution of Foraminifera revealed by comparison of ribosomal DNA sequences and the fossil record', *Molecular Biology and Evolution*, 14: 498–505.

Pawlowski, J., Bolivar, I., Fahrni, J., de Vargas, C. and Bowser, S.S. (1999) 'Molecular evidence that *Reticulomyxa filosa* is a freshwater naked foraminifer', *Journal of Eukaryotic Microbiology*, 46: 612–17.

Pawlowski, J., Fahrni, J.F., Brykczynska, U., Habura, A. and Bowser, S.S. (2002) 'Molecular data reveal high taxonomic diversity of allogromiid Foraminifera in Explorers Cove (McMurdo Sound, Antarctica)', *Polar Biology*, 25: 96–105.

Philippe, H., Chenuil, A. and Adoutte, A. (1994) 'Can the Cambrian explosion be inferred through molecular phylogeny?', in M. Akam, P.W.H. Holland, P.W. Ingham and G.A. Wray (eds) *Development 1994 Supplement*, Cambridge, MA: The Company of Biologists Limited, pp. 15–25.

Ragan, M.A., Bird, C.J., Rice, E.L., Gutell, R.R., Murphy, C.A. and Singh, R.K. (1994) 'A molecular phylogeny of the marine red algae (Rhodophyta) based on the nuclear small-subunit rRNA gene', *Proceedings of the National Academy of Sciences, USA*, 91: 7276–80.

Rambaut, A. and Bromham, L. (1998) 'Estimating divergence dates from molecular sequences', *Molecular Biology and Evolution*, 15: 442–8.

Robinson-Rechavi, M. and Huchon, D. (2000) 'RRTree: relative-rate tests between groups of sequences on a phylogenetic tree', *Bioinformatics*, 16: 296–7.

Rodríguez-Trelles, F., Tarrío, R. and Ayala, F.J. (2001) 'Xanthine dehydrogenase (XDH): episodic evolution of a "neutral" protein', *Journal of Molecular Evolution*, 53: 485–95.

Ross, C.A. and Ross, J.R.P. (1991) 'Paleozoic Foraminifera', *Biosystems*, 25: 39–51.

Smith, A.B., Lafay, B. and Christen, R. (1992) 'Comparative variation of morphological and molecular evolution through geologic time: 28S ribosomal RNA versus morphology in echinoids', *Philosophical Transactions of the Royal Society, London*, B338: 365–82.

Sogin, M. (1997) 'History assignment: when was the mitochondrion founded?', *Current Opinion in Genetics and Development*, 7: 792–9.

Sorhannus, U. (1996) 'Higher ribosomal RNA substitution rates in Bacillariophyceae and Dasycladales than in Mollusca, Echinodermata and Actinistia–Tetrapoda', *Molecular Biology and Evolution*, 13: 1032–8.

Van de Peer, Y. and De Wachter, R. (1997) 'Evolutionary relationships among the eukaryotic crown taxa taking into account site-to-site rate variation in 18S rRNA', *Journal of Molecular Evolution*, 45: 619–30.

Van de Peer, Y., Baldauf, S.L., Doolittle, W.F. and Meyer, A. (2000) 'An updated and comprehensive rRNA phylogeny of (crown) Eukaryotes based on rate-calibrated evolutionary distances', *Journal of Molecular Evolution*, 51: 565–76.

Vossbrinck, C.R., Maddox, J.V., Friedman, S., Debrunner-Vossbrinck, B.A. and Woese, C.R. (1987) 'Ribosomal RNA sequence suggests Microsporidia are extremely ancient Eukaryotes', *Nature*, 326: 411–14.

Wallis, M. (2000) 'Episodic evolution of protein hormones: molecular evolution of pituitary prolactin', *Journal of Molecular Evolution*, 50: 465–73.

—— (2001) 'Episodic evolution of protein hormones in mammals', *Journal of Molecular Evolution*, 53: 10–18.

Wallis, O.C., Zhang, Y.P. and Wallis, M. (2001) 'Molecular evolution of GH in primates: characterisation of the GH genes from slow loris and marmoset defines an episode of rapid evolutionary change', *Journal of Molecular Endocrinology*, 26: 249–58.

Wilson, P.A. and Norris, R.D. (2001) 'Warm tropical ocean surface and global anoxia during the mid-Cretaceous period', *Nature*, 412: 425–9.

Winker, S. and Woese, C.R. (1991) 'A definition of the domains Archaea, Bacteria and Eucarya in terms of small subunit ribosomal RNA characteristics', *Systematic and Applied Microbiology*, 14: 305–10.

Chapter 7

Dating the origin of land plants

Charles H. Wellman

ABSTRACT

Ascertaining the time of origin of land plants has been a long standing scientific concern. Initial estimates were based on the early land plant megafossil record, and suggested a late Silurian origin. However, the megafossil record is notoriously poor, and subsequent work on the early land plant dispersed microfossil record (spores and phytodebris) indicates an earlier origin, in the Llanvirn (Ordovician). The disparity between the plant megafossil/microfossil record is believed to be largely a consequence of the low preservation potential of the earliest land plants: these are considered to have been bryophyte-like, and most likely lacked recalcitrant tissues, such as those containing lignin, that facilitate preservation as megafossils. More recently, other techniques have been employed to shed light on the time of origin of land plants. The geochemical record has been examined for signals reflecting global change considered to have been concomitant with the developmental of a significant terrestrial vegetation. Such techniques are, however, fraught with difficulties. Assessing the extent and pattern of global change associated with the development of terrestrial vegetation is largely conjectural. Furthermore, the relationships between global change and geochemical signals are often poorly understood. Another approach has been the utilization of molecular clock techniques. Recent molecular clock analyses suggest that land plants diverged significantly earlier than expected based on fossil evidence. However, concern has been expressed that many of the currently utilized molecular clock techniques are flawed, and indeed that the molecular clock does not exist. None the less, research into molecular clocks is still in its infancy, and there is hope that, ultimately, more reliable techniques will appear that will engender greater confidence in results. It is argued here that currently the most reliable estimate for the time of origin of land plants remains the minimum estimate of Llanvirn (Ordovician) provided by the early land plant dispersed microfossil record. In future this benchmark may well be breached as new fossil finds and new techniques (possibly refined molecular clock techniques) come to light.

Introduction

The origin of land plants was one of the most important events in the history of life on Earth. The invasion of the land by plants initiated the development of complex and substantial terrestrial ecosystems. This had profound effects on the environment, with the newly evolved flora having direct impact on the composition of the atmosphere, rates of weathering and erosion, and consequently the nature of soils and sedimentation patterns. The actual timing of the origin of land plants has been a long standing scientific concern. Initial investigations focused on the fossil record (which has in itself proved contentious). More recently different lines of enquiry have been explored: clues from environmental change associated with the invasion of the land and evidence from molecular clocks. This contribution critically reviews evidence for the timing of the origin of land plants, and attempts to explain divergent results from the different lines of enquiry.

What are land plants and what was the invasion of the land?

Land plants (embryophytes) and their evolutionary relationships

'Land plants' constitute the Kingdom Embryophyta and are the most conspicuous component of modern terrestrial ecosystems. They are widespread over the land surface, except in the driest deserts and coldest upland and polar regions. They represent a phenomenal biomass, and are relatively diverse with at least 300 000 extant species. All of these species are multicellular organisms that are photosynthetic (except for rare parasitic forms that have lost their photosynthetic capabilities) and inhabit terrestrial environments (except for rare secondarily aquatic forms). The embryophytes possess numerous autapomorphies (see below), including the fact that they all develop from a multicellular embryo enclosed in maternal tissue (with the embryo nutritionally and developmentally dependent on the maternal tissue for at least some time during early development). Clearly the embryophytes are one of the major kingdoms of eukaryote life and constitute the vast bulk of terrestrial biomass.

There is very strong evidence that the embryophytes are monophyletic, and essentially evolved as an adaptive response to the invasion of the land (i.e. their origin is intimately related to their invasion of the terrestrial environment). Evidence that the embryophytes are a monophyletic group is convincing and is supported by morphological/anatomical, ultrastructural, biochemical, and molecular data (reviewed in Graham 1993; Kenrick and Crane 1997; Graham and Gray 2001). It is generally accepted that the charophycean green algae are the closest extant relatives to the embryophytes, and that the first embryophytes evolved from ancestors that would be classified with modern charophycean green algae (Graham 1993; Graham and Gray 2001; Karol et al. 2001).

It is the opinion of many workers that, when we consider the origin of land plants, we are essentially concerning ourselves with the origin of the embryophytes. However, equating the origin of land plants with that of embryophytes can be problematic with certain grey areas (as is often the case when attempting to define the origin of major clades). One must bear in mind that equating land plants with embryophytes excludes: (i) terrestrial green algae; (ii) possible organisms on the charophycean green algae to embryophyte lineage that may have possessed some, but not all, embryophyte characters, and may even have lived in a predominantly sub-aerial habitat; (iii) potential 'failed land plant attempts' that may have evolved from within the green algae independently, but ultimately became extinct. Graham (1993) warned against equating land plants with embryophytes, noting the possibilities that aquatic algae might have evolved an embryo prior to invading the land or may have invaded the land before evolving an embryo. These possibilities must be borne in mind, whilst recalling that all extant land plants belong with the monophyletic group Embryophyta that evolved once and once only.

Despite the fact that there is overwhelming evidence that the embryophytes are monophyletic, relationships among the major embryophyte groups are unresolved. It is generally accepted that the bryophytes are basal and diverged earlier than the tracheophytes (vascular plants). However, relationships among the extant bryophyte groups (liverworts, hornworts, and mosses) are controversial. Most recent analyses suggest that the bryophytes are paraphyletic with respect to the vascular plants, with a moss/vascular plant sister group relationship, with either the liverworts or hornworts as the earliest divergent group (Figure 7.1). However, this subject is contentious, and many different phylogenetic schemes have been proposed (reviewed by Kenrick and Crane 1997; Kenrick 2000).

The evolutionary transformation from aquatic algae to terrestrial land plants

Based on the knowledge that the embryophytes are monophyletic and evolved from a charophycean green algal ancestor, it is possible to compare extant charophycean green algae with purported early divergent embryophytes (liverworts or hornworts), and draw some general conclusions regarding the evolutionary innovations involved in the origin of embryophytes and the likely nature of the most primitive forms. Recently, Graham and Gray (2001) have undertaken just such an exercise. They suggest that the ancestors of embryophytes were multicellular charophycean green algae that inhabited shallow freshwater environments. They consider that, due to the unpredictability of such environments (i.e. their propensity to dry out), these organisms would have evolved characteristics that facilitated survival when water bodies dried out exposing them to the harsh subaerial environment. By carefully considering characters present in extant charophycean green algae and early divergent embryophytes, Graham and Gray were able to hypothesize which features were required in order to make the transition from the aqueous to subaerial environment, and identify which of these features were plesiomorphic and which are autapomorphies of the embryophytes (presumably many of which are essential acquisitions required for life in a subaerial environment). Furthermore, they were able to identify potential pre-adaptive features among the charophycean green algae.

Graham and Gray (2001) suggested that in order to survive in the subaerial environment, the earliest embryophytes required characteristics that would: (i) prevent decay during periods of dormancy, drought avoidance, and perhaps also short-term desiccation; and (ii) enable reproduction in conditions of low moisture content. They identified a number of such characteristics. These include plesiomorphic features common to both charophycean green algae and bryophytes (resistant compounds in vegetative cell walls and cytoplasmic desiccation resistance) as well as autapomorphic characters of the bryophytes (histogenetic meristems, placental transfer tissues, and reproductive propagules enclosed in a sporopollenin wall [spores]). These autapomorphies, in many cases, can be identified as evolving from features already present in extant charophycean green algae but used in a different context (i.e. pre-adaptive).

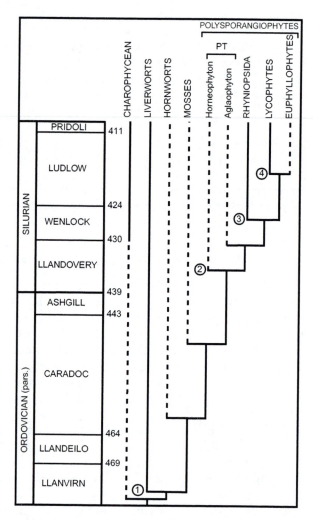

Figure 7.1 Evolutionary tree for basal land plants. The stratigraphical column is from Harland *et al.* (1990). Note the traditional use of the Llanvirn stage, as utilized throughout this contribution. Phylogenetic relationships based on Kenrick (2000, figure 1a). Nodes calibrated to the stratigraphical column using: (1) the earliest dispersed spores, believed to derive from liverworts; (2) the time when trilete spores first become abundant in the fossil record (true trilete spores possibly represent a synapomorphy of polysporangiophytes – i.e. the clade nested within *Horneophyton*) (Gray 1985; Wellman and Gray 2000); (3) the earliest rhyniophytoid megafossils (these most likely belong with the Rhyniopsida, but possibly represent protracheophytes [PT]); (4) the earliest lycophyte megafossils (evidence summarized in Edwards and Wellman 2001). Solid lines represent fossil record, dashed lines represent range extensions.

For example, the sporopollenin layer in the zygote wall of extant *Coleochaete* has been identified as pre-adaptive with respect to the sporopollenin spore wall of land plants (Blackmore and Barnes 1987; Graham 1993; Wellman, in press). When considering mutations that enabled green algal ancestors to survive in the subaerial environment as land plants, Knoll and Bambach (2000, p. 8) note that 'principal among these mutations must have been one involving a simple change in the timing of gene expression for sporopollenin biosynthesis from just after zygote formation to after meiosis and spore formation . . . such a mutation, arguably lethal in water, would have provided protection against desiccation and harmful radiation at a critically vulnerable phase of the lifecycle'.

Terrestrial life before the land plants

When considering the origin of land plants it is important to consider the terrestrial biota (aquatic and subaerial) present before they originated. Prior to the origin of the embryophytes the land surface was not barren as is often stated, but was almost certainly inhabited with what is often, rather disparagingly, referred to as an algal scum. This biota would have included photosynthesizing organisms (probably photolithotrophs such as cyanobacteria and green algae – unicellular and multicellular) and possibly also parasites, grazers, and decomposers (viruses, bacteria, archaea, and possibly also fungi) that exploited the primary producers whilst either living or dead (Raven 1997; Boucot and Gray 2001; Raven and Edwards 2001). Evidence for this early terrestrial biota occurs in the form of fossil remains and geochemical signals (Keller and Wood 1993; Horodyski and Knauth 1994; Rye and Holland 2000). This primitive terrestrial biota would have contributed to the formation of biologically active soils (reviewed in Boucot and Gray 2001; Raven and Edwards 2001) and it is quite possible that it could have supported multicellular animal life (possibly various arthropods, annelids etc.). Thus the earliest land plants would have invaded a landscape with rudimentary soils, but already inhabited by a variety of other organisms.

Graham and Gray (2001) suggest that the algal ancestors of the embryophytes inhabited shallow freshwater aquatic environments that were ephemeral and characterized by unpredictable water availability. In addition to their propensity to dry out, other unfavourable characteristics probably included variation in temperature and possibly also salinity. They termed such harsh environments 'environments of unpredictable unfavourableness'. It should also be borne in mind that subaerial environmental conditions would most likely have been very different from those today, in terms of atmospheric composition and solar luminosity (see Graham 1993).

How long did the evolutionary transformation take?

Of major relevance to the problem of defining land plants and pinpointing the exact timing of their origin is the controversial question of how rapidly they evolved. Was the transformation from an aquatic green alga to a subaerial embryophyte relatively slow or rapid? Were the characteristics that define embryophytes, and permit subaerial existence, acquired more-or-less simultaneously or sequentially? Such thorny questions are at the very root of the long standing debate on the tempo and mode of evolution and are not easily addressed. There is a major disjunction between putative charophycean green algal ancestors and embryophytes (based on characters of their extant relatives), and the evolutionary transformation must have been considerable. However, from a theoretical perspective at least, this evolutionary transformation could have been relatively rapid.

The 'environments of unpredictable unfavourableness' of Graham and Gray (2001) would have exerted immense evolutionary pressures on multicellular green algae inhabiting them. The major problem would have been periods of desiccation associated with the water bodies drying out. In order to survive, the algae must have: (i) evolved characters (morphological and biochemical) that permitted desiccation tolerance; and/or (ii) produced desiccation resistant resting cysts or spores that either awaited the water body to refill or were transported (subaerially) into another suitable aquatic environment. Whether one, or most likely both, of these strategies was adopted, the immense evolutionary pressures would quite possibly have promoted rapid acquisition of favourable characteristics permitting subaerial survival. It is difficult to know if the plants evolved these characters sequentially (in which case the organisms may have been facultatively subaerial initially, with subaerial existence later becoming more permanent), or whether these characters were acquired more-or-less simultaneously. I consider the latter more likely, because evolutionary pressures conferring advantage upon organisms that could survive subaerial conditions for longer periods of time would have been immense, promoting rapid evolution of characters enabling subaerial exposure.

It should be noted that a prerequisite for reproduction in, and hence successful colonization of, the subaerial environment is the production of abundant spores that are easily dispersed subaerially (i.e. small) and well protected (i.e. with a resistant sporopollenin wall). As soon as such spore-producing plants evolved, their palaeogeographical spread would have been rapid because spores are ideally suited to long distance subaerial dispersal by wind and most likely had the ability to self-fertilize, therefore readily establishing palaeogeographically widespread founder populations (reviewed in Gray 1985).

Fossil evidence

Introduction

Traditionally, fossil evidence has been the principle tool utilized in research into the timing of the origin of land plants. Initially such research was based exclusively on land plant megafossils. These are relatively complete remains, and early land plant megafossils have been studied since the 1850s. More recently, since the late 1950s, attention has been paid to the early land plant dispersed microfossil record. This consists of spores (subaerially dispersed by plants during life) and phytodebris (fragmentary cuticles and tubular remains produced on disarticulation of a plant).

Early land plant megafossil record

The fossil record for early land plant megafossils is relatively poor and highly biased. This is primarily a reflection of the low preservation potential of the primitive land plants, particularly the earliest bryophytes that almost certainly lacked recalcitrant lignified tissues. To date there are only approximately 25 reported plant megafossil assemblages from the Silurian. These are all from the late Silurian, and probably post-date the acquisition of lignin (i.e. the origin of the polysporangiophytes) (see Figure 7.1). Some workers (e.g. Kenrick 2000) have suggested that representational bias is important. That is, plant megafossils are poorly known pre-Lower Devonian due to the paucity of non-marine strata available for exploration. It should be noted, however, that plants are easily transported into marine environments where conditions are often favourable for preservation. While plants are most commonly preserved in terrestrial environments, it is not uncommon for them to occur in marine deposits. All of the known late Silurian plant megafossil assemblages are from marine deposits (i.e. the plants are allochthonous and have been transported into the marine environment). Convincing plant megafossils simply have not been recovered below this level, despite extensive exploration that has included rare examples of non-marine Ordovician–Silurian strata that appear suitable for plant megafossil preservation. Thus I consider that the plant megafossil record begins once plants had evolved recalcitrant tissues that enhanced preservation potential. Earlier plants were of low preservation potential and most likely will only occur as megafossils in cases of exceptional preservation (if any exist and can be located). The early land plant megafossil record has been reviewed most recently by Edwards and Wellman (2001).

Convincing early land plant megafossils are known from the late Wenlock (late Silurian) (Edwards et al. 1983), and consist of rhyniophytoid plants. These diminutive plants comprise a naked, bifurcating axis with terminal sporangia. Bifurcating axes, that probably represent plants of a similar grade of organization, but lack preserved sporangia, are known from the Llandovery (Schopf et al. 1966). The enigmatic vascular plant, *Pinnatiramosus qianensis*, of Wenlock (late Silurian) age from southwest China is remarkable in its peculiar morphology and advanced anatomy (Geng 1986; Cai et al. 1996). Recently, however, Edwards et al. (2001) have suggested that these fossils may represent roots of Permian plants growing on (and infiltrating) a Silurian substrate.

Most of the reported Silurian land plant assemblages are dominated by simple rhyniophytoids. These are plants comprising a bifurcating axis (sometimes bearing stomata) with terminal sporangia (sometimes containing *in situ* spores). However, none possess a well preserved stele, and thus the presence of tracheids and vascular status cannot be confirmed. Consequently they are termed rhyniophytoid, rather than included in the vascular rhyniophyte plants, that are proven from earliest Devonian deposits (Edwards et al. 1992). Although simple, Silurian rhyniophytoid plants exhibit extensive diversity in terms of variations within *in situ* spores and sporangial characters (e.g. Fanning et al. 1988). Some assemblages contain more complex plants, including lycophytes (summarized in Edwards and Wellman 2001).

Early land plant microfossil record

In contrast to the early land plant megafossil record, the dispersed microfossil record is excellent. These microfossils represent organs shed during the lifetime of the plants (spores) or fragments of recalcitrant tissue released following disarticulation of the plant (cuticles and tubular structures). Spores have an excellent fossil record because: (i) they have high preservation potential because they have an outer wall composed of sporopollenin, an extremely recalcitrant macromolecule; (ii) they are produced in

Table 7.1 Estimated number of spores per sporangium in selected early land plants (calculated by dividing the volume of the sporangium by the volume of a spore unit of mean diameter and assuming 74 per cent close packing)

Taxon	Reference	Locality age	Spore unit	Units per sporangium
Specimen B	Habgood (2000)	North Brown Clee Hill Lochkovian	Permanent tetrad	840 tetrads
NMW98.23G.1	Edwards et al. (1999)	Ludford Lane Pridoli	Permanent tetrad	20 520 tetrads
Culullitheca richardsonii	Wellman et al. (1998b)	North Brown Clee Hill Lochkovian	Permanent dyad	2690 dyads
Fusitheca fanningiae	Wellman et al. (1998b)	North Brown Clee Hill Lochkovian	Permanent dyad	1220 dyads
Type A	Wellman et al. (1998a)	Ludford Lane Pridoli	Hilate cryptospore	31 520 singles
Type B	Wellman et al. (1998a)	Ludford Lane Pridoli	Hilate cryptospore	9090 singles
Cooksonia pertoni	Fanning et al. (1988)	Perton Lane Pridoli	Trilete spore	86 100 singles
Cooksonia pertoni	Fanning et al. (1988)	North Brown Clee Hill Lochkovian	Trilete spore	64 740 singles

vast quantities (Table 7.1 presents estimated spore production per sporangium for selected early land plants); and (iii) they have a huge dispersal potential and are easily transported into environments favourable for preservation (e.g. lacustrine and river floodplain environments and shallow marine clastic deposits). There are numerous reports of early land plant spore assemblages with excellent spatial and temporal coverage (reviewed by Steemans 1999; Wellman and Gray 2000; Edwards and Wellman 2001). Similarly, fragmented cuticles and tubular structures that are small and produced in large quantities may be transported large distances (primarily by water) into environments favourable to preservation, and are composed of recalcitrant materials that are easily preserved.

Dispersed spores

As early as 1959 trilete spores, that undoubtably derive from land plants, were reported from deposits of Llandovery (early Silurian) age (Hoffmeister 1959), and it was demonstrated that there was an increase in trilete spore diversity throughout the Silurian (Richardson and Lister 1969; Richardson and Ioannides 1973). These spores were uncontroversially accepted as deriving from embryophytes, as similar spores are known from extant embryophytes and *in situ* from early land plant megafossils (see review in Gray 1985; Wellman and Gray 2000).

Somewhat later, in a ground-breaking insight, Gray and Boucot (1971) realized that spores with rather peculiar morphology (united in permanent tetrads) recovered from early Silurian deposits were derived from land plants. The first occurrence of such tetrads, and hence the earliest evidence for land plants, was subsequently extended back into the Caradoc (Ordovician) (Gray *et al.* 1982) and then the Llanvirn

(Ordovician) (Vavrdova 1984; McClure 1988; Gray 1993; Strother *et al.* 1996). Clearly such an observation suggests a significant disparity between the first reported micro-fossil and megafossil evidence for early land plants. It is likely that this is a reflection of preservational bias. It is not until lignified tissues with high preservation potential evolved (possibly coincident with the evolution of vascular plants) that the fossil record of megafossils begins.

Subsequent research has demonstrated that there are a number of distinct non-trilete spore morphotypes that dominate the early land plant dispersed spore record. These have been collectively termed cryptospores (see Richardson *et al.* 1984; Richardson 1988, 1996; Strother 1991; Steemans 2000). They include monads, dyads, and tetrads, that are either naked or enclosed within an envelope, and the separated products of dyads (hilate cryptospores). Initially the affinities of cryptospores was controversial, with many authors disputing a land plant origin. Subsequent research, however, has provided compelling support for land plant affinities (recently reviewed in Wellman and Gray 2000). More precise relationships have been suggested for some of the morphotypes, with Gray (1985, 1991) arguing persuasively that permanent tetrads derived from bryophyte-like, or more precisely, liverwort-like, plants.

At this juncture it is helpful to consider the type of spore record one might expect from the earliest land plants. There is strong evidence that bryophytes are the most basal land plants. Consequently an examination of the spores of extant bryophytes might provide a search image for the type of spores expected of the earliest land plants. Gray and subsequent workers (Gray 1985; Wellman and Gray 2000) have compared the spores comprising the early dispersed spore fossil record with those of extant bryophytes. While there are some obvious similarities regarding morphology and ultrastructure, it should be borne in mind that many extant bryophytes produce rather non-descript spores that occur singly and are small and thin walled. The latter appear to have no counterpart in the early dispersed spore fossil record, even though taxon composition has been described in its entirety for certain Ordovician and Silurian spore assemblages from non-marine deposits (i.e. if such spores were a component of the spore assemblage they would have been observed). In addition, dispersed spore assemblages from the Llanvirn to Llandovery (an interval of some 40 myr duration) are very similar in composition, exhibiting virtual stasis in terms of morphological/ taxonomic composition (Wellman 1996), and the spore morphologies possessed by a great many extant bryophytes simply are not present. Interestingly there are very few dispersed spores recognized as bryophyte in origin throughout the entire dispersed spore fossil record (reviewed in Traverse 1988). Either such spores are not preserved (perhaps they are too thin walled) and/or are a derived feature of bryophytes. I con-sider the latter explanation a distinct possibility, and contend that the Ordovician–Lower Silurian spore record reflects 'ancestral' spore morphologies.

Evidence for the affinities of cryptospores is derived from four main sources: (i) occurrence (i.e. depositional environment) of the dispersed fossil spores; (ii) inferences based on comparison with the spores of extant land plants (size and morphology); (iii) studies of land plant fossils preserving *in situ* spores; and (iv) analysis of spore wall ultrastructure. This evidence is summarized below.

It has long been noted that cryptospores are distributed in a similar range of depo-sitional environments in which the spores/pollen of extant land plants occur, and with similar abundances. Their occurrence in continental and nearshore marine deposits (with abundances usually decreasing offshore) is wholly consistent with them representing the subaerially released spores of land plants, that were transported to their sites of deposition through the actions of wind and water. However, while there are numerous examples of spore assemblages derived from continental deposits from the Devonian, few examples exist for the Ordovician–Silurian interval (see Wellman and Gray 2000). These findings are almost certainly an artefact of the stratigraphical record: the Ordovician–early Silurian was a time of persistently high sea levels and continental deposits are rare, with those that do exist often possessing geological characteristics unsuitable for the preservation of palynomorphs (e.g. unsuitable lithologies and/or high thermal maturity).

Comparisons with the spores of extant and fossil land plants demonstrate that cryptospores are similar in terms of size, gross morphology, and possession of a thick sporopollenin spore wall (regarded as a synapomorphy for embryophytes). Sporopollenin walls may have multiple functions (Blackmore and Barnes 1987; Graham and Gray 2001; Wellman, in press), but almost certainly the primary one is to protect propagules during transport following subaerial release. Thus the possession of such walls in early land plant spores provides evidence that they were functionally similar to their modern counterparts. Furthermore, the small size of early land plant spores is within the range of subaerially dispersed spores produced by extant free-sporing plants.

Based largely on analogy with the reproductive propagules of extant embryophytes, Gray (1985, 1991) has argued persuasively that permanent tetrads are a primitive character in embryophytes and that such tetrads derive from land plants of a bryophyte, most likely liverwort, grade of organization. She noted that among extant free-sporing embryophytes only liverworts regularly produce permanent tetrads as mature spores, some of which are contained within an envelope similar to those enclosing certain fossil spore tetrads, but that a tetrad regularly occurs in the spore ontogeny of embryophytes. The affinities of monads and dyads are more equivocal, primarily because such morphologies do not have an obvious modern counterpart, either in mature spores or in spore ontogeny (reviewed in Wellman et al. 1998a,b). Dyads rarely occur in extant (non-angiosperm) embryophytes, and only through meiotic abnormalities (Wellman et al. 1998a,b). The abundance of dyads in early land plant spore assemblages indicates that they were commonly produced and are therefore probably not the products of meiotic abnormalities. Their occurrence is most comfortably explained by invoking successive meiosis, with separation occurring following the first meiotic division and sporopollenin deposition on the products of the second division. It has been noted that monads, dyads, and tetrads often have identical envelopes, and some authors have suggested that they are closely related, perhaps even deriving from a single species (Johnson 1985; Richardson 1988, 1992; Strother 1991; Hemsley 1994; Taylor 2001).

Studies of in situ spores provide the only direct link between the dispersed spore and plant megafossil records, and are critical to our understanding of the affinities of dispersed spore types. Unfortunately, however, plant megafossils are practically unknown until the late Silurian and hence there are no in situ spore records for the first 50 million or so years of the early land plant dispersed spore record. However, there are a handful of exceptionally well-preserved plant megafossil assemblages from the latest Silurian–earliest Devonian in which in situ spores are preserved in sufficient detail to enable detailed analysis. Such exceptional preservation is known from two main localities: Ludford Lane (Pridoli, late Silurian) and north Brown Clee Hill (Lochkovian, Early Devonian) from the Welsh Borderland, UK, and the record of *in situ* early land plant spores is based primarily on material from these localities (see reviews by Fanning *et al.* 1991; Edwards 1996, 2000; Edwards and Richardson 1996; Edwards and Wellman 2001). At both localities extremely small plant fragments (sometimes referred to as mesofossils) are preserved as relatively uncompressed coalifications that preserve exquisite cellular detail (Edwards 1996).

At the Ludford Lane and north Brown Clee Hill localities, the vast majority of *in situ* spores are trilete. Rare specimens, however, contain cryptospores. In fact most cryptospore morphotypes have now been recovered *in situ* (naked and envelope-enclosed permanent tetrads and dyads, and hilate monads). However, it is unclear if the parent plants represent relict populations, and provide a true reflection of earlier cryptospore-producing plants, or if the cryptospores are plesiomorphic in more advanced plants, or perhaps even arose due to convergence (Gray 1991; Edwards 2000). It must be borne in mind that the fossils occur some 65 million years after the earliest reported cryptospores from the Llanvirn (Ordovician).

Interpretation of the parent plants is not always straightforward as the mesofossils are fragmentary. Usually only terminal parts of the axes (+/− sporangia) are preserved, and cellular detail is variable. Furthermore, many of the rhyniophytoid plants preserve unusual character combinations, confusing considerations of affinities. However, trilete spores have been recovered from the rhyniophyte *Cooksonia pertoni* (Fanning *et al.* 1988) which is demonstrably a true tracheophyte (Edwards *et al.* 1992), and the vast majority of trilete spore producers appear to have constituted plants with bifurcating axes, with terminal sporangia, and often possessing stomata. Interestingly, some dyads and tetrads derive from plants with bifurcating axes/sporangia (Edwards *et al.* 1995a, 1999; Wellman *et al.* 1998a), a character not represented among extant bryophytes (see Edwards 2000). Another interesting observation is the presence of stomata on plants containing *in situ* hilate cryptospores (Edwards 2000; Habgood 2000). Stomata are absent from liverworts, but present in most hornworts, mosses, and vascular plants (although losses are not uncommon in these groups, and are generally considered to be related to ecological factors and functional requirements) (see review in Kenrick and Crane 1997).

It is well established that analysis of spore wall ultrastructure characters can be extremely useful when attempting to ascertain the phylogenetic relationships of extant land plants, and similar research has been extrapolated back in time and is now routinely undertaken on fossil spores (e.g. Kurmann and Doyle 1994). Such research is also of paramount importance in studies of spore wall development. Recently there has been a surge of interest in wall ultrastructure in early land plant spores, and it is hoped that exploitation of this potentially rich data source will provide characters useful in ascertaining the affinities of these ancient plants, and shed light on the nature of spore wall development.

To date, studies of wall ultrastructure in early land plant spores are in their infancy, and two principal lines of inquiry have been explored. Some of the earlier cryptospores have been studied based on analysis of isolated dispersed spores from the Late Ordovician (Ashgill) to early Silurian (Llandovery) of Ohio, USA (Taylor 1995a,b, 1996, 1997, 2000). Later cryptospores and trilete spores have been studied based on analysis of *in situ* spores exceptionally preserved in mesofossils from the latest Silurian (Ludford Lane) and earliest Devonian (north Brown Clee Hill) localities (Rogerson *et al.* 1993; Edwards *et al.* 1995b, 1996a, 1999; Wellman *et al.* 1998a,b; Wellman 1999). Studies on late Silurian–Early Devonian dispersed spores remain an unexploited, but potentially extremely useful, data source. None the less, many early land plant spore morphotypes have now been ultrastructurally examined. Taylor has studied early examples of naked and envelope-enclosed tetrads (Taylor 1995b, 1996, 1997) and naked and envelope-enclosed dyads (Taylor 1995a, 1996, 1997). Later tetrads (Edwards *et al.* 1999), dyads (Wellman *et al.* 1998a) and hilate monads (Wellman *et al.* 1998b) have also been examined, as have a variety of trilete spore taxa (Rogerson *et al.* 1993; Edwards *et al.* 1995b, 1996a; Wellman 1999).

In terms of ascertaining phylogenetic relationships, findings to date are rather difficult to interpret, with no clear patterns emerging regarding the relationships between different mesofossil taxa, *in situ* spore morphology, and wall ultrastructure. This is most likely a consequence of the frailty of the database, as studies are extremely limited to date, and there are major gaps in our knowledge. However, interpretation of wall ultrastructure in early embryophyte spores is also problematic due to a number of technical and theoretical factors (summarized in Wellman *et al.* 1998a,b; Edwards and Wellman 2001). None the less, these problems are not insurmountable. If the database continues to improve at its current rate phylogenetic 'noise' to 'signal' ratio may improve. It is anticipated that, in the future, spore wall ultrastructure will play an increasingly important role in phylogenetic analysis of early land plants. Taylor (2001) recently summarized his findings and proposed a tentative hypothesis for evolutionary relationships among early cryptospore producers. He suggested that at least two separate lineages occur, but stressed that the phylogenetic relationships between these groups and more recent land plants remain uncertain, although he has suggested possible liverwort affinities for some of the dyads (e.g. Taylor 1995a).

In conclusion, spores provide an important source of information on the phylogenetic relationships of early land plants. This is particularly true for the earliest land plants that left no megafossil record, as we are dependent on the dispersed microfossil record as our only source of information. Analysis of these earliest land plant spores suggests that the producers included bryophytes (Gray 1985, 1991), and possibly also stem-group embryophytes. It is anticipated that our understanding of early land plant spores will continue to improve as further localities are discovered, and more work is undertaken on these and pre-existing localities, particularly if these searches were to turn up megafossils of the spore producers. Additionally, it is probable that some of the identified gaps in knowledge will be filled following further research on *in situ* spores and spore wall ultrastructure, which may shed further light on the evolutionary relationships of early land plants.

Embryophyte spores from the Cambrian?

Recently it has been suggested that the cryptospore fossil record may extend as far back as the Middle Cambrian. Strother and Beck (2000) described an intriguing assemblage of palynomorphs from the Middle Cambrian Bright Angel Shale from Arizona, USA. The palynomorphs have a relatively thick wall, are arranged as monads and polyads (containing two, three, four, and more than four units), and are often enclosed within a thin envelope. To a certain extent they resemble cryptospores derived from embryophytes, but they differ in that the walls are thinner and they are loosely arranged, lacking the rigidity, regular arrangement, and symmetry of embryophyte cryptospores. Strother and Beck acknowledge that 'there are no spores in the Bright Angel assemblage that are convincingly like spores from any known embryophytes' (Strother and Beck 2000: p. 417) and 'there is no direct evidence that the fossils from the Bright Angel Shale represent either embryophytes or their evolutionary ancestors' (Strother and Beck 2000: p. 419). This view is supported by Taylor (1999) who presented preliminary results of an analysis of wall ultrastructure in the palynomorphs from the Bright Angel Shale, noting that the walls are quite different from those reported from embryophyte cryptospores.

Strother and Beck interpreted the Bright Angel Shale palynomorphs as deriving from either freshwater aquatic or subaerial 'cryptogams' that colonized a very early terrestrial landscape. This is based on their interpretation that the muddy component of the depositional system (incorporating the palynomorphs) was derived from freshwater sources washed into estuaries (Strother 2000). Evidence for this includes sedimentological structures indicative of intertidal to supratidal conditions, and the fact that normal marine palynomorphs and marine megafossils are absent.

Strother (2000) briefly mentioned a further Middle Cambrian palynomorph assemblage containing similar cryptospore-like fossils from the Rogersville Shale from Tennessee, USA. These palynomorphs are more similar to embryophyte cryptospores (personal observation) but still lack their rigidity and symmetry. More evidence (e.g. further TEM analysis) is required before they can be accepted as indicative of embryophytes, and until such time I consider it prudent to regard the earliest fossil evidence for embryophytes to be Llanvirn cryptospores. I see no reason why the embryophyte cryptospore record could not extend back beyond the Llanvirn, but more proof is required, as it is quite possible that these palynomorphs derive from organisms (inhabiting terrestrial aquatic, or even subaerial habitats) unrelated to the embryophytes.

Dispersed phytodebris

Phytodebris includes fragmentary cuticles and tubular structures that are produced during plant disarticulation and transported by water into sedimentary environments. Their affinities have long been controversial (see Banks 1975, and Gray and Boucot 1977, for an early debate on this subject). The relevance of these fragmentary remains to the study of the earliest land plants has recently been reviewed by Gray (1985), Gensel et al. (1991), Wellman and Gray (2000), and Edwards and Wellman (2001).

Many post-Early Silurian dispersed cuticles clearly derive from embryophytes, and some even possess structures such as stomata (Gensel et al. 1991; Edwards and Wellman 1996). Earlier forms, however, are more controversial. It has been suggested that some represent a cuticular covering of nematophytes (Lang 1937). These enigmatic 'plants' have a unique tubular anatomy (possibly with a cuticular covering). There is evidence to suggest that they possibly represent pathogens or decomposers (Edwards et al. 1996b, 1998; Edwards and Richardson 2000), and it has been suggested that at least some have fungal affinities (Hueber 2001). More recently, it has been suggested that some of these cuticles may derive from early embryophytes (i.e. bryophytes). Experiments were conducted in which extant bryophytes were exposed to high temperature acid hydrolysis in order to study the fragments that survived (i.e. recalcitrant tissues that one would expect to survive in the fossil record) (Kroken *et al.* 1996; Graham and Gray 2001; Kodner and Graham 2001). These authors noted similarities between some of the extant bryophyte remains and fossil cuticles, and suggested that some of the early dispersed cuticles may derive from primitive embryophytes. This is a distinct possibility and requires further examination. However, it should be noted that geochemical studies by Edwards *et al.* (1996b) showed that some of the cuticles were markedly different in chemical composition from coeval ones derived from axial higher land plants, suggesting a fundamentally different affinity.

The dispersed tubular structures are even more problematic. They exhibit a variety of forms and occur in complex associations, and almost certainly derive from diverse sources. Some probably represent fungal hyphae (Sherwood-Pike and Gray 1985; Wellman 1995). Others appear to derive from nematophytes (Gray 1985; Burgess and Edwards 1991; Wellman 1995; Wellman and Gray 2000). Recently, following experimental disaggregation of extant bryophytes (see above), a similarity between some of the surviving extant bryophyte remains and some of the fossil tubular structures was noted, and it has been suggested that some of the fossils may represent fragments of ancient bryophytes (Kroken *et al.* 1996; Graham and Gray 2001; Kodner and Graham 2001). I am yet to be convinced by this argument owing to problems regarding the size and symmetry of the remains. However, few extant bryophytes have been examined to date, and this interesting avenue of research should be explored further.

Dispersed phytodebris does not extend the embryophyte fossil record back beyond the Llanvirn. The earliest fragments of cuticle are reported from the Caradoc (Gray *et al.* 1982), although the age of these deposits has been questioned (Richardson and McGregor 1986). It is not until the Llandovery that such cuticle becomes abundant. Tubular structures, on the other hand, extend back to the Llanvirn where they co-occur with cryptospores. As noted by Graham and Gray (2001), the differences in the stratigraphical ranges of cuticles and tubular structures suggests that they have a different origin.

Conclusions

The embryophyte fossil record, whilst not always easy to interpret, provides tangible evidence for the timing of the origin of land plants. However, it must be interpreted with care, and the infamous vagaries of the fossil/stratigraphical record taken into account. Clearly, only minimum age estimates for groups are possible (i.e. preservation of the first fossilized remains following the origin of embryophytes), and this is dependent upon the evolution of structures likely to leave a fossil record (disregarding any hitherto undiscovered exceptionally preserved floras), a suitable stratigraphical record preserving relevant fossils, and the actual discovery and correct interpretation of such fossils.

The earliest generally accepted evidence for embryophytes is dispersed cryptospore assemblages from the Llanvirn (Ordovician). We must judge to what extent this reliably represents the timing of the origin of land plants. There is convincing evidence that they derive from embryophytes of bryophytic grade (see discussion above). However, we must consider some potential problems: (i) Were there earlier embryophytes that did not produce fossilizable parts? (ii) Is there an earlier, but unrecognized, fossil record for embryophytes? (iii) Is there an earlier, but undiscovered, fossil record for embryophytes?

The first problem addresses the possibility that the earliest embryophytes did not leave a fossil record because they lacked any recalcitrant parts with high fossilization potential. This is unlikely because some of the features considered to be essential for successful colonization of the subaerial environment would almost certainly have included parts composed of recalcitrant materials with high fossilization potential. Spores are a case in point. It has been argued (Wellman, in press) that successful exploitation of the subaerial environment by plants was impossible without spores enclosed within a resistant sporopollenin wall. It follows that such spores must have evolved coincidently with (or possibly before) the invasion of the land, and populations of spore-bearing plants would have spread rapidly due to the wide dispersal potential of such spores.

The second problem addresses the possibility that there is a pre-Llanvirn fossil record of embryophytes but we simply do not recognize these fossils as deriving from land plants. For example, could the Middle Cambrian palynomorphs described by Strother and Beck (2000) derive from embryophytes? At the heart of this problem is the ability to recognize embryophytes unequivocally. We need to recognize structures that clearly belong to embryophytes (i.e. autapomorphies). Sporopollenin-walled spores are considered to be an embryophyte autapomorphy and consequently are a good example. However, the onus is on the researcher to identify unequivocal evidence for embryophyte affinities. Hence, when identifications are uncertain (as is the case with the Middle Cambrian material) it is best to err on the side of caution.

The third problem addresses the possibility that pre-Llanvirn embryophytes, capable of, or indeed, producing fossils existed, but we simply have not discovered them. This is unlikely. In the past half century a vast literature has accumulated based on a massive quantity of palynological research undertaken on Precambrian and Early Palaeozoic deposits, that is both stratigraphically and palaeogeographically extensive. Considering the amount of pre-Llanvirn sediments palynologically analysed, it can be argued that if the embryophytes were widespread we would have discovered them (unless we simply do not recognize them – see above). Thus we are only likely not to have discovered pre-Llanvirn fossil-producing embryophytes if either the rock record for the relevant time is missing or inadequate, or the embryophytes comprised small populations that were palaeogeographically restricted. The former is unlikely as nearshore marine deposits (that should contain dispersed microfossils) are common in pre-Llanvirn successions and have been exhaustively examined for embryophyte remains. It can be argued that the latter is also unlikely, based on the assumption that in order to reproduce and survive successfully, the earliest subaerial embryophytes would have produced abundant, small, sporopollenin-coated spores that were easily dispersed over large distances. One would expect such plant populations to spread rapidly as founder populations were established by widely dispersed spores (bear in mind these spores are believed to have been able to self-fertilize; Gray, 1985). Consequently, it is anticipated that the earliest embryophytes would have rapidly colonized new areas soon after their inception and quickly become widespread. Hence we might expect the appearance of their spores in the fossil record soon after they had first evolved.

In conclusion, I consider that the Llanvirn benchmark for the origin of embryophytes is the most reliable provided to date. This is based primarily on the following assumptions: (i) plants could not have invaded the land until they had evolved spores with a resistant sporopollenin wall that were produced in vast numbers and easily transported large distances; (ii) due to the large transportation potential of their spores, palaeogeographical spread of the earliest land plants would have been rapid; and (iii) the spores would have left a rich fossil record, which should have been identified owing to the intensity of palynological work undertaken on Precambrian and Lower Palaeozoic deposits.

Environmental evidence

It is generally accepted that the invasion of the land by plants would have had a profound effect on the environment of planet Earth. Scientists have identified a number of potential ways in which such global change could be picked up in the fossil/stratigraphical record (summarized in Retallack 2000), and these signals serve as a proxy for the appearance of significant terrestrial vegetation. Most of these methods rely on the assumption that increased weathering, associated with the development of a substantial vegetation, would fertilize the oceans with nutrients, and hence drive up isotopic values such as $\partial^{13}C$, $\partial^{18}O$ and $^{87}Sr/^{86}Sr$. There are, however, a number of potential pitfalls associated with these methods. These essentially relate to the fact that a number of independent factors can cause shifts in isotopic values, and these can be difficult to disentangle. The nature of $^{87}Sr/^{86}Sr$ ratios provides a good example of such problems.

It has been suggested that the $^{87}Sr/^{86}Sr$ ratio in marine carbonate rocks acts as a proxy for continental silicate weathering (e.g. Edmond 1992). This model assumes that ^{87}Sr is mainly derived from the weathering of silicate minerals in continental rocks, and that it accumulates in marine carbonate rocks that are the major sink for strontium. Because it is assumed that continental weathering will have increased rapidly following the development of a substantial terrestrial flora, it follows that this event may be recorded in the $^{87}Sr/^{86}Sr$ record. In fact $^{87}Sr/^{86}Sr$ values begin to rise quite significantly from a low in the Middle Cambrian, and it has been suggested that this might be related to the appearance of a significant flora comprising land plants (summarized in Retallack 2000).

However, a number of problems have been identified that render interpretation of continental silicate weathering based on $^{87}Sr/^{86}Sr$ ratios dubious (e.g. Blum 1997; Edmond and Huh 1997; Quade et al. 1997; Blum et al. 1998; Broecker and Sanyal 1998; Boucot and Gray 2001). Chief among these is the possibility that the strontium content of marine carbonates may derive from a variety of sources of different age, either through reworking of pre-existing ^{87}Sr-rich deposits or because the strontium derives from dissolved calcite of metamorphic origin. Furthermore, increases in continental silicate weathering may relate to factors other than development of a significant flora, perhaps increasing during glacial intervals or periods of enhanced tectonic activity and related uplift. Consequently estimates of increasing continental weathering (and relating this to the origin of land plants) based on evidence from $^{87}Sr/^{86}Sr$ ratios must be treated with caution.

Molecular clock evidence

Graham (1993) considered the timing of the origin of land plants and noted that 'in the future it may be possible to use the ticking of molecular clocks (Zuckerkandl and Pauling, 1965) in the nucleic acids of modern organisms to obtain information on the time of origin of one or more possible embryophyte clades'. However, Graham warned that 'a prerequisite would be information on relative rates of nucleotide substitution in green algae and "lower" land plants having similar generation times, and many sequences would be needed in order to reduce the chance of stochastic error in estimation of divergence time (Li and Grauer, 1991)'. Recently, Muse (2000) reviewed current knowledge of rates and patterns of nucleotide substitution in plants. His judgement is fairly gloomy regarding prospects for a workable molecular clock for plants. Three of his main conclusions are: (i) the three plant genomes vary extensively in both synonymous and non-synonymous substitution rates; (ii) genes within each genome display a wide variety of synonymous and non-synonymous rates; and (iii) there does not appear to be a time-calibrated molecular clock, at least at higher taxonomic levels.

None the less, scientists have begun to experiment with the use of molecular clock techniques to estimate divergence times among the plant kingdom (e.g. lycopsids: Wikström and Kenrick 2001; angiosperms: Sanderson and Doyle 2001; Wikström *et al.* 2001) and there has even been an attempt to estimate the time of the origin of land plants. Heckman *et al.* (2001) used molecular clock techniques to estimate when land plants first appeared, and came up with a minimum estimate of approximately 700 Ma, some 225 million years earlier than suggested by the fossil record. Clearly this disparity with fossil-based estimates requires attention.

Heckman *et al.* obtained divergence time estimates for a green alga (*Chlamydomonas*) versus embryophytes, and bryophytes (the moss *Physcomitrella*) versus tracheophytes (various angiosperms). They analysed non-chloroplast nuclear protein sequences which permitted animal-based calibration. Calibration was undertaken using multiple external calibrations from older divergences among animal phyla and kingdoms (plants, animals, fungi) derived from an analysis of 75 nuclear proteins calibrated with the vertebrate fossil record (Wang *et al.* 1999). Times were estimated using two techniques: the multigene and average-distance approach. Their results suggest that the green algae/land plant divergence was 1061 Ma ± 109 myr and the bryophyte/tracheophyte divergence was 703 Ma ± 45 myr. They suggest that the latter provides a minimum molecular clock estimate for the colonization of land by plants.

One must address the large disparity between ages derived from fossil and molecular clock evidence. Either the fossils or molecular clock (or both) are wrong. I am of the opinion that the fossil record is more likely to provide an accurate time (see discussion above), although it is clear that fossils can only provide minimum ages. There are a number of potential pitfalls with the techniques adopted by Heckman *et al.* which may explain the anomalous result. These include: (i) their technique assumes that the proteins of animals, fungi, and plants evolved at the same rates (many workers dispute this); and (ii) calibration is indirect (essentially using the animal–plant–fungus, nematode–chordate and nematode–arthropod, and arthropod–chordate divergences, obtained from the previous analysis of Wang *et al.* 1999), and ultimately calibrated using the vertebrate fossil record – in fact a datum at 310 Ma (indirect calibration is likely to enhance errors, such as range extension, when applied to different datasets, particularly when applied over vast tracts of geological time as is the case in this analysis). I consider that the molecular clock estimate for the colonization of land by plants provided by Heckman *et al.* (2001) is far too old, primarily as a consequence of the factors outlined above.

Molecular clock techniques are consistently providing divergence estimates far greater than expected based on fossil evidence, and concern has been expressed that many of the currently utilized molecular clock techniques are flawed, and indeed that there may be no molecular clock (e.g. Rodríguez-Trelles *et al.* 2001). There are a number of potential sources of error when using molecular clock techniques to estimate divergence times, and these are often enhanced when considering increasingly older events. One of the major problems is rate variation, which is now an acknowledged fact and has a number of different causes (see Muse 2000; Sanderson and Doyle 2001). None the less, attempts are being made to compensate for rate variation. For example, Sanderson (1997) has developed an approach termed 'non-parametric rate smoothing' (NPRS). This method accepts rate variation, as opposed to rate constancy, but assumes that rate changes are autocorrelated (i.e. immediate descendants inherit rate change from the ancestral lineage). It should be noted, however, that there is little evidence to support the assumption that rate variation is autocorrelated. None the less, studies utilizing NPRS do appear to be reducing the discrepancy between fossil- and molecular clock-based estimates (e.g. angiosperms; Wikström *et al.* 2001).

Research into the molecular clock is still in its infancy, and it would be extremely valuable if it could be accurately utilized. There are a number of potential problems, chief among them rate variation, and these are particularly problematic when considering divergences that occurred deep in geological time. Hopefully as more research is undertaken, more reliable methods to compensate for rate variation will appear, and we can have more confidence in the results. Until such time I consider it premature to rely solely on data from the molecular clock.

Conclusions

It is concluded that the most reliable date for the origin of embryophytes is the minimum age of Llanvirn (Ordovician) provided by the earliest occurrence of dispersed microfossils (specifically spores) that are widely accepted as deriving from land plants. Other lines of evidence, namely geochemistry and the molecular clock, are considered to be unreliable at present, largely due to problems with the actual techniques and in interpreting the data. It is suggested that a widespread and abundant early land plant dispersed spore record would develop soon after the origin of land plants, as these structures, which are vital for the invasion of land, have high fossilization potential, would have been produced in vast numbers, and would have been widely distributed (in addition to palaeogeographical spread of the founder populations being rapid).

Acknowledgements

The research reported in this chapter was supported by the NERC research grant NER/B/S/2001/00211. I would like to thank Dr Paul Kenrick for an insightful review of an earlier draft of the manuscript.

References

Banks, H.P. (1975) 'The oldest vascular land plants: a note of caution', *Review of Palaeobotany and Palynology*, 20: 13–25.

Blackmore, S. and Barnes, S.H. (1987) 'Embryophyte spore walls: origin, homologies and development', *Cladistics*, 3: 185–95.

Blum, J.D. (1997) 'The effect of Late Cenozoic glaciation and tectonic uplift on silicate weathering rates and the $^{87}Sr/^{86}Sr$ record', in W.F. Ruddiman (ed.) *Tectonic Uplift and Climate Change*, New York: Plenum.

Blum, J.D., Gazis, C.A., Jacobson, A.D. and Chamberlin, C.P. (1998) 'Carbonate versus silicate weathering in the Raikhot watershed within the High Himalayan Crystalline Series', *Geology*, 26: 411–14.

Boucot, A.J. and Gray, J. (2001) 'A critique of Phanerozoic climatic models involving changes in the CO_2 content of the atmosphere', *Earth Science Reviews*, 56: 1–159.

Broecker, W.S. and Sanyal, A. (1998) 'Does atmospheric CO_2 police the rate of chemical weathering', *Global Biogeochemical Cycles*, 12: 403–8.

Burgess, N.D. and Edwards, D. (1991) 'Classification of uppermost Ordovician to Lower Devonian tubular and filamentous macerals from the Anglo–Welsh Basin', *Botanical Journal of the Linnean Society*, 106: 41–66.

Cai, C., Ouyang, S., Wang, Y., Fang, Z., Rong, J., Geng, L. and Li, X. (1996) 'An Early Silurian vascular plant', *Nature*, 379: 592.

Edmond, J.M. (1992) 'Himalayan tectonics, weathering processes, and the strontium isotope record in marine limestones', *Science*, 258: 1594–7.

Edmond, J.M. and Huh, Y. (1997) 'Chemical weathering yields from basement and orogenic terrains in hot and cold climates', in W.F. Ruddiman (ed.) *Tectonic Uplift and Climate Change*, New York: Plenum, pp. 329–51.

Edwards, D. (1996) 'New insights into early land ecosystems: a glimpse of a Lilliputian world', *Review of Palaeobotany and Palynology*, 90: 159–74.

—— (2000) 'The role of mid-Palaeozoic mesofossils in the detection of early Bryophytes', *Philosophical Transactions of the Royal Society, London*, B355: 733–55.

Edwards, D. and Richardson, J.B. (1996) 'Review of *in situ* spores in early land plants', in J. Jansonius and D.C. McGregor (eds) *Palynology: Principles and Applications*, Salt Lake City, Utah: American Association of Stratigraphic Palynologists Foundation, Publishers Press, pp. 391–407.

—— (2000) 'Progress in reconstructing vegetation on the Old Red Sandstone Continent: two *Emphanisporites* producers from the Lochkovian of the Welsh Borderland', in P.F. Friend and B.P.J. Williams (eds) *New Perspectives on the Old Red Sandstone*, Geological Society, London, Special Publications, 180: 355–70.

Edwards, D. and Wellman, C.H. (1996) 'Older plant macerals (excluding spores)', in J. Jansonius and D.C. McGregor (eds) *Palynology: Principles and Applications*, Salt Lake City, Utah: American Association of Stratigraphic Palynologists Foundation, Publishers Press, pp. 383–7.

—— (2001) 'Embryophytes on land: The Ordovician to Lochkovian (Lower Devonian) record', in P.G. Gensel and D. Edwards (eds) *Plants Invade the Land*, New York: Columbia University Press, pp. 3–28.

Edwards, D., Feehan, J. and Smith, D.G. (1983) 'A late Wenlock flora from Co. Tipperary, Ireland. *Botanical Journal of the Linnean Society*, 86: 19–36.

Edwards, D., Davies, K.L. and Axe, L. (1992) 'A vascular conducting strand in the early land plant *Cooksonia*', *Nature*, 357: 683–5.

Edwards, D., Duckett, J.G. and Richardson, J.B. (1995a) 'Hepatic characters in the earliest land plants', *Nature*, 374: 635–6.

Edwards, D., Davies, K.L., Richardson, J.B. and Axe, L. (1995b) 'The ultrastructure of spores of *Cooksonia pertoni*', *Palaeontology*, 38: 153–68.

Edwards, D., Davies, K.L., Richardson, J.B., Wellman, C.H. and Axe, L. (1996a) 'Ultrastructure of *Synorisporites downtonensis* and *Retusotriletes* cf. *coronadus* in spore masses from the Prídolí of the Welsh Borderland', *Palaeontology*, 39: 783–800.

Edwards, D., Abbott, G.D. and Raven, J.A. (1996b) 'Cuticles of early land plants: a palaeoecophysiological evaluation', in G. Kersteins (ed.) *Plant Cuticles – an Integrated Functional Approach*, Oxford: BIOS Scientific Publishers, pp. 1–31.

Edwards, D., Wellman, C.H. and Axe, L. (1998) 'The fossil record of early land plants and interrelationships between primitive embryophytes: too little too late?' in J.W. Bates, N.W. Ashton and J.G. Duckett (eds) *Bryology for the 21st Century*, Leeds, UK: Maney Publishing and the British Bryological Society, pp. 15–43.

—— (1999) 'Tetrads in sporangia and spore masses from the Upper Silurian and Lower Devonian of the Welsh Borderland', *Botanical Journal of the Linnean Society*, 130: 111–15.

Edwards, D., Li Cheng-Sen, Wang Yi and Bassett, M.G. (2001) '*Pinnatiramosus*: the ultimate Chinese puzzle?' *Abstracts of The Palaeontological Association 45th Annual Meeting, Geological Museum University of Copenhagen*, p. 13.

Fanning, U., Richardson, J.B. and Edwards, D. (1988) 'Cryptic evolution in an early land plant', *Evolutionary Trends in Plants*, 2: 13–24.

—— (1991) 'A review of *in situ* spores in Silurian land plants', in S. Blackmore and S.H. Barnes (eds) *Pollen and Spores, Patterns of Diversification, Systematics Association Special Volume 44*, Oxford: Clarendon Press, pp. 25–47.

Geng, B. (1986) Anatomy and morphology of *Pinnatiramosus*, a new plant from the Middle Silurian (Wenlockian) of China', *Acta Botanica Sinica*, 28: 664–70.

Gensel, P.G., Johnson, N.G. and Strother, P.K. (1991) 'Early land plant debris (Hooker's "waifs and strays"?)', *Palaios*, 5: 520–47.

Graham, L.E. (1993) *'Origin of Land Plants'*, New York: John Wiley & Sons, Inc.

Graham, L.E. and Gray, J. (2001) 'The origin, morphology and ecophysiology of early embryophytes: neontological and paleontological perspectives', in P.G. Gensel and D. Edwards (eds) *Plants Invade the Land*, New York: Columbia University Press, pp. 140–58.

Gray, J. (1985) 'The microfossil record of early land plants: advances in understanding of early terrestrialization, 1970–1984', *Philosophical Transactions of the Royal Society, London*, B309: 167–95.

—— (1991) '*Tetrahedraletes, Nodospora*, and the "cross" tetrad: an accretion of myth', in S. Blackmore and S.H. Barnes (eds) *Pollen and Spores, Patterns of Diversification, Systematics Association Special Volume 44*, Oxford: Clarendon Press, pp. 49–87.

—— (1993) 'Major Paleozoic land plant evolutionary bio-events', *Palaeogeography, Palaeoclimatology, Palaeoecology*, 104: 153–69.

Gray, J. and Boucot, A.J. (1971) 'Early Silurian spore tetrads from New York: earliest New World evidence for vascular plants?', *Science*, 173: 918–21.

—— (1977) 'Early vascular land plants: proof and conjecture', *Lethaia*, 10: 145–74.

Gray, J., Massa, D. and Boucot, A.J. (1982) 'Caradocian land plant microfossils from Libya', *Geology*, 10: 197–201.

Habgood, K.S. (2000) 'Two cryptospore-bearing land plants from the Lower Devonian (Lochkovian) of the Welsh Borderland', *Botanical Journal of the Linnean Society*, 133: 203–27.

Harland, W.B., Armstrong, R.L., Cox, A.V., Craig, L.E., Smith, A.G. and Smith, D.G. (1990) *'A Geologic Timescale 1989'*, Cambridge: Cambridge University Press.

Heckman, D.S., Geiser, D.M., Eidell, B.R., Stauffer, R.L., Kardos, N.L. and Hedges, S.B. (2001) 'Molecular evidence for the early colonization of land by fungi and plants', *Science*, 293: 1129–33.

Hemsley, A.R. (1994) 'The origin of the land plant sporophyte: an interpolation scenario', *Biological Reviews*, 69: 263–73.

Hoffmeister, W.S. (1959) 'Lower Silurian plant spores from Libya', *Micropalaeontology*, 5: 331–4.

Horodyski, R.J. and Knauth, L.P. (1994) 'Life on land in the Precambrian', *Science*, 263: 494–8.

Hueber, F.M. (2001) 'Rotted wood–alga–fungus: the history and life of *Prototaxites* Dawson 1859', *Review of Palaeobotany and Palynology*, 116: 123–58.

Johnson, N.G. (1985) 'Early Silurian palynomorphs from the Tuscarora Formation in central Pennsylvania and their paleobotanical and geological significance', *Review of Palaeobotany and Palynology*, 45: 307–60.

Karol, K.G., McCourt, R.M., Cimino, M.T. and Delwiche, C.F. (2001) 'The closest living relative of land plants', *Science*, 294: 2351–3.

Keller, C.K. and Wood, B.D. (1993) 'Possibility of chemical weathering before the advent of vascular land plants', *Nature*, 364: 223–5.

Kenrick, P. (2000) 'The relationships of vascular plants', *Philosophical Transactions of the Royal Society, London*, B355: 847–55.

Kenrick, P.K. and Crane, P.R. (1997) '*The Origin and Early Diversification of Land Plants*', Washington: Smithsonian Institution Press.

Knoll, A.H. and Bambach, R.K. (2000) 'Directionality in the history of life: diffusion from the left wall or repeated scaling of the right?', *Paleobiology*, 26 (supplement): 1–14.

Kodner, R.B. and Graham, L.E. (2001) 'High-temperature, acid-hydrolyzed remains of *Polytrichum* (Musci, Polytrichaceae) resemble enigmatic Silurian–Devonian tubular microfossils', *American Journal of Botany*, 88: 462–6.

Kroken, S.B., Graham, L.E. and Cook, M.E. (1996) 'Occurrence and evolutionary significance of resistant cell walls in charophytes and bryophytes', *American Journal of Botany*, 83: 1241–54.

Kurmann, M.H. and Doyle, J.A. (1994) '*Ultrastructure of Fossil Spores and Pollen: Its Bearing on Relationships Among Fossil and Living Groups*', Kew: The Royal Botanic Gardens.

Lang, W.H. (1937) 'On the plant-remains from the Downtonian of England and Wales', *Philosophical Transactions of the Royal Society, London*, B227: 245–91.

Li, W.-H. and Grauer, D. (1991) '*Fundamentals of molecular evolution*', Sunderland, Massachusetts: Sinauer.

McClure, H.A. (1988) 'The Ordovician–Silurian boundary in Saudi Arabia', *Bulletin of the British Museum (Natural History), Geology Series*, 43: 155–63.

Muse, S.V. (2000) 'Examining rates and patterns of nucleotide substitutions in plants', *Plant Molecular Biology*, 42: 25–43.

Quade, J., Roe, L., DeCelles, P.G. and Ojha, T.P. (1997) 'The late Neogene $^{87}Sr/^{86}Sr$ record of lowland Himalayan rivers', *Science*, 276: 1828–31.

Raven, J.A. (1997) 'The role of marine biota in the evolution of terrestrial biota: gases and genes', *Biogeochemistry*, 39: 139–64.

Raven, J.A. and Edwards, D. (2001) 'Roots: evolutionary origins and biogeochemical significance', *Journal of Experimental Botany*, 52: 381–401.

Retallack, G.J. (2000) 'Ordovician life on land and Early Palaeozoic global change', in R.A. Gastaldo and W.A. Dimichele (eds) *Phanerozoic Terrestrial Ecosystems, Paleontological Society Papers*, 6: 21–45.

Richardson, J.B. (1988) 'Late Ordovician and Early Silurian cryptospores and miospores from northeast Libya', in A. El-Arnauti, B. Owens and B. Thusu (eds) *Subsurface Palynostratigraphy of Northeast Libya*, Benghazi, Libya: Garyounis University Publications, pp. 89–109.

—— (1992) 'Origin and evolution of the earliest land plants', in W.J. Schopf (ed.) *Major Events in the History of Life*, Boston: Jones and Bartlett Publishers, pp. 95–118.

Richardson, J.B. (1996) 'Lower and middle Palaeozoic records of terrestrial palynomorphs', in J. Jansonius and D.C. McGregor (eds) *Palynology: Principles and Applications*, Salt Lake City, Utah: American Association of Stratigraphic Palynologists, Publishers Press, pp. 555–74.

Richardson, J.B. and Ioannides, N. (1973) 'Silurian palynomorphs from the Tanezzuft and Acacus Formations, Tripolitania, North Africa', *Micropalaeontology*, 19: 257–307.

Richardson, J.B. and Lister, T.R. (1969) 'Upper Silurian and Lower Devonian spore assemblages from the Welsh Borderland and South Wales', *Palaeontology*, 12: 201–52.

Richardson, J.B. and McGregor, D.C. (1986) 'Silurian and Devonian spore zones of the Old Red Sandstone Continent and adjacent regions', *Geological Survey of Canada Bulletin*, 364: 1–79.

Richardson, J.B., Ford, J.H. and Parker, F. (1984) 'Miospores, correlation and age of some Scottish Lower Old Red Sandstone sediments from the Strathmore region (Fife and Angus)', *Journal of Micropalaeontology*, 3: 109–24.

Rodríguez-Trelles, F., Tarrío, R. and Ayala, F.J. (2001) 'Erratic overdispersion of three molecular clocks: GPDH, SOD, and XDH', *Proceedings of the National Academy of Sciences, USA*, 98: 11405–10.

Rogerson, E.C.W., Edwards, D., Davies, K.L. and Richardson, J.B. (1993) 'Identification of *in situ* spores in a Silurian *Cooksonia* from the Welsh Borderland', *Special Papers in Palaeontology*, 49: 17–30.

Rye, R. and Holland, H.D. (2000) 'Life associated with a 2.76 Ga ephemeral pond?: Evidence from Mount Roe #2 paleosol', *Geology*, 28: 483–6.

Sanderson, M.J. (1997) 'A nonparametric approach to estimating divergence times in the absence of rate constancy', *Molecular Biology and Evolution*, 14: 1218–31.

Sanderson, M.J. and Doyle, J.A. (2001) 'Sources of error and confidence intervals in estimating the age of angiosperms from *RBCL* and 18S RDNA data', *American Journal of Botany*, 88: 1499–516.

Schopf, J.M., Mencher, E., Boucot, A.J. and Andrews, H.N. (1966) 'Erect plants in the Early Silurian of Maine', *United States Geological Survey Professional Paper*, 550-D: 69–75.

Sherwood-Pike, M.A. and Gray, J. (1985) 'Silurian fungal remains: probable records of the Class Ascomycetes', *Lethaia*, 18: 1–20.

Steemans, P. (1999) 'Paléodiversification des spores et des cryptospores de l'Ordovicien au Dévonien inférieur', *Geobios*, 32: 341–52.

—— (2000) 'Miospore evolution from the Ordovician to Silurian', *Review of Palaeobotany and Palynology*, 113: 189–96.

Strother, P.K. (1991) 'A classification schema for the cryptospores', *Palynology*, 15: 219–36.

—— (2000) 'Cryptospores: the origin and early evolution of the terrestrial flora', in R.A. Gastaldo and W.A. Dimichele (eds) *Phanerozoic Terrestrial Ecosystems, The Paleontological Society Papers*, 6: 3–20.

Strother, P.K. and Beck, J.H. (2000) 'Spore-like microfossils from Middle Cambrian strata: expanding the meaning of the term cryptospore', in M.M. Harley, C.M. Morton and S. Blackmore (eds) *Pollen and Spores: Morphology and Biology*, Kew: Royal Botanic Gardens, pp. 413–24.

Strother, P.K., Al-Hajri, S. and Traverse, A. (1996) 'New evidence for land plants from the lower Middle Ordovician of Saudi Arabia', *Geology*, 24: 55–8.

Taylor, W.A. (1995a) 'Spores in earliest land plants', *Nature*, 373: 391–2.

—— (1995b) 'Ultrastructure of *Tetrahedraletes medinensis* (Strother and Traverse) Wellman and Richardson, from the Upper Ordovician of southern Ohio', *Review of Palaeobotany and Palynology*, 85: 183–7.

—— (1996) 'Ultrastructure of lower Paleozoic dyads from southern Ohio', *Review of Palaeobotany and Palynology*, 92: 269–79.

—— (1997) 'Ultrastructure of lower Paleozoic dyads from southern Ohio II: *Dyadospora murusattenuata*, functional and evolutionary considerations', *Review of Palaeobotany and Palynology*, 97: 1–8.

—— (1999) 'Preliminary analysis of early land plant spores and cryptospores from the Cambrian through the upper Silurian', *XVI International Botanical Congress (St Louis) Abstract 2003*.

—— (2000) 'Spore wall development in the earliest land plants', in M.M. Harley, C.M. Morton and S. Blackmore (eds) *Pollen and Spores: Morphology and Biology*, Kew: Royal Botanic Gardens, pp. 425–34.

—— (2001) 'Evolutionary hypothesis of cryptospore producing plants based on wall ultra-structure', in D.K. Goodman and R.T. Clarke (eds) *Proceedings of the IX International Palynological Congress, Texas, USA, 1996*, Dallas, Texas, USA: American Association of Stratigraphic Palynologists Foundation, pp. 11–15.

Traverse, A. (1988) *'Paleopalynology'*, Boston: Unwin Hyman.

Vavrdova, M. (1984) 'Some plant microfossils of possible terrestrial origin from the Ordovician of Central Bohemia', *Vestnik Ustredniho Ustavu Geologickeho*, 3: 165–70.

Wang, D.Y.C., Kumar, S. and Hedges, S.B. (1999) 'Divergence time estimates for the early history of animal phyla and the origin of plants, animals and fungi, *Proceedings of the Royal Society, London*, B266: 163–71.

Wellman, C.H. (1995) ' "Phytodebris" from Scottish Silurian and Lower Devonian continental deposits', *Review of Palaeobotany and Palynology*, 84: 255–79.

—— (1996) 'Cryptospores from the type area of the Caradoc Series in southern Britain', *Special Papers in Palaeontology*, 55: 103–36.

—— (1999) 'Sporangia containing *Scylaspora* from the Lower Devonian of the Welsh Borderland', *Palaeontology*, 42: 67–81.

—— (in press) 'Origin, function and development of the spore wall in early land plants', in Poole, I. and Hemsley, A.R. (eds) *Evolutionary Physiology at the Sub Plant Level*, London: Academic Press.

Wellman, C.H. and Gray, J. (2000) 'The microfossil record of early land plants', *Philosophical Transactions of the Royal Society, London*, B355: 717–32.

Wellman, C.H., Edwards, D. and Axe, L. (1998a) 'Permanent dyads in sporangia and spore masses from the Lower Devonian of the Welsh Borderland', *Botanical Journal of the Linnean Society*, 127: 117–47.

—— (1998b) 'Ultrastructure of laevigate hilate spores in sporangia and spore masses from the Upper Silurian and Lower Devonian of the Welsh Borderland', *Philosophical Transactions of the Royal Society, London*, B353: 1983–2004.

Wikström, N. and Kenrick, P. (2001) 'Evolution of Lycopodiaceae (Lycopsida): estimating divergence times from *rbcl* gene sequences by use of nonparametric rate smoothing', *Molecular Phylogenetics and Evolution*, 19: 177–86.

Wikström, N., Savolainen, V. and Chase, M.W. (2001) 'Evolution of the angiosperms: calibrating the family tree', *Proceedings of the Royal Society, London*, B268: 2211–20.

Zuckerkandl, E. and Pauling, L. (1965) 'Evolutionary divergence and convergence in proteins', in V. Bryson and H.J. Vogel (eds) *Evolving Genes and Proteins*, New York: Academic Press, pp. 97–116.

Chapter 8

Angiosperm divergence times: congruence and incongruence between fossils and sequence divergence estimates

Niklas Wikström, Vincent Savolainen and Mark W. Chase

ABSTRACT

The documentation of derived angiosperm lineages from increasingly older geological deposits, and growing evidence of considerable diversity in flower, seed, and pollen morphology in the mid-Cretaceous both imply that the timing of early angiosperm cladogenesis may be older than our current fossil-based estimates indicate. An alternative to fossils for calibrating the phylogenetic tree comes from divergence in DNA sequence data. Here, we report on an analysis using non-parametric rate smoothing and a three gene dataset covering c. 75 per cent of all angiosperm families recognized in recent classifications. The results provide an initial hypothesis of angiosperm diversification times; by using an internal calibration point, an independent evaluation of angiosperm and eudicot origins is accomplished. Results are compared with fossil-based estimates of both magnolids and eudicot divergence times, and possible directions of future analyses are discussed.

Introduction

Flowering plants (angiosperms) have dominated terrestrial ecosystems since the Late Cretaceous (Crane 1987), and their estimated 250 000–300 000 living species represent an overwhelming majority of extant land plants. Our understanding of angiosperm origin and diversification has, however, been hampered by a number of problems. The early fossil record was, until recently, comparatively poorly understood and insufficiently known (Crane *et al.* 1995) and, up until the 1960s, angiosperms were widely held to have originated in the Late Palaeozoic or possibly early Mesozoic (Axelrod 1952, 1970; Thomas 1957; Eames 1959; Takhtajan 1969). The assignment of early fossil leaves to putatively derived angiosperm lineages led to this conclusion, but later work, based not only on leaves but also on pollen, flowers, and fruits, changed this view dramatically, pushing the angiosperm origin into the Early Cretaceous (Doyle 1969; Doyle and Hickey 1976; Hughes 1976; Hickey and Doyle 1977; Doyle 1978; Friis and Skarby 1981; Muller 1981; Friis 1984; Friis *et al.* 1988). From a neobotanical perspective, resolving the relationships among extant lineages, as well as establishing a rooting point for the angiosperm clade, have both been problematic. However, in the late 1980s and early 1990s several landmark papers appeared providing the skeleton of a more rigorous hierarchical framework in which the fossil data could be interpreted (Doyle and Donoghue 1986, 1987; Donoghue and Doyle 1989a,b; Chase *et al.* 1993; Doyle *et al.* 1994). Following this progress, a more or less coherent picture of an Early Cretaceous origin followed by a rapid diversification of early angiosperm lineages has appeared, and Crane *et al.* (1995) described the Valanginian (135–141 Ma) appearance of angiosperms (through putative magnolid pollen), the Barremian–Aptian boundary (125 Ma) appearance of eudicots (based on their unique triaperturate pollen), and the appearances of hamamelids and rosids in the Albian–early Cenomanian (97–112 Ma) as an orderly sequence, and a pattern that any claim of a pre-Cretaceous angiosperm origin must confront.

During the last five years, we have seen additional progress on angiosperm phylogenetics, and some of the patterns emerging prove more difficult to fit into the coherent sequence described by Crane *et al.* (1995). In the palaeobotanical community, increasing morphological diversity in seeds, pollen, and fruits has been documented from comparatively old geological deposits (Friis *et al.* 1999, 2000, 2001). A second development is that derived angiosperm lineages are being documented from increasingly older geological deposits. Crepet and Nixon (1998), for example, documented Clusiaceae from Turonian (88–90 Ma) deposits of New Jersey, Keller *et al.* (1996) and Herendeen *et al.* (1999) documented Actinidiaceae from Campanian (74–83 Ma) and Santonian (83–87 Ma) deposits. Herendeen *et al.* (1999) suggested a possible affinity to Apiaceae/Araliaceae for one of their Santonian fossils, and Basinger and Dilcher (1984) documented a possible Rosaceae/Rhamnaceae from the early Cenomanian (94–97 Ma) of Nebraska.

In parallel to this development, the neobotanical community has seen an explosion in the amount of molecular data addressing the relationships of extant lineages, and the hierarchical framework has been transformed from a mere skeleton into a more rigorous one with most extant families represented, and relationships receiving an increasing amount of support (Soltis *et al.* 1997, 1999, 2000; Qiu *et al.* 1999; Chase *et al.* 2000; Qiu 2000; Savolainen *et al.* 2000a,b). The full impact of the palaeobotanical development can only be appreciated by considering the patterns emerging from these molecular analyses, and Figure 8.1 illustrates this by indicating the putative positions of some key fossil taxa on a phylogram representing one of the more than 8000 most parsimonious trees resulting from the analyses by Soltis *et al.* (1999). Groups recently documented from Cenomanian–Campanian deposits are highly derived and nested well inside rosids and asterids, and the comparatively high levels of sequence divergence separating the Cenomanian–Campanian fossils (taxa 5–8; 74–97 Ma) and the fossil taxa (taxa 1–4) constituting the orderly sequence of Crane *et al.* (1995) indicate one of three possible conclusions. Either the putatively rapid and explosive diversification of early angiosperms was accompanied by much more rapid molecular change than subsequently has occurred, or cladogenesis in early angiosperms took place earlier than our current fossil-based estimates indicate, or the assignment of these fossils to derived angiosperm families is incorrect.

By accepting this argument, that the levels of sequence divergence on our phylograms indicate problems, we rely in one way or other on embracing the concept of molecular clocks (Zuckerhandl and Pauling 1962, 1965). However, all methods, whether they use rigorous or more relaxed clock assumptions for inferring time from divergence in DNA sequence data, suffer from their own problems. Some are associated with small datasets and stochastic errors (Hillis *et al.* 1996; Sanderson and Doyle 2001) and others with an inability to infer rate changes correctly over the tree (Sanderson

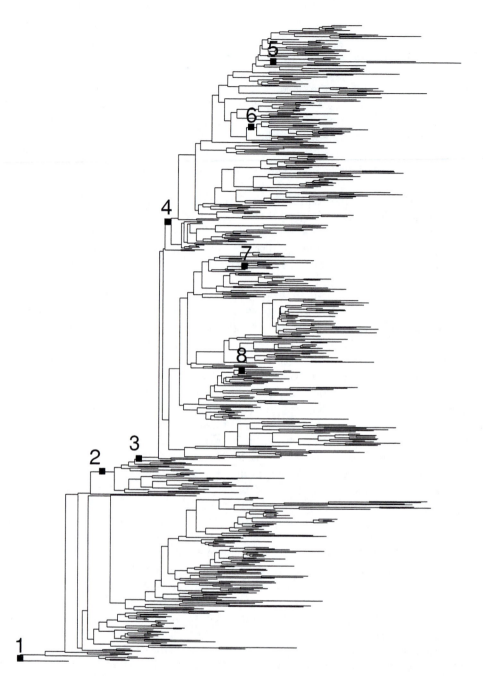

Figure 8.1 Cladogram with branch lengths representing one of the more than 8000 most parsimonious trees from the analyses by Soltis *et al.* (1999). Branch lengths are calculated using parsimony with ACCTRAN optimization. Putative positions of fossil taxa (1–4) constituting the orderly sequence of Crane *et al.* (1995) are indicated: (1) Valanginian–Heuterivian (132–141 Ma) pollen generally accepted as the first angiosperm fossil record; (2) Barremian (125 Ma) record of tricolpate pollen; (3) Albian (97–112 Ma) records of 'lower Hamamelididae'; and (4) Cenomanian (90–97 Ma) records of early 'rosids'. In addition, the putative position of more recently documented fossil taxa (5–8) are indicated: (5) Turonian (88–90 Ma) Clusiaceae documented by Crepet and Nixon (1998); (6) Cenomanian (94–97 Ma) Rhamnaceae/Rosaceae documented by Basinger and Dilcher (1984); (7) Santonian (83–87 Ma) Araliaceae/Apiaceae documented by Herendeen *et al.* (1999); and (8) Santonian (83–87 Ma) Actinidiaceae documented by Herendeen *et al.* (1999).

1997, 1998). Furthermore, until recently they all relied on the assumption that sequences evolve roughly at constant rates by enforcing a rigorous clock assumption on our analyses. Sanderson (1997) developed a different approach, the non-parametric rate smoothing (NPRS), which allows rates to change over the tree but assumes that such changes are autocorrelated, effectively meaning that changes in rates are assumed to be inherited from an ancestral lineage by its immediate descendants. The method smooths local changes in rates over the tree using an optimization algorithm, and searches for the solution that minimizes the inferred rate changes (Sanderson 1997). As a case study, we have used NPRS to estimate divergence times in angiosperms using a three-gene dataset based on plastid *rbcL* and *atpB* exons and nuclear 18S rDNA covering 560 angiosperm taxa (Soltis *et al.* 1999, 2000). Our primary aim was to provide an initial hypothesis of angiosperm divergence times based on sequence divergence data representing a majority of extant families, and by using an internal calibration point, an independent evaluation of angiosperm and eudicot origins is accomplished. Results are compared with fossil-based estimates extracted from Magallón and Sanderson (2001) for magnoliids and from Magallón *et al.* (1999) for eudicots; possible directions for future analyses are discussed.

Material and methods

Data

Nucleotide sequence data covering a majority of all flowering plant families have over the last decade been assembled for three different loci, *rbcL* (Chase *et al.* 1993, 2000; Savolainen *et al.* 2000b) and *atpb* (Savolainen *et al.* 2000a) from the plastid genome, and 18S rDNA (Soltis *et al.* 1997) from the nucleus. These efforts recently culminated in a comprehensive phylogenetic analysis including 560 angiosperms and seven outgroup gymnosperm taxa (Soltis *et al.* 1999, 2000). In total, these analyses included representatives of about 75 per cent of all families recognized in the most up to date classification (APG 1998). We used their complete dataset to calculate branch lengths on one of the more than 8000 most parsimonious trees obtained by Soltis *et al.* (1999, 2000); the tree used corresponds to that reported in their 'B series' of figures. The seven outgroup taxa used by Soltis *et al.* (1999, 2000) were initially included to obtain branch length estimates for the first ingroup branching point but were subsequently removed from the analyses. Branch lengths were estimated using both parsimony (accelerated and delayed transformations; ACCTRAN and DELTRAN, respectively), and maximum likelihood methods. The HKY85 model of sequence evolution (Hasegawa *et al.* 1985) was used in the likelihood estimates, and transition/transversion ratios as well as nucleotide frequencies were estimated from the data. Calculations were done using PAUP 4.0b4a (Swofford 1998).

NPRS analyses

The non-parametric rate smoothing analyses were done using the r8s program (Sanderson 1997). To prevent the optimization algorithm from converging on a local optimum, the searches were started at five different initial time estimates (num_time_guesses=5) and were restarted three times for each guess (num_restarts=3). Three consecutive analyses were done using the different branch length estimates from ACCTRAN, DELTRAN, and maximum likelihood optimizations. No minimum age constraints were enforced during the analyses.

A bootstrap resampling procedure was employed to estimate errors arising from the stochastic nature of the substitution process (Efron and Tibshirani 1993; Sanderson 1997). One hundred bootstrap replicates of the dataset were constructed using the seqboot program (Felsenstein 1993), and branch lengths were estimated on our single tree using ACCTRAN optimization for each replicate and input to the r8s program. Bootstrap estimates of standard errors for each node were calculated for the age distribution estimates obtained (Efron and Tibshirani 1993).

Calibrating relative ages

The output from NPRS analyses is a set of relative ages, and a calibration point has to be selected to convert these ages into absolute times. In most cases, this is done with reference to the fossil record. Important considerations for choosing the calibration point include: (1) terminal nodes should be avoided to minimize any effects of poor taxon sampling; (2) the fossil taxon should indisputably be part of the group defined by the selected node; (3) the age of the fossil taxon should as closely as possible represent the actual divergence time for the selected node; and (4) relationships of the selected group to other taxa should be well supported by the bootstrap/ jackknife.

Based on the occurrence of *Protofagacea* (Herendeen *et al.* 1995) and *Antiquacupula* (Sims *et al.* 1998) in the Campanian and Late Santonian of Georgia, we chose to fix the split between the Fagales and Cucurbitales (node CALL; Figure 8.5) in the late Santonian (84 Ma). Several floral features indicate that they are both part of the Fagales lineage, and both have flowers and fruits born in a typical Fagales cupule (Herendeen *et al.* 1995; Sims *et al.* 1998). Evaluating their precise relationships is, however, complicated by uncertainties surrounding the origin of the cupule. Analyses based on both morphological and molecular data indicate that the Fagaceae *sensu lato* are paraphyletic and that the Fagaceae *sensu stricto* and Notofagaceae form two separate lineages (Nixon 1989; Chase *et al.* 1993; Manos *et al.* 1993; Manos 1997; Manos and Steele 1997). The cupule must therefore either have originated twice or originated once in the Fagales lineage and subsequently been lost in the lineage leading to the Betulaceae, Casuarinaceae, Juglandaceae, and Myricaceae. Although our choice of calibration point (Fagales–Cucurbitales split) could be seen as too conservative, it allows us to control the direction of any errors incorporated through the calibration, and we can be confident that we are underestimating the true age of our calibration point.

Results

Results are presented in the form of chronograms (Figures 8.3–8.6), calibrated against the geological timescale (Harland *et al.* 1990), focusing on taxa for which fossil-based estimates were reported by Magallón *et al.* (1999) and Magallón and Sanderson (2001). In addition, Table 8.1 gives details on both the fossil-based and molecular estimates. Molecular estimates are given for all three analyses using

Table 8.1 Comparison between estimates from our analyses and fossil-based estimates extracted from Magallón et al. (1999) and Magallón and Sanderson (2001)

Taxon	Estimated age	ACCTRAN (Ma)	DELTRAN (Ma)	ML (Ma)	Bootstrap estimated error (myr)	Specific fossil-based age (Ma)	Implied fossil-based age (Ma)
Aristolochiaceae–Lactoridaceae	Aptian–Hauterivian (node 545)	132	133	122	6	91 (Lactoridaceae)	
Lactoridaceae	Santonian–Albian (node 547)	85	107	97	5	91	
Calycanthales[a]	Albian–Aptian (node 537)	111	114	108	7	109	
Ceratophyllaceae	Valanginian–Kimmeridgian (node 5)	155	154	140	7	60	125 (Eudicots)
Chloranthaceae	Valanginian–Callovian (node 417)	158	154	141	7	121 (Hedyosmum)	123 (Winteraceae)
Hedyosmum	Aptian–Barremian (552)	132	131	121	8	121	
Illiciales	Hauterivian–Tithonian (node 554)	146	148	**133**	7	93 (Illiciaceae)	
Illiciaceae	Cenomanian–Albian (555)	102	108	93	11	93	
Laurales	Albian–Aptian (node 537)	111	114	108	7	109 (Hernandiaceae)	
Hernandiaceae	Turonian–Cenomanian (node 539)	96	89	95	7	109	
Magnoliales[a]	Barremian–Hauterivian (node 552)	132	131	121	8	109	
Magnoliales	Barremian–Hauterivian (node 552)	132	131	121	8	98 (Magnoliaceae)	
Magnoliaceae	Cenomanian (node 527)	92	95	93	7	98	
Nymphaeales	Kimmeridgian–Bajocian (node 2)	171	168	153	8	121 (Nymphaeaceae)	125 (Eudicots)
Nymphaeaceae	Albian–Hauterivian (node 557)	122	133	105	8	121	
Winteraceae[a]	Albian (node 532)	99	105	99	6	123	
Alismatiflorae	Albian (node 515)	112	110	107	6	68 (Cymodoceaceae)[b]	
Arales	Albian–Aptian (node 514)	124	115	111	6	106 (Epipremnum)[b]	
Arecaceae[a]	Cenomanian–Albian (node 425)	99	91	92	5	84	
Cyperales	Lutetian–Ypresian (node 433)	43	50	52	3	45 (Cyperaceae)	
Cyperaceae	Chattian–Bartonian (node 434)	28	38	39	2	45	
Poales	Lutetian (node 436)	45	48	49	3	68 (Poaceae)	
Poaceae	Rupelian–Lutetian (node 437)	35	43	44	3	68	
Typhales	Danian (node 431)	63	62	64	4	77	
Zingiberales	Maastrichtian–Campanian (node 439)	81	73	79	4	84 (Zingiberaceae)	
Zingiberaceae	Chattian–Bartonian (node 449)	42	25	43	3	84	
Ranunculid clade	Barremian–Tithonian (node 6)	147	144	131	6	69 (Menispermaceae)	118 (Trochodendrales)[c]
Menispermaceae	Albian–Aptian (node 407)	103	113	103	6	69	
Nelumbonaceae	Barremian–Valanginian (node 396)	135	137	125	6	100	108 (Platanaceae)
Platanaceae	Albian–Aptian (node 397)	113	117	108	7	108	
Proteaceae	Albian–Aptian (node 397)	113	117	108	7	97	108 (Platanaceae)
Sabiaceae	Barremian–Valanginian (node 8)	140	140	128	5	69	118 (Trochodendrales)[c]

Table 8.1 (Cont'd)

Taxon	Estimated age	ACCTRAN (Ma)	DELTRAN (Ma)	ML (Ma)	Bootstrap estimated error (myr)	Specific fossil-based age (Ma)	Implied fossil-based age (Ma)
Buxaceae	Aptian (node 393)	122	124	113	6	104	
Trochodendrales[c]	Aptian–Hauterivian (node 10)	135	134	123	5	118 (Tetracentraceae)[c]	
Tetracentraceae[c]	Cenomanian–Albian (node 392)	106	100	95	7	118	
Caryophyllid clade	Albian (node 360)	105	111	104	4	83 (Amaranthaceae)	
Amaranthaceae	Chattian–Bartonian (node 382)	28	40	38	2	83	
Saxifragoids	Albian–Aptian (node 14)	121	119	111	4	89 (saxifragaleans)	
Saxifragaleans	Campanian–Cenomanian (node 191)	91	78	88	5	89	
Geraniaceae	Maastrichtian–Santonian (node 182)	71	85	79	4	89	
Expanded Capparales	Santonian–Turonian (node 133)	90	89	85	4	8	
Capparales	Ypresian (node 167)	54	54	52	3	89 (Capparales)	
Sapindales	Campanian–Santonian (node 134)	82	84	80	4	89	
Rutaceae	Lutetian (node 141)	47	47	45	4	67 (Rutaceae, Aceraceae)	
Expanded Malvales	Campanian–Santonian (node 134)	82	84	80	4	67	69 (Malvales)
Bombacaceae	Chattian–Rupelian (node 156)	28	31	29	4	69 (Bombacaceae)	
Myrtales	Albian (node 17)	107	104	100	3	69	
Combretaceae	Campanian (node 122)	79	78	75	4	84 (Combretaceae)	
Cucurbitales	Our calibration point	–	–	–	–	84	84 (Fagales)
Cucurbitaceae	Maastrichtian (node 97)	65	66	66	2	58 (Cucurbitaceae)	
Urticales	Maastrichtian (node 104)	67	65	67	3	58	69 (Urticales)
Celtidoideae	Ypresian–Thanetian (node 110)	55	56	57	4	69 (Celtidoideae)	
Rosaceae	Campanian (node 103)	76	76	76	3	69	
Prunoideae	Chattian–Rupelian (node 119)	29	31	35	4	44 (Prunoideae)	
Higher Hamamelididae	Our calibration point	–	–	–	–	44	
Normapolles clade	Thanetian–Danian (node 89)	61	60	61	4	84 (Normapolles)	
Polygalaceae	Maastrichtian (node 85)	66	68	67	3	84	68 (Polygalaceae)
Fabaceae	Campanian (node 82)	79	78	74	3	68	
Expanded Cunoniaceae	Coniacian–Cenomanian (node 21)	91	89	88	3	56–65	
Elaeocarpaceae	Danian–Maastrichtian (node 67)	66	65	64	4	58 (Elaeocarpaceae)	
Malpighiales	Coniacian–Cenomanian (node 21)	91	89	88	3	58 (Euphorbiaceae)	
Euphorbiaceae	Maastrichtian (node 63)	69	71	71	3	58	
Cornalean clade	Albian–Aptian (node 315)	114	112	106	5	69 (mastixioid taxa)	
Mastixioid taxa	Coniacian–Cenomanian (node 348)	87	90	92	6	69	
Ericalean clade	Albian–Aptian (node 315)	114	112	106	5	89 (Ericaceae s. lat.)	89 (ericalean clade)

Taxon	Age span (node)					Fossil age	Other fossil
Ericaceae s. lat.	Ypresian (node 323)	50	58	56	5	89	
Ilex clade	Albian (node 214)	107	103	99	5	69 (Ilex)	
Ilex	Lutetian–Ypresian (node 253)	49	55	52	5	69	
Apiales	Santonian–Turonian (node 225)	90	89	85	5	69 (Araliaceae)	69 (Araliaceae)
Araliaceae	Bartonian–Lutetian (node 229)	41	45	43	4	69	
Dipsacales	Santonian–Turonian (node 217)	88	90	85	5	53 (Caprifoliaceae)	
Caprifoliaceae	Ypresian–Thanetian (node 219)	56	58	54	4	53	
Asterales	Cenomanian–Albian (node 215)	101	98	94	5	29 (Menyanth., Gooden.)	
Menyanthaceae	Maastrichtian (node 238)	69	69	65	4	29	
Goodeniaceae	Bartonian (node 244)	39	42	40	4	29	
Garrya clade	Albian (node 254)	107	105	100	5	46 (*Eucommia*)	
Eucommia	Campanian–Santonian (node 313)	80	84	81	5	46	
Boraginales	Campanian (node 258)	81	77	78	4	53	53 (Boraginales)
Solanales	Campanian–Santonian (node 257)	86	82	82	4	53 (Convolvulaceae)	
Convolvulaceae	Maastrichtian (node 295)	66	65	65	4	53	
Gentianales	Santonian–Turonian (node 256)	89	86	83	4	53 (Apocynaceae, Rubiaceae)	
Apocynaceae	Lutetian–Ypresian (node 306)	45	53	53	4	53	
Rubiaceae	Danian (node 304)	64	61	63	4	53	
Lamiales	Maastrichtian (node 259)	74	71	74	4	37 (Oleaceae)	
Oleaceae	Ypresian–Danian (node 260)	63	55	64	4	37	
Santalales	Albian–Aptian (node 359)	113	118	111	4	53 (Olacaceae)	83 (Caryophyllid clade)
Olacaceae	Santonian–Cenomanian (node 385)	85	97	87	5	53	
Dilleniaceae	Albian (node 360)	105	111	104	4	53	
Vitis-Leeaceae	Albian–Aptian (node 15)	117	115	108	4	58	84 (Fagales)
Gunneraceae	Albian–Aptian (node 391)	115	118	108	5	89	

The fossil-based estimates are, contrary to their use, assumed to provide a minimum age for the split between the taxon in question and its sister group. Column 1 lists the taxa according to their usage, and columns 2–5 list our age estimates (node numbers correspond to those on the chronograms, Figures 8.3–8.6). The age span given in column 2 results from the three consecutive analyses using ACCTRAN and DELTRAN optimization and maximum likelihood for calculating branch lengths. The specific ages, resulting from each analysis are reported in columns 3–5. Column 6 indicates the bootstrap estimates of standard error, and columns 7–8 list fossil-based estimates extracted from Magallón et al. (1999) and Magallón and Sanderson (2001). The specific fossil-based age (column 7) corresponds to that given by Magallón et al. (1999) and Magallón and Sanderson (2001) and also indicates on what taxon the age estimate was based. Column 8 lists instances when the topology implies that some other fossil provides a better minimum age estimate for a node. In addition, further comparisons are made whenever the fossil-based estimate was based on a less inclusive taxon. The age for the Aristolochiaceae–Lactoridaceae clade, for example, was based on Lactoridaceae pollen (Zavada 1987). We have therefore included comparisons with both the Aristolochiaceae–Lactoridaceae and Lactoridaceae clades.

[a] The fossil-based age concerns the stem-group and not the crown-group as for remaining taxa (Magallón and Sanderson 2001).

[b] We are unable to compare the fossil-based estimate for the less inclusive taxon with a molecular estimate since the taxon was not included in the molecular dataset.

[c] Magallón et al. (1999) indicated an Aptian (124.5–112 Ma) occurrence for *Populus potomacensis*, but the original documentation of this taxon indicated the Albian (112–97 Ma; Doyle and Hickey 1976).

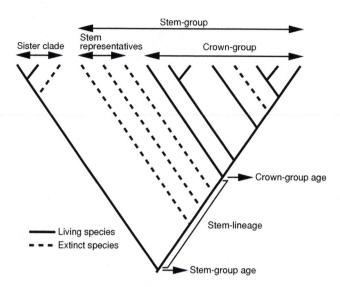

Figure 8.2 Schematic drawing adopted from Magallón and Sanderson (2001) explaining our use of crown-groups, stem-groups, and their ages. The crown-group age corresponds to the first phylogenetic split within the crown-group and only fossils demonstrated to be nested within the crown-group can provide a minimum age estimate for this group. The stem-group age may be considerably older and corresponds to the split between the crown-group in question and its extant sister group. Any representative, though preferably a stem one, may provide a minimum age estimate for the stem-group. Also note that a minimum age estimate for the sister clade may represent the best available age estimate for the stem-group.

parsimony (ACCTRAN and DELTRAN) and maximum likelihood for calculating branch lengths, and also lists the bootstrap estimates of standard error. Bootstrap estimates of standard error, as well as the chronograms presented (Figures 8.3–8.6) are based on the analysis using parsimony (ACCTRAN optimization) for calculating branch lengths, and the chronograms have been generalized to a level enabling us to present graphically the comparison between our molecular estimates and the fossil-based estimates (Magallón *et al.* 1999; Magallón and Sanderson 2001). Complete details of the results, documenting age estimates for all included taxa, have been presented elsewhere (Wikström *et al.* 2001).

Discussion

Crown-groups versus stem-groups

Understanding the difference between crown-groups, stem-groups, and how fossils provide minimum age estimates for these is of crucial importance when comparing our molecular estimates with fossil-based ones. The two terms were initially introduced by Jefferies (1979) but his stem-group definition only referred to the stem representatives (Figure 8.2), implying the existence of paraphyletic groups (Doyle and Donoghue 1993; Smith 1994). We have therefore adopted the usage of Magallón and Sanderson (2001) including both stem representatives and the crown-group in the stem-group definition, and our usage of stem-group corresponds with the total-group concept of Jefferies (1979). Following the definitions of Magallón and Sanderson (2001), a crown-group is the least inclusive monophyletic group that includes all the extant members of a clade (Figure 8.2). The crown-group may include extinct represent-atives, but only those that diverged after the origin of the most recent common ancestor of all extant representatives. The stem-group is the most inclusive mono-phyletic group containing the extant members of a clade, but no other extant taxa, plus all the extinct lineages that diverge from the lineage leading to the crown-group (Figure 8.2). The crown-group age corresponds to the first phylogenetic split within the crown-group, and only fossils demonstrated to be nested within the crown-group can provide a minimum age estimate for the crown-group. The stem-group age may be considerably older and corresponds to the split between the crown-group in question and its extant sister group (Magallón and Sanderson 2001), and any repres-entative, though preferably a stem-taxon, may provide a minimum age estimate for the stem-group (Figure 8.2). Also note that an age estimate for the origin of the sister clade may provide the best available estimate for the stem-group age, and this is what the implied age column in Table 8.1 indicates.

Interpreting incongruence

We need to consider what different types of incongruence might represent while comparing our molecular-based estimates with those based on fossils. The most easily interpreted type is the case where the fossil evidence indicates an older age than our corresponding molecular estimate. In this case, either the assignment of the fossil taxon to the group in question is wrong, or else the molecular estimate fails to infer the taxon age correctly. There are examples for both these cases. The fossil Amaranthaceae from the Santonian–Campanian (Collinson *et al.* 1993), for ex-ample, has been confirmed to be something other than Amaranthaceae (Friis, personal communication), and one might suspect that some other assignments such as the Maastrichtian Bombacaceae (Muller 1981) also fall into this category. There are, however, several examples in which solid fossil evidence indicates that our molecu-lar estimates are underestimating the taxon age, and in particular, this pattern is seen among some of the more derived groups (see below for a more detailed account of sources of error in the molecular estimates).

The reverse pattern, where molecular estimates indicate older ages than the available fossil evidence, is less easily interpreted. In some cases there may be good reasons to assume that the available fossil information only provides a poor estimate for the timing of divergence, but in others one might consider the molecular estimate to be wrong, overestimating the 'true' taxon age. Extending the temporal range of a taxon further back in time does imply the existence of a gap in the fossil record, but given that fossils only provide minimum age estimates (Doyle and Donoghue 1993), we should expect a gap of some size to exist. The question then is, when is a gap large enough to indicate incongruence? In the zoological community, at least two different methods have been suggested for evaluating whether the fossil record is com-plete enough to discard the existence of such gaps (Marshall 1998; Foote *et al.* 1999), but to our knowledge, neither one has been used by the palaeobotanical community.

It would, for example, be interesting to apply these methods to the origin of eudicots. Intuitively, it is often argued that the fossil record of triaperturate pollen (indicating the eudicot origin) is good enough to discard the possibility of anything but a minor gap, but it would perhaps be worthwhile to demonstrate this using either one of these methods.

Origins of angiosperms and eudicots

The crown-group of extant angiosperms is resolved to have originated in the Early–Middle Jurassic (158–179 Ma), and eudicots are indicated as Late Jurassic–mid-Cretaceous (131–147 Ma). Despite the conservative age estimate for our calibration point, these estimates are older then nearly all fossil-based estimates.

Claims of a pre-Cretaceous angiosperm diversification have been made before based both on fossil evidence (Cornet and Habib 1992; Cornet 1993), and molecular clock estimates (Ramshaw et al. 1972; Martin et al. 1989, 1993; Wolfe et al. 1989; Brandl et al. 1992; Goremykin et al. 1997). Such claims, however, have generally been rejected, and Crane et al. (1995) considered the appearances of angiosperms in the Valanginian, eudicots around the Barremian-Aptian boundary, and rosids and hamamelids in the Early Cenomanian (Figure 8.1) as an orderly sequence, and one that argues against a pre-Cretaceous origin. From a molecular point of view, it is not the sequence of appearances that poses a problem, but the ages themselves. The fossil evidence indicates that the time intervals separating basal branches are short and that major angiosperm lineages diverged within a comparatively short time span (Hickey and Doyle 1977; Lidgard and Crane 1988; Crane and Lidgard 1989; Taylor and Hickey 1990; Crane et al. 1995). Nevertheless, we see a substantial amount of nucleotide change on those branches (Figure 8.1) and, in our molecular-based estimates, angiosperm and eudicot origins are pushed further back in time than the fossil-based estimates (Figure 8.3).

If claims of a pre-Cretaceous angiosperm diversification need to confront the orderly sequence of appearance seen in the fossil record, claims of a Cretaceous diversification need to confront the long branch lengths observed on our molecular phylogenetic trees (Soltis et al. 1999, 2000). There may, of course, be alternative explanations for those long branch lengths. They may be incorrectly inferred, and true branch lengths might be considerably shorter. Such an explanation, however, would have serious consequences with respect to our phylogenetic analyses implying that support for basal branches is based on spurious and incorrectly inferred evidence. Other, non-molecular, lines of evidence corroborating these phylogenetic analyses indicate that this is unlikely (Nandi et al. 1998; Doyle and Endress 2000). Alternatively, if these branch lengths are correct, then the inferred rates may not be, resulting in estimates of the time intervals between cladogenic events that are too large. The rapid morphological diversification of early angiosperms must then have been accompanied by more, or equally, rapid molecular change, and it also implies that rates of DNA change accelerated and slowed down in several lineages in a correlated manner over time. This seems highly unlikely but, if it did occur, it would appear to imply historical environmental changes as the cause (the only cause that could simultaneously affect multiple lineages occurring worldwide would have to be global environmental change). A pattern with both rapid morphological and molecular change, would,

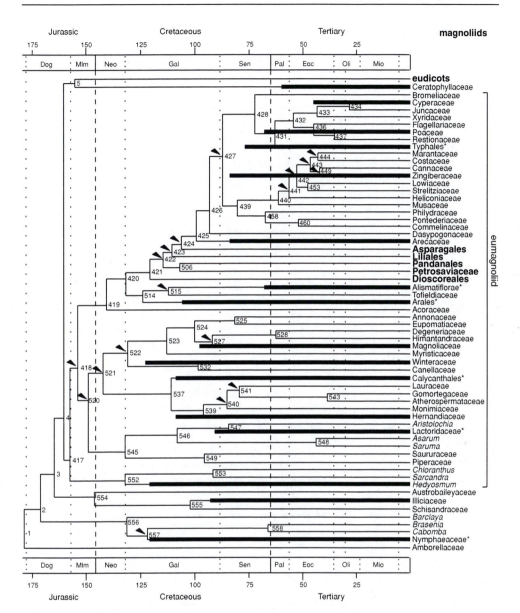

Figure 8.3 Chronogram calibrated against the geological timescale (Harland *et al.* 1990) focusing on magnoliids. Fossil-based estimates extracted from Magallón and Sanderson (2001) are indicated with thick bars. The chronogram is based on the analysis using ACCTRAN optimization for calculating branch lengths, and has been generalized, usually down to family level, to facilitate a graphical comparison between our molecular-based estimates and the fossil-based estimates (Magallón and Sanderson 2001). Non-generalized chronograms, indicating age estimates for all included taxa, have been presented elsewhere (Wikström *et al.* 2001), and node numbers correspond to those given there. With few exceptions (taxa marked with *) taxon names follow those used by Soltis *et al.* (1999) and arrows indicate nodes that received less than 50 per cent jackknife support in their phylogenetic analyses (Soltis *et al.* 1999, 2000). Taxa marked with * have been generalized and named to correspond with the usage of Magallón and Sanderson (2001).

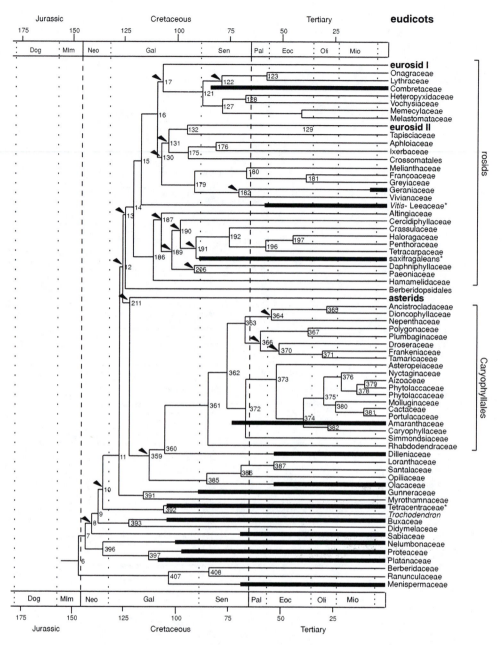

Figure 8.4 Chronogram calibrated against the geological timescale (Harland *et al.* 1990) focusing on eudicots. Fossil-based estimates extracted from Magallón *et al.* (1999) are indicated with thick bars. The chronogram is based on the analysis using ACCTRAN optimization for calculating branch lengths, and has been generalized, usually down to family level, to facilitate a graphical comparison between our molecular-based estimates and the fossil-based estimates (Magallón *et al.* 1999). Non-generalized chronograms, indicating age estimates for all included taxa, have been presented elsewhere (Wikström *et al.* 2001), and node numbers correspond to those given there. With few exceptions (taxa marked with *) taxon names follow those used by Soltis *et al.* (1999) and arrows indicate nodes that received less then 50 per cent jackknife support in their phylogenetic analyses (Soltis *et al.* 1999, 2000). Taxa marked with * have been generalized and named to correspond with the usage of Magallón *et al.* (1999).

however, contrast with that seen in groups that have diversified more recently such as the Asterales and Lamiales. Here, we see no apparent correlation between rapid diversification and morphological change on the one hand, and molecular change on the other (Soltis *et al.* 2000; Magallón and Sanderson 2001), and it seems illogical for more recent patterns to be qualitatively different from older ones. Bateman (1999) tried to address this issue of correlated or non-correlated change of morphological and molecular characters by looking at architectural radiations on volcanic islands, and this island approach may provide a way to address this issue at a more general level. Furthermore, comparisons between our molecular-based estimates and those based on fossils for 'basal' angiosperm and eudicot lineages indicate a reasonable amount of congruence between the two (Figures 8.3–8.4), and if we accept fossil-based estimates such as the Barremian–Aptian split between the Nymphaeaceae and Cabombaceae (Friis *et al.* 2001), the Barremian–Aptian crown-group origin of the Chloranthaceae (Friis *et al.* 1999), and the early Albian split between the Platanaceae and Proteaceae (Crane and Herendeen 1996), the crown-group origins of angiosperms and eudicots become, almost by necessity, older than our current fossil-based estimates indicate.

Fossil-based estimates within magnolids and eudicots

In assessing absolute diversification rates in angiosperms, Magallón *et al.* (1999) and Magallón and Sanderson (2001) compiled and summarized the available fossil evidence for the diversification times of major angiosperm lineages. We have used their summaries to compare our molecular estimates with those based on fossil evidence. One complicating factor is that their ages mostly refer to crown-groups, and there is no reasonable way to deal with crown-group ages for the molecular estimates. In the molecular analyses, the thoroughness of taxon sampling varies considerably between different clades, and this complicates a direct comparison with their crown-group estimates. We have therefore treated all their estimates as stem-group ages. However, for most of their crown-group ages, they have either specified a less inclusive taxon that their estimate was based on, or provided supporting references for their age estimates. By using either the less inclusive taxon they specified, or referring to the original literature, we have extended our comparisons to encompass also the less inclusive taxa. Their estimate of the Aristolochiaceae–Lactoridaceae clade, for example, was based on the occurrence of Lactoridaceae pollen from the Turonian of SW Africa (Zavada 1987). We have thus not only compared their Aristolochiaceae–Lactoridaceae clade estimate with our estimate for the split between the Aristolochiaceae–Lactoridaceae clade and its sister group, but also included the comparison with our Lactoridaceae age (Table 8.1). The comparisons with the less inclusive taxa are generally the more appropriate ones, and are also the ones represented in the chronograms (Figures 8.3–8.6).

'Basal' nodes

If we treat the fossil-based estimates in a conservative way, only comparing our estimates with the more inclusive groups, such as the Aristolochiaceae–Lactoridaceae clade (Table 8.1), the molecular-based ages are uniformly older than the

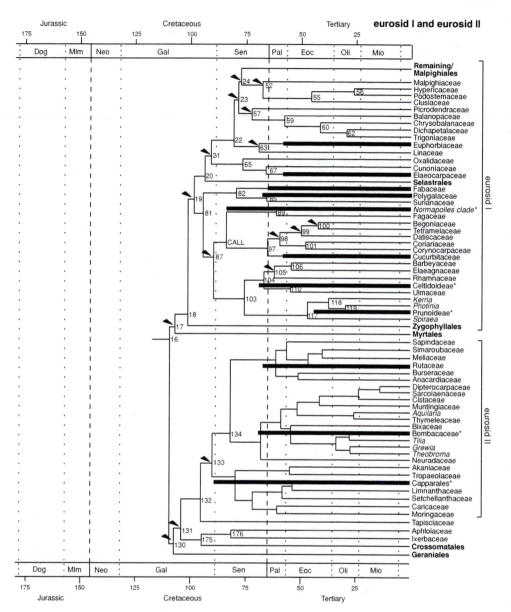

Figure 8.5 Chronogram calibrated against the geological timescale (Harland *et al.* 1990) focusing on eurosid I and eurosid II. Fossil-based estimates extracted from Magallón *et al.* (1999) are indicated with thick bars. The chronogram is based on the analysis using ACCTRAN optimization for calculating branch lengths, and has been generalized, usually down to family level, to facilitate a graphical comparison between our molecular-based estimates and the fossil-based estimates (Magallón *et al.* 1999). Non-generalized chronograms, indicating age estimates for all included taxa, have been presented elsewhere (Wikström *et al.* 2001), and node numbers correspond to those given there. With few exceptions (taxa marked with *) taxon names follow those used by Soltis *et al.* (1999) and arrows indicate nodes that received less then 50 per cent jackknife support in their phylogenetic analyses (Soltis *et al.* 1999, 2000). Taxa marked with * have been generalized and named to correspond with the usage of Magallón *et al.* (1999).

corresponding fossil-based estimates. However, the fact that the fossil-based estimates concern crown-groups and not stem-groups (Magallón *et al.* 1999; Magallón and Sanderson 2001) implies that this comparison may not be appropriate, and there is considerably more congruence if we instead compare our estimates with the less inclusive groups, such as Lactoridaceae (Table 8.1, Figures 8.3–8.6).

This congruence, between molecular and fossil-based estimates of 'basal' angiosperm divergence times, should perhaps be a concern for those who reject the idea of pushing the angiosperm and eudicot origins back in time (Crane *et al.* 1995). The means by which ages of groups such as the Nymphaeaceae, crown-group Chloranthaceae (*Hedyosmum*), and Illiciaceae (Figure 8.3) as well as the Platanaceae, Tetracentraceae, and Buxaceae (Figure 8.4) are resolved in the molecular analyses ultimately depend on how the NPRS analyses have resolved the evolutionary rates among those 'basal' nodes. If the analysis is doing a reasonable job for these groups, it would seem illogical to conclude that it does so poorly with respect to the angiosperm and eudicot origins. The fossil-based arguments (Crane *et al.* 1995) imply that a rapid morphological expansion of early angiosperm lineages was accompanied by equally rapid molecular evolution. However, looking at the tree (Figure 8.1), such rapid molecular evolution along the spine of the tree has passed without any trace among living early branching taxa. NPRS analyses use an optimization scheme to resolve changes in evolutionary rates, and without any visible trace of such changes, the angiosperm and eudicot origins are being pushed back in time.

Terminal nodes

The molecular estimates tend to underestimate ages for more terminal nodes in the tree. This is true if we compare our estimates with the fossil-based estimates from Table 8.1 and Figures 8.3–8.6 (e.g. Zingiberaceae, Poaceae, Cyperaceae, Rutaceae, Araliaceae), and also if we extend our comparison to other more terminal nodes with reliable fossil-based estimates. Examples include the Moraceae, Salicaceae, and Aceraceae (Collinson *et al.* 1993), and the list could no doubt be expanded through more comprehensive comparisons.

One possible explanation for this general pattern could be that we are underestimating the true age for our calibration point. Uncertainties surrounding both the precise relationships of *Protofagaceae* and *Antiquacupula* within the Fagales clade, and the documentation of *Normapolles* pollen from corresponding or even older ages than our calibration point (Sims *et al.* 1999) may be seen to support such a view. However, recalibrating our tree, fixing the *Normapolles* clade (Figure 8.5) in the late Santonian, would increase all our ages by about 37 per cent pushing the eudicot origin into the Early Jurassic and the angiosperm origin into the Triassic. Considering the fossil record, these ages seem unlikely and maybe the incongruence seen among more terminal nodes simply indicates that the NPRS analyses fail to resolve the evolutionary rates in a reasonable way. A second, and partial, explanation for this pattern relates to the resolution of homoplasy and how this resolution is affected by taxon sampling (Sanderson 1990). For homoplastic characters, parsimony only provides a lower bound on the number of changes, and the inferred positions and numbers of those changes are affected by the thoroughness of the taxon sampling. Sanderson (1990), for example, demonstrated that decreased taxon sampling often

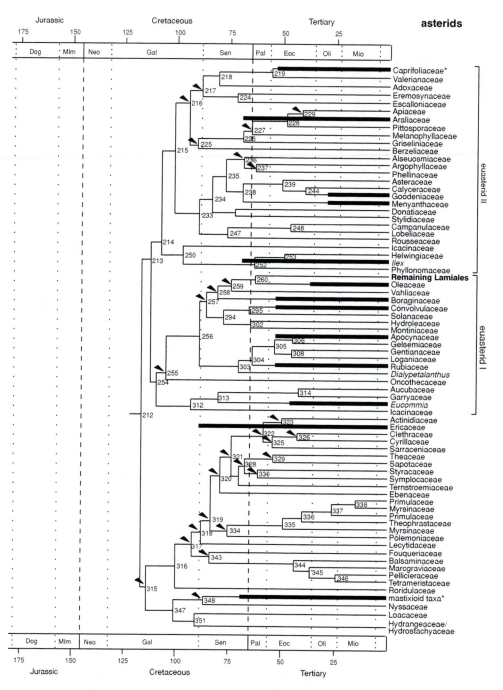

Figure 8.6 Chronogram calibrated against the geological timescale (Harland *et al.* 1990) focusing on asterids. Fossil-based estimates extracted from Magallón *et al.* (1999) are indicated with thick bars. The chronogram is based on the analysis using ACCTRAN optimization for calculating branch lengths, and has been generalized, usually down to family level, to facilitate a graphical comparison between our molecular-based estimates and the fossil-based estimates (Magallón *et al.* 1999). Non-generalized chronograms, indicating age estimates for all included taxa, have been presented elsewhere (Wikström *et al.* 2001), and node numbers correspond to those given there. With few exceptions (taxa marked with *) taxon names follow those used by Soltis *et al.* (1999), and arrows indicate nodes that received less than 50 per cent jackknife support in their phylogenetic analyses (Soltis *et al.* 1999, 2000). Taxa marked with * have been generalized and named to correspond with the usage of Magallón *et al.* (1999).

leads to a dramatic decrease in the estimates of branch lengths. In the terminal clades in our tree, taxon sampling is sparse and families such as Poaceae, Cyperaceae, Rutaceae, and Araliaceae are only represented by a few accessions. We would expect an extended sample of these groups to have the effect of pushing their age estimates closer in line with the fossil-based estimates.

Sources of error

Errors in analyses of this kind arise from several sources, including: (1) noise introduced from the stochastic nature of the substitution process; (2) rate variations that invalidate the assumptions of the method; (3) the calibration point obtained from the fossil record; and (4) use of an incorrect tree. There is, however, no reasonable way of quantifying the effect they have all taken together. Sanderson and Doyle (2001) recently addressed this issue when they analysed the effect of various factors on point estimates for the age of angiosperms. They also conducted a series of resampling experiments designed to provide a statistical estimate of the relative magnitude of errors due to these factors. Their analyses clearly indicated that several factors may introduce large errors in our divergence time estimates (Sanderson and Doyle 2001). However, all their point estimates of the angiosperm age were based on strict molecular clock models, and it is not clear how methods trying to deal with unequal rates such as NPRS would have been affected by the factors analysed.

Following Sanderson (1997, 1998), we estimated errors introduced from the stochastic nature of the substitution process using a bootstrap resampling procedure (Efron and Tibshirani 1993). The bootstrap estimates of standard error resulting from these analyses are comparatively small (on average c. 5 myr), indicating that stochastic errors can be reduced by including sufficient data (Wikström *et al.* 2001).

Perhaps a more serious source of error involves rate variations and an inability to infer shifts in substitution rates correctly (Sanderson 1997). If any amount or any type of change is allowed, the estimates from NPRS analyses will be associated with large errors (Sanderson 1997). There is simply no way to avoid making assumptions about the nature of both rate changes and rates themselves. The NPRS approach allows substitution rates to change and assumes that such changes are autocorrelated, but there is not much empirical support for this assumption (but see Harvey *et al.* 1991, for a discussion of autocorrelation and heritability of cladogenesis). A proper evaluation would require knowledge of absolute rates, which itself requires knowledge of absolute divergence times (Springer 1995). The assumption is, however, intuitively reasonable, and a different examination of its validity could perhaps be accomplished by looking at how rate changes are inferred and by trying to corroborate these changes through different kinds of analyses.

Our choice of calibration point may incorporate errors into our analyses, and as discussed above, our choice may be too conservative. If *Antiquacupula* and *Protofagaceae* were shown to have a more derived position within the Fagales clade, or if we were to recalibrate our results using *Normapolles* pollen records, all our estimates would become older.

How the timing of a group divergence is ultimately resolved depends on correctly inferring its relationship to other groups, and this is probably not the case for some of the groups in our phylogenetic analysis. The tree used simply corresponds to one out of more than 8000 most parsimonious trees reported by Soltis *et al.* (1999, 2000), but evaluating the amount of uncertainty this places on all our estimates is not possible. This simply has to be looked at on a group-by-group basis. It is worth noting that the great majority of groups are consistently resolved and receive at least some jackknife support; in particular the spine of the tree is consistently resolved, and the variability that generates the greater than 8000 equally shortest trees occurs within terminal groups that are not an issue here (Soltis *et al.* 1999, 2000). Relationships within some of the more derived groups such as the Malpighiales and Lamiales, are resolved in the tree used here, but many of these receive less than 50 per cent jackknife support (Soltis *et al.* 1999, 2000). Resolving the relationships differently within these groups will probably have limited consequences on the timings for the more inclusive groups (Malpigiales, Lamiales). We have also indicated on the chronograms (Figures 8.3–8.6) nodes with less than 50 per cent jackknife support.

Future analyses

The analyses conducted are unconstrained, including no fossil-based age constraints; this permits us to evaluate how the molecular data on their own resolve angiosperm diversification. One way forward from the current analyses might involve including constraints, and the type of analyses (NPRS) allows for either minimum- or maximum-age constraints to be enforced during the analyses. Enforcing constraints affects the actual analyses and forces the results to fall within the boundaries specified. Although after such an inclusion we can no longer independently evaluate the fossil-based estimates, such an approach may provide not only a way of improving the actual estimates, but also a way of evaluating how evolutionary rates would be resolved, should we accept the current fossil-based estimates. An analysis including fossil-based constraints would, however, require a detailed and critical evaluation of the available fossil information, which is clearly beyond the scope of our current work.

Existing methods for estimating divergence time cannot combine data (necessary if stochastic errors are to be further reduced) and at the same time take different rate characteristics into account, much as early likelihood models used for phylogeny reconstruction were all simple and without such capabilities. Perhaps work such as this will promote not only an evaluation of the assumptions used in NPRS analyses but also further developments, so that we can look forward to corresponding improvements in age estimation analyses over those we have seen in the development of likelihood models for phylogeny reconstruction. By using the available fossil information as age constraints, analyses of this kind have the advantage of providing ways to estimate the time of origin for all the groups lacking a decent fossil record. They also force our estimates into a more rigorous hierarchical framework, and without such a framework, the full implications of documenting derived groups from successively older geological deposits become less clear. The continued incongruence between the results presented here and fossil-based estimates such as the Turonian Clusiaceae (Crepet and Nixon 1998), the Campanian and Santonian Actinidiaceae (Keller *et al.* 1996; Herendeen *et al.* 1999), the Santonian Apiaceae/Araliaceae (Herendeen *et al.* 1999), and the Cenomanian Rosaceae/Rhamnaceae (Basinger and Dilcher 1984) indicate that we may still underestimate the timing for early angiosperm diversification, or that there are historical patterns to changes in the rate of molecular evolution that we still know little about. This last possibility should be evaluated, and the least we should do is to quantify the consequences of accepting the fossil-based estimates.

Acknowledgements

This research was supported by a Marie Curie Fellowship of the European Community Programme Improving Human Research Potential and the Socio-Economic Knowledge under contract number HPMFCT-2000-00631 to N. Wikström.

References

APG (1998) 'An ordinal classification for the families of flowering plants', *Annals of the Missouri Botanical Garden*, 85: 531–53.

Axelrod, D.I. (1952) 'A theory of angiosperm evolution', *Evolution*, 6: 29–60.

—— (1970) 'Mezozoic paleogeography and early angiosperm history', *Botanical Review*, 36: 277–319.

Basinger, J.F. and Dilcher, D.L. (1984) 'Ancient bisexual flowers', *Science*, 224: 511–13.

Bateman, R. (1999) 'Architectural radiations cannot be optimally interpreted without morphological and molecular phylogenies', in M.H. Kurmann and A.R. Hemsley (eds) *The Evolution of Plant Architecture*, Kew: Royal Botanic Gardens, pp. 221–50.

Brandl, R., Mann, W. and Sprintzl, M. (1992) 'Estimation of the monocot–dicot age through tRNA sequences from the chloroplast', *Proceedings of the Royal Society, London*, B249: 13–17.

Chase, M.W., Soltis, D.E., Olmstead, R.G., Morgan, D., Les, D.H., Mishler, B.D., Duvall, M.R., Price, R.A., Hills, H.G., Qiu, Y.-L., Kron, K.A., Rettig, J.H., Conti, E., Palmer, J.D., Manhart, J.R., Sytsma, K.J., Michaels, H.J., Kress, J.W., Karol, K.G., Clark, D.W., Hedren, M., Gaut, B.S., Jansen, R.K., Kim, K.-J., Wimpee, C.F., Smith, J.F., Furnier, G.R., Strauss, S.H., Xiang, Q.-Y., Plunkett, G.M., Soltis, P.S., Swensen, S.M., Williams, S.E., Gadek, P.A., Quinn, C.J., Equiarte, L.E., Golenberg, E., Learn Jr., G.H., Graham, S.W., Barrett, S.C.H., Dayanandan, S. and Albert, V.A. (1993) 'Phylogenetics of seed plants: An analysis of nucleotide sequences from the plastid gene *rbcL*', *Annals of the Missouri Botanical Garden*, 80: 528–80.

Chase, M.W., Soltis, D.E., Soltis, P.S., Rudall, P.J., Fay, M.F., Hahn, W.H., Sullivan, S., Joseph, J., Molvray, M., Kores, P.J., Givnish, T.J., Sytsma, K.J. and Pires, J.C. (2000) 'Higher-level systematics of the monocotyledons: an assessment of current knowledge and a new classification', in K.L. Wilson and D.A. Morrison (eds) *Monocots: Systematics and Evolution*, Collingwood, Victoria, Australia: CSIRO Publishing, pp. 3–16.

Collinson, M.E., Boulter, M.C. and Holmes, P.L. (1993) 'Magnoliophyta ('Angiospermae')', in M.J. Benton (ed.) *The Fossil Record 2*, London: Chapman & Hall, pp. 809–41.

Cornet, B. (1993) 'Dicot-like leaf and flowers from the Late Triassic tropical Newark Supergroup rift zone, U.S.A.', *Modern Geology*, 19: 81–99.

Cornet, B. and Habib, D. (1992) 'Angiosperm-like pollen from the ammonite-dated Oxfordian (Upper Jurassic) of France', *Review of Palaeobotany and Palynology*, 71: 269–94.

Crane, P.R. (1987) 'Vegetational consequences of the angiosperm diversification', in E.M. Friis, W.G. Chaloner and P.R. Crane (eds) *The Origins of Angiosperms and their Biological Consequences*, Cambridge: Cambridge University Press, pp. 107–44.

Crane, P.R. and Herendeen, P.S. (1996) 'Cretaceous floras containing angiosperm flowers and fruits from eastern North America', *Review of Palaeobotany and Palynology*, 90: 319–37.

Crane, P.R. and Lidgard, S. (1989) 'Angiosperm diversification and paleolatitudinal gradients in Cretaceous floristic diversity', *Science*, 246: 675–8.

Crane, P.R., Friis, E.M. and Pedersen, K.R. (1995) 'The origin and early diversification of angiosperms', *Nature*, 374: 27–33.

Crepet, W.L. and Nixon, K.C. (1998) 'Fossil Clusiaceae from the Late Cretaceous (Turonian) of New Jersey and implications regarding the history of bee pollination', *American Journal of Botany*, 85: 1122–33.

Donoghue, M.J. and Doyle, J.A. (1989a) 'Phylogenetic analysis of angiosperms and the relationships of Hamamelidae', in P.R. Crane and S. Blackmore (eds) *Evolution, Systematics, and Fossil History of the Hamamelidae. Volume 1. Introduction and 'Lower' Hamamelidae*, Systematics Association Special Volume 40, Oxford: Clarendon Press, pp. 17–45.

—— (1989b) 'Phylogenetic studies of seed plants and angiosperms based on morphological characters', in B. Fernholm, K. Bremer and H. Jörnvall (eds) *The Hierarchy of Life: Molecules and Morphology in Phylogenetic Analysis*, Amsterdam: Elsevier Science Publishers, pp. 181–93.

Doyle, J.A. (1969) 'Cretaceous angiosperm pollen of the Atlantic Coastal Plain and its evolutionary significance', *Journal of the Arnold Arboretum*, 50: 1–35.

—— (1978) 'Origin of angiosperms', *Annual Review of Ecology and Systematics*, 9: 365–92.

Doyle, J.A. and Donoghue, M.J. (1986) 'Seed plant phylogeny and the origin of angiosperms: an experimental cladistic approach', *The Botanical Review*, 52: 321–431.

—— (1987) 'The origin of angiosperms: a cladistic approach', in E.M. Friis, W.G. Chaloner and P.R. Crane (eds) *The Origins of Angiosperms and their Biological Consequences*, Cambridge: Cambridge University Press, pp. 17–49.

—— (1993) 'Phylogenies and angiosperm diversification', *Paleobiology*, 19: 141–67.

Doyle, J.A. and Endress, P.K. (2000) 'Morphological phylogenetic analysis of basal angiosperms: comparison and combination with molecular data', *International Journal of Plant Science*, 161: S121–53.

Doyle, J.A. and Hickey, L.J. (1976) 'Pollen and leaves from the mid-Cretaceous Potomac Group and their bearing on the early angiosperm evolution', in C.B. Beck (ed.) *Origin and Early Evolution of Angiosperms*, Columbia: Columbia University Press, pp. 139–206.

Doyle, J.A., Donoghue, M.J. and Zimmer, E.A. (1994) 'Integration of morphological and ribosomal RNA data on the origin of angiosperms', *Annals of the Missouri Botanical Garden*, 81: 419–50.

Eames, A.J. (1959) 'The morphological basis for a Paleozoic origin of the angiosperms', *Recent Advances in Botany*, 1: 721–5.

Efron, B. and Tibshirani, R.J. (1993) *An Introduction to the Bootstrap*, New York: Chapman & Hall.

Felsenstein, J. (1993) *PHYLIP (Phylogeny Inference Package) version 3.5c*, University of Washington, Seattle: Distributed by the author.

Foote, M., Hunter, J.P., Janis, C.M. and Sepkoski, J.J. (1999) 'Evolutionary and preservational constraints of origins of biologic groups: divergence times of eutherian mammals', *Science*, 283: 1310–14.

Friis, E.M. (1984) 'Preliminary report of Upper Cretaceous angiosperm reproductive organs from Sweden and their level of organization', *Annals of the Missouri Botanical Garden*, 71: 403–18.

Friis, E.M. and Skarby, A. (1981) 'Structurally preserved angiosperm flowers from the Upper Cretaceous of southern Sweden', *Nature*, 291: 485–6.

Friis, E.M., Crane, P.R. and Pedersen, K.R. (1988) 'The reproductive structures of Cretaceous Platanaceae', *Biologiske Skrifter*, 31: 1–55.

—— (1999) 'Early angiosperm diversification: The diversity of pollen associated with angiosperm reproductive structures in Early Cretaceous floras from Portugal', *Annals of the Missouri Botanical Garden*, 86: 259–96.

Friis, E.M., Pedersen, K.R. and Crane, P.R. (2000) 'Reproductive structures and organization of basal angiosperms from the Early Cretaceous (Barremian or Aptian) of western Portugal', *International Journal of Plant Science*, 161: S169–82.

—— (2001) 'Fossil evidence of water lilies (Nymphaeales) in the Early Cretaceous', *Nature*, 410: 357–60.

Goremykin, V.V., Hansman, S. and Martin, W.F. (1997) 'Evolutionary analysis of 58 proteins encoded in six completely sequenced chloroplast genomes: revised molecular estimates of two seed plant divergence times', *Plant Systematics and Evolution*, 206: 337–51.

Harland, W.B., Armstrong, R.L., Cox, A.V., Craig, L.E., Smith, A.G. and Smith, D.G. (1990) *A Geologic Time Scale 1989*, Cambridge: Cambridge University Press.

Harvey, P.H., Nee, S., Mooers, A.O. and Partridge, L. (1991) 'These hierarchical views of life: phylogenies and metapopulations', in R.J. Berry, T.J. Crawford and G.M. Hewitt (eds) *Genes in Ecology: the 33rd Symposium of the British Ecological Society*, Oxford: Blackwell, pp. 123–37.

Hasegawa, M., Kishino, H. and Yano, T. (1985) 'Dating the human–ape splitting by a molecular clock of mitochondrial DNA', *Journal of Molecular Evolution*, 21: 160–74.

Herendeen, P.S., Crane, P.R. and Drinnan, A.N. (1995) 'Fagaceous flowers, fruits, and capsules from the Campanian (Late Cretaceous) of central Georgia, U.S.A.', *International Journal of Plant Science*, 156: 93–116.

Herendeen, P.S., Magallón-Puebla, S., Lupia, R., Crane, P.R. and Kobylinska, J. (1999) 'A preliminary conspectus of the Allon Flora from the Late Cretaceous (Late Santonian) of central Georgia, U.S.A.', *Annals of the Missouri Botanical Garden*, 86: 407–71.

Hickey, L.J. and Doyle, J.A. (1977) 'Early Cretaceous fossil evidence for angiosperm evolution', *The Botanical Review*, 43: 3–104.

Hillis, D.M., Mable, B.K. and Moritz, C. (1996) 'Applications of molecular systematics: the state of the field and a look to the future', in D.M. Hillis, C. Moritz, and B.K. Mable (eds) *Molecular Systematics*, 2nd edn, Sunderland, Massachusetts: Sinauer Associates, pp. 515–43.

Hughes, N.F. (1976) *Palaeobiology of Angiosperm Origins*, Cambridge: Cambridge University Press.

Jefferies, R.P.S. (1979) 'The origin of chordates – a methodological essay', in M.R. House (ed.) *The Origin of Major Invertebrate Groups*, Systematics Association Special Volume 12, pp. 443–7.

Keller, J.A., Herendeen, P.S. and Crane, P.R. (1996) 'Fossil flowers of the Actinidiaceae from the Campanian (Late Cretaceous) of Georgia', *American Journal of Botany*, 83: 528–41.

Lidgard, S. and Crane, P.R. (1988) 'Quantitative analyses of the early angiosperm radiation', *Nature*, 331: 344–6.

Magallón, S. and Sanderson, M.J. (2001) 'Absolute diversification rates in angiosperm clades', *Evolution*, 55: 1762–80.

Magallón, S., Crane, P.R. and Herendeen, P.S. (1999) 'Phylogenetic pattern, diversity, and diversification of eudicots', *Annals of the Missouri Botanical Garden*, 86: 297–372.

Manos, P.S. (1997) 'Systematics of *Nothofagus* (Nothofagaceae) based on rDNA spacer sequences (ITS): taxonomic congruence with morphology and plastid sequences', *American Journal of Botany*, 84: 1137–55.

Manos, P.S. and Steele, K.P. (1997) 'Phylogenetic analyses of "higher" Hamamelididae based on plastid sequence data', *American Journal of Botany*, 84: 1407–19.

Manos, P.S., Nixon, K.C. and Doyle, J.J. (1993) 'Cladistic analyses of restriction site variation within the chloroplast DNA inverted repeat region of selected Hamamelididae', *Systematic Botany*, 18: 551–62.

Marshall, C.R. (1998) 'Determining stratigraphic ranges', in S.K. Donovan and C.R.C. Paul (eds) *The Adequacy of the Fossil Record*, Chichester: John Wiley, pp. 23–54.

Martin, W., Gierl, A. and Saedler, H. (1989) 'Molecular evidence for pre-Cretaceous angiosperm origins', *Nature*, 339: 46–8.

Martin, W., Lydiate, D., Brinkmann, H., Forkmann, G., Saedler, H. and Cerff, R. (1993) 'Molecular phylogenies in angiosperm evolution', *Molecular Biology and Evolution*, 10: 140–62.

Muller, J. (1981) 'Fossil pollen records of extant angiosperms', *The Botanical Review*, 47: 1–146.

Nandi, W.I., Chase, M.W. and Endress, P.K. (1998) 'A combined cladistic analysis of angiosperms using *rbcL* and non-molecular data sets', *Annals of the Missouri Botanical Garden*, 85: 137–212.

Nixon, K.C. (1989) 'Origins of Fagaceae', in P.R. Crane and S. Blackmore (eds) *Evolution, Systematics and Fossil History of the Hamamelidae, Volume 2 "Higher" Hamamelidae*, Oxford: Oxford University Press, pp. 23–43.

Qiu, Y.-L. (2000) 'Phylogeny of basal angiosperms: analyses of five genes from three genomes', *International Journal of Plant Science*, 161: S3–27.

Qiu, Y.-L., Lee, J., Bernasconi-Quadroni, F., Soltis, D.E., Soltis, P.S., Zanis, M., Chen, Z., Savolainen, V. and Chase, M.W. (1999) 'The earliest angiosperms: Evidence from mitochondrial, plastid and nuclear genomes', *Nature*, 402: 404–7.

Ramshaw, J.A.M., Richardson, D.L., Meatyard, B.T., Brown, R.H., Richardson, M., Thompson, E.W. and Boulter, D. (1972) 'The time of origin of the flowering plants determined by using amino acid sequence data of cytochrome c', *New Phytologist*, 71: 773–9.

Sanderson, M.J. (1990) 'Estimating rates of speciation and evolution: a bias due to homoplasy', *Cladistics*, 6: 387–91.

—— (1997) 'A nonparametric approach to estimating divergence times in the absence of rate constancy', *Molecular Biology and Evolution*, 14: 1218–31.

—— (1998) 'Estimating rate and time in molecular phylogenies: beyond the molecular clock', in D.E. Soltis, P.S. Soltis and J.J. Doyle (eds) *Molecular Systematics of Plants II: DNA sequencing*, Norwell, Massachusetts: Kluwer Academic Publishers, pp. 242–64.

Sanderson, M.J. and Doyle, J.A. (2001) 'Sources of error and confidence intervals in estimating the age of angiosperms from *rbcL* and 18S rDNA data', *American Journal of Botany*, 88: 1499–516.

Savolainen, V., Chase, M.W., Morton, C.M., Hoot, S.B., Soltis, D.E., Bayer, C., Fay, M.F., de Bruijn, A., Sullivan, S. and Qiu, Y.-L. (2000a) 'Phylogenetics of flowering plants based upon a combined analysis of plastid *atpB* and *rbcL* gene sequences', *Systematic Biology*, 49: 306–62.

Savolainen, V., Fay, M.F., Albach, D.C., Backlund, A., van der Bank, M., Cameron, K.M., Johnson, S.A., Lledó, M.D., Pintaud, J.-C., Powell, M., Sheahan, M.C., Soltis, D.E., Soltis, P.S., Weston, P., Whitten, W.M., Wurdack, K.J. and Chase, M.W. (2000b) 'Phylogeny of the eudicots: a nearly complete familial analysis based on *rbcL* gene sequences', *Kew Bulletin*, 55: 257–309.

Sims, H.J., Herendeen, P.S. and Crane, P.R. (1998) 'New genus of fossil Fagaceae from the Santonian (Late Cretaceous) of central Georgia, U.S.A.', *International Journal of Plant Science*, 159: 391–404.

Sims, H.J., Herendeen, P.S., Lupia, R.A., Christopher, R.A. and Crane, P.R. (1999) 'Fossil flowers with *Normapolles* pollen from the Late Cretaceous of southeastern North America', *Review of Palaeobotany and Palynology*, 106: 131–51.

Smith, A.B. (1994) *Systematics and the Fossil Record: Documenting Evolutionary Patterns*, Oxford: Blackwell.

Soltis, D.E., Soltis, P.S., Nickrent, D.L., Johnson, L.A., Hahn, W.J., Hoot, S.B., Sweere, J.A., Kuzoff, R.K., Kron, K.A., Chase, M.W., Swensen, S.M., Zimmer, E.A., Chaw, S.-M., Gillespie, L.J., Kress, W.J. and Sytsma, K.J. (1997) 'Angiosperm phylogeny inferred from 18S ribosomal DNA sequences', *Annals of the Missouri Botanical Garden*, 84: 1–49.

Soltis, D.E., Soltis, P.S., Chase, M.W., Mort, M.E., Albach, D.C., Zanis, M., Savolainen, V., Hahn, W.H., Hoot, S.B., Fay, M.F., Axtell, M., Swensen, S.M., Nixon, K.C. and Farris, J.S. (2000) 'Angiosperm phylogeny inferred from a combined dataset of 18S rDNA, *rbcL*, and *atpB* sequences', *Botanical Journal of the Linnean Society*, 133: 381–461.

Soltis, P.S., Soltis, D.E. and Chase, M.W. (1999) 'Angiosperm phylogeny inferred from multiple genes: A research tool for comparative biology', *Nature*, 402: 402–4.

Springer, M. (1995) 'Molecular clocks and the incompleteness of the fossil record', *Journal of Molecular Evolution*, 41: 531–8.

Swofford, D.L. (1998) *PAUP*. Phylogenetic analyses using parsimony (*and other methods). Version 4*, Sunderland, Massachusetts: Sinauer Associates.

Takhtajan, A. (1969) *Flowering Plants: Origin and Dispersal*, Edinburgh: Oliver.

Taylor, D.W. and Hickey, L.J. (1990) 'An Aptian plant with attached leaves and flowers: implications for angiosperm origin', *Science*, 247: 702–4.

Thomas, H.H. (1957) 'Plant morphology and the evolution of the flowering plants', *Proceedings of the Linnean Society of London*, 168: 125–33.

Wikström, N., Savolainen, V. and Chase, M.W. (2001) 'Evolution of the angiosperms: calibrating the family tree', *Proceedings of the Royal Society, London*, B268: 2211–20.

Wolfe, K.H., Gouy, M., Yang, Y.-W., Sharp, P.M. and Li, W.-H. (1989) 'Date of the monocot–dicot divergence estimated from chloroplast DNA sequence data', *Proceedings of the National Academy of Sciences, USA*, 86: 6201–5.

Zavada, M.S. (1987) 'First fossil evidence for the primitive angiosperm family Lactoridaceae', *American Journal of Botany*, 74: 1590–4.

Zuckerhandl, E. and Pauling, L. (1962) 'Molecular disease, evolution, and genetic heterogeneity', in M. Kasha and B. Pullman (eds) *Horizons in Biochemistry*, New York: Academic Press, pp. 189–225.

—— (1965) 'Evolutionary divergence and convergence', in V. Bryson and H.J. Vogel (eds) *Evolving Genes and Proteins*, New York: Academic Press, pp. 97–166.

Chapter 9

The limitations of the fossil record and the dating of the origin of the Bilateria

Graham E. Budd and Sören Jensen

ABSTRACT

The origin of the bilaterian phyla is one of the classic areas of conflict between molecular dates and the fossil record. Here we examine possible sources of bias in the fossil record, and conclude that on several grounds, not least broad stratigraphic and phlyogenetic congruence, it is hard to see the Cambrian fossil record as representing anything but a broadly 'real-time' event. Refinements of molecular estimates of the dating of bilaterian origins have increasingly cut the gap between fossil and molecular estimates, and a rapprochement must be hoped for in the next few years. Further refinements of our understanding of this event must increasingly focus on better phylogenies, and a consideration of the ecology and biogeography of the event.

Introduction

The apparent conflict between molecular and fossil evidence for the origin of the Bilateria is one that is increasingly being treated in a more mature way than the protagonists simply gainsaying the data of their opponents. The pressure that has been brought to bear on both sides has led to a critical examination of both types of data. Whilst the fossil record has been criticized in this process, estimates of the degree of uncertainty introduced by its undeniable imperfections have been much more difficult to obtain. In other words, the validity of the logical step from 'the fossil record is flawed' to 'the fossil record can tell us nothing about timing of the origin of animals' needs to be carefully scrutinized. The purpose of this present chapter is to lay out briefly the various reasons for thinking the fossil record of the Cambrian and terminal Proterozoic might be considerably misrepresentative of the true timing of the events it is purported to depict, together with attempts that have been made to circumvent them, and to examine them critically. Finally, we offer some further ways in which the record could be tested, and outline some of the latest evidence suggesting that the fossil and molecular conflict might soon be resolved.

The problems of the Cambrian fossil record

The so-called 'Cambrian explosion' is, together with the extinction events at the end of the Cretaceous, the most well-known biological event in Earth history (cf. Conway Morris 1998a; Knoll and Carroll 1999; Budd and Jensen 2000; Zhuravlev and

Riding 2000; Erwin and Davidson 2002 for recent reviews). Whilst it has been most notable for the rather sudden appearance of invertebrate macrofossils in the record, it is increasingly being recognized as a very complex interwoven network of events, really a revolution in how the biological world was organized (e.g. Butterfield 2000; Zhuravlev and Riding 2000). As a result, research has been expanded to consider the context in which this event took place, typically the period of 100 million years termed 'Geon 5' (600–500 Ma: see Hofmann 1990). New aspects that have been considered recently include changes in the flora, represented especially by the enigmatic acritarch record (Butterfield 1997; Vidal and Moczydlowska-Vidal 1997), and the importance of environmental change (e.g. the evidence for extensive glaciations that in its extreme constitutes the 'Snowball Earth' hypothesis; Kirschvink 1992; Hoffmann et al. 1998).

This broadening perspective has led to the inescapable conclusion that the Cambrian explosion, considered in its widest sense, is a truly important Earth history event. As such, there can be no possibility that the events of this time are, overall, an illusion created by the vagaries of preservation in the rock record. It is therefore considerably ironic that the macroinvertebrate fossil record, which led researchers as far back as Buckland and Darwin to worry about the problem, is the one under most scrutiny today. Darwin struggled with the problem of the Cambrian explosion (which according to his stratigraphic understanding, took place at the base of the Silurian) and recognized the potentially grave problems it presented for his theory of evolution by natural selection of tiny incremental changes (Darwin 1985, ch. 9). Indeed, his discussion of the problem in *The Origin of Species* is well worth reading for its amusingly modern tone. For Darwin, there was no possibility that the fossil record as known could truly represent the history of life. There simply must have been vast periods of time in which the world 'swarmed with living creatures' (Darwin 1985, p. 313; originally published in 1859). Yet his efforts to explain why these creatures had not been found were lame, and the consequent weakness represents a genuine tear in the argumentative fabric of *The Origin of Species* as a whole.

Almost 150 years later, Darwin's problem remains as stark as ever. Despite extensive searching, the rock record of the Precambrian (as now defined) has not once yielded a single truly convincing body fossil of bilaterian affinity. The only possible exceptions to this rule are provided by the so-called Ediacaran fossils (the most comprehensive overall introduction in English, tellingly, remains Glaessner 1984; more recent overviews include Gehling 1991, Runnegar and Fedonkin 1992, Runnegar 1995, Narbonne 1998, and Grazdhankin and Seilacher 2002). In recent years, two important facts have become apparent about these highly enigmatic fossils that Glaessner and many others considered as straightforward members of extant bilaterian phyla. The first is that they are uniformly younger than previously thought. Glaessner (1984), for example, estimates the oldest Ediacaran fossils at 650–660 Ma, but more recent dating shows that the oldest Ediacaran fossils are not much older than 565 Ma, and that diverse Ediacaran assemblages are younger than about 555 Ma (Martin et al. 2000). Despite the recent fixing of the Precambrian–Cambrian boundary stratigraphically (Landing 1994), there remains considerable uncertainty about basal Cambrian correlation, and the likelihood that some Cambrian rocks contain 'Ediacaran' fossils (Jensen et al. 1998) means that rocks generally thought to be Precambrian because of their (Ediacaran) fossil content may in actuality be Cambrian. The result of this revision is that Ediacaran fossils appear temporally to be very close to the Cambrian faunas, and may have a considerable overlap with them. Secondly, opinion today is hardening that none of these taxa can be unequivocally assigned to crown-group Bilateria. The candidate most recently favoured by some as a crown bilaterian, *Kimberella* (Fedonkin and Waggoner 1997) merely shows relatively complex morphology that is coupled with bilateral symmetry, compatible with being a stem-group bilaterian or even, according to some theoretical reconstructions, a stem-group eumetazoan (Jägersten 1972). New morphological details of Ediacaran fossils are continually emerging (e.g. Dzik and Ivantsov 2002), but these have not necessarily clarified the affinities of the organisms in question.

Finally, it should be noted that rocks slightly older than Ediacaran age, the Doushantuo Formation of South-West China, now dated at around 599 Ma (Barfod *et al.* 2002) have been claimed to yield embryos of bilaterian affinity (Chen *et al.* 2000; Xiao 2002), although this claim has been subject to searching criticism (e.g. Xiao *et al.* 2000). Reports of segmented, annelid-like taxa in the Doushantuo Formation (e.g. Chen and Xiao 1991) and elsewhere (e.g. Russia: Gnilovskaya *et al.* 2000) are not compelling, and are more likely to be the remmants of multicellular algae (Xiao *et al.* 2000). Nevertheless, the presence of cnidarian-like taxa does seem possible, especially with the recent publication of further morphological details (Chen *et al.* 2002), including taxa of apparent tabula-like structure; although presumably an algal affinity cannot be definitively ruled out. It is worth noting in this context the great diversity of algae in these sediments (Steiner 1994; Xiao *et al.* 2000). If these taxa are stem- or crown-group cnidarians, then the implication would be that at least stem-group bilaterians would have evolved at this stage. Nevertheless, the unmineralized nature of the most prominent tabula-like form gives pause for thought (Xiao *et al.* 2000); raising questions about when mineralization in 'corals' took place. Tabulates are clearly not crown-group corals and, although of cnidarian grade (e.g. Copper 1985), could conceivably be not particularly closely related to extant corals. As Xiao *et al.* (2000) suggest, these taxa might even be stem-group eumetazoans.

To summarize: the Precambrian fossil record has, to a certain extent, fulfilled Darwin's prediction, although not in a way that he would have imagined. Many fossils are known, including eukaryotes (e.g. Butterfield 2000). Furthermore, very close to the end of the terminal Proterozoic, many large body fossils appear, but they do not demonstrate conclusive evidence of being crown-group bilaterians. The expected Darwinian pattern of a deep fossil history of the bilaterians, potentially showing their gradual development, stretching hundreds of millions of years into the Precambrian, has singularly failed to materialize. Since the body fossil record has been so unyielding, attention has shifted in recent years to the trace fossil record. Trace fossils have certain advantages over body fossils: they can be, and often are, preserved in sediments, such as sandstones, in which body fossils are usually absent or fragmentary; they are not (normally) subject to transport or to other types of biostratinomic alteration; and their preservation is not dependent on the generating organism possessing hard parts, for example. Indeed, the low degree of bioturbation in the Proterozoic and earliest Cambrian, both in terms of intensity and depth, arguably resulted in sediment properties that were particularly conducive to preservation of very shallow tiers (Droser *et al.* 2002).

Nevertheless, it could be argued that, although advantageous, the trace fossil record does suffer from some less favourable secular variation. For example, it might be argued that the transition from the microbial mat-dominated sediment surfaces of the late Proterozoic to more bioturbated soft sediment bottoms of the Phanerozoic (Seilacher 1999) might bias the record in favour of Phanerozoic traces. While it has been argued that mats may have been important in structuring early benthic communities (Seilacher 1999; Bottjer *et al.* 2000), these mats can hardly be seen as leathery, impenetrable objects: many lifestyles of macro-organisms are thought to have involved constant disruption and utilization of mats. This would include 'mat-stickers' penetrating the mats, and 'under-mat miners' burrowing just beneath the mats (Seilacher 1999). If so, the presence of the mats might actually enhance trace fossil preservation (Budd and Jensen 2000); although it should be again stressed that many Precambrian bedding planes show no signs of having once been covered by biomats (e.g. they have rippled surfaces).

Two candidates for extremely old trace fossils have recently been described. The first (Seilacher *et al.* 1998), from India, was reported as being some 1 Ga, a date that at the time of description coincided well with molecular estimates of the divergence of the Bilateria, and is now considered to date from more than 1.6 Ga (Rasmussen *et al.* 2002a; Ray *et al.* 2002). This description has not, it is fair to say, been widely accepted. The sediments from which it was described are full of sedimentary structures such as mud cracks and, as detailed in Budd and Jensen (2000), these traces show some suspicious features such as an irregular, crinkly appearance and tapering terminations; all of which are suggestive of inorganic origin. The other candidate is from the Stirling Range of South Australia (Rasmussen *et al.* 2002b). These millimetric structures are from rocks that are dated at between 1.2 and 2 Ga. They consist of parallel-sided ridges preserved in sandstones, sometimes showing a distinct narrowing towards a rounded termination. The authors interpret these structures, with some hesitation, as the surficial traces made by a probable soft-bodied worm, citing an earlier study (Collins *et al.* 2000) as evidence that secreted mucus can act as a glue to stick displaced sediment grains together. The problem with this interpretation, apart from any doubts about the preserved morphology (e.g. the narrowing of the terminations would be highly unusual in a trail), is primarily a taphonomic one. By their nature, these traces would have been made on a sandy surface, and whether or not binding mucus was produced, preservation of this level in the fossil record must be considered to be doubtful in general. Indeed, a rigourist school of thought would argue that all locomotion trace fossils preserved in the marine fossil record are essentially undertracks – impressions made on an interface within the sediment, not on its surface (Seilacher 1957). On these grounds alone, the interpretation of these structures as made by a bilaterally symmetrical worm moving across the surface of the sediment must be strongly questioned. Even more problematic is how the traces as found are preserved. The authors describe them as being preserved in sharp positive hyporelief on the base of sandstones (i.e. as positive protrusions from the underside of the sandstone). This is a very common mode of preservation of trace fossils, where an overlying sediment type casts impressions in an underlying one. However, the structures under discussion are meant to be not casts but sediment accumulations glued by mucus. In order for these originally positive-relief structures to be preserved on the underside of the overlying bed, they must somehow have become transferred from one bed to another. Even if this is possible, they still apparently show the same composition of the bed they now reside in; and they do not show any sign of being compressed.

A more complex taphonomic history for these structures is hinted at in a recent discussion of their nature (Conway Morris *et al.* 2002). Here, the original authors suggest that the original mucus-bound strings consisted of mud, and have been later replaced/cast by sand. Such a scenario would require the mucus strings to be partly covered by more mud, and then during compaction of the upper layers, to collapse more than this mud (because of decay of the mucus within). Even if this scenario had any evidential support, it seems highly implausible that the end result would be the sharply defined ridges as now preserved. All of these difficulties render the likelihood of these structures representing true trace fossils extremely low. The simple fact is that whatever the true status of these and other candidate early trace fossils, none of them can be straightforwardly accepted as trace fossils; they all require a certain or high amount of special pleading: in itself a significant point.

The above points, we think, make it clear that whatever the resolution of the misfit between the fossil record and molecular evidence for the origin of animals, it does not come about through a misunderstanding of the known fossil record. The other options are to investigate what our reasonable expectations of the fossil record should be, and to re-examine the molecular evidence. Perhaps we should not be too surprised about the mismatch.

Telling the true time from the fossil record

The most persistent and telling critics of the fossil record have, perhaps surprisingly, been palaeontologists. Hence, Smith (1999) and Smith and Peterson (2002) have pointed out important features of the fossil record that, at least potentially, greatly weaken its reliability. These include the facies dependence of fossils; the reliance on outcrops of the right age and type; the variability of preservation rate through time; and the problems of identifying basal members of clades. More importantly, a considerable amount of effort is being expended in quantifying what this weakening of fossil reliability means (e.g. Strauss and Sadler 1989; Marshall 1990, 1997; Foote *et al.* 1999; Tavaré *et al.* 2002). The simplest, and most unrealistic, approach is to consider the actual fossil finds during an interval to be Poisson distributed, and use the calculated mean to place confidence limits on the exponentially distributed 'gaps' that extend above and below the known range. Bayesian inference (Strauss and Sadler 1989) and likelihood estimates (Huelsenbeck and Rannala 1997) of true range can also be made (see Huelsenbeck and Rannala 1998 for discussion). Other ways of relaxing the Poisson distribution criterion would presumably include modelling fossil finds with a gamma distribution, although as Huelsenbeck and Rannala (1998) point out, this sort of refinement can hardly be done justice to by the general imperfection of the stratigraphic record.

An important critique of these distribution-based analyses is that they fail to take into account important facts that particularly apply to the beginnings (although not always to the ends) of taxon ranges. At the origin of a clade, it is standard dogma that there can be only a small founder population, and that as the history of the clade unfolds, it will diversify (although what this word means here requires some attention, as is discussed below). Clearly, the chance of finding the first member, or even the first few members of a clade, is vanishingly small. On the other hand, at the acme of the clade, when diversity is highest, the chances of finding a fossil must be greatly enhanced. If a clade is abruptly terminated by an extinction caused by something like a meteorite impact, a similar argument will not apply to the end of the range of the clade. Simply looking at preservation rates from the period of time when a clade has already diversified and then extrapolating them back in time is thus likely to underestimate considerably the true time of origin.

If the above point is accepted, then in order to model the potential likelihood or probability of finding possible fossils below their lowest known occurrence, some modelling of the earliest history of the clade is necessary (Foote *et al.* 1999; Tavaré *et al.* 2002). Although this approach has been applied to taxa such as mammals and primates, it is potentially limited by two problems. The first is that it is difficult to decide what sort of model is appropriate for the (unknown) diversity increase. Although on *a priori* grounds a diversity-dependent model is usually chosen, such choice seems to beg the question. For what one is trying to estimate in the first place is the time lag of diversification that took place before the first fossil is found: yet it is this very variable that must be (tacitly) assumed by one's choice of diversification model. Tavaré *et al.* (2002) attempt to circumvent this problem by modelling their diversification curve from the known data and then extrapolating backwards, assuming at the very least that the curve represents the monotonous continuation of the initial diversification event. True diversification events, on the other hand, may be much more complex involving various types of positive feedback, lags, and so on (Erwin 2001).

Another potentially much more serious problem is the use of diversity as a proxy for 'rate of fossilization'. Consider the example of the planktonic graptolite *Rhabdinopora*. This graptolite probably represents a single invasion of the planktonic realm by the benthic dendroid graptoloids. It seems, as far as can be ascertained, to appear virtually simultaneously in huge numbers all over the North Atlantic Province (Cooper *et al.* 1998). However, its diversity, a complex of subspecies, is tiny. If one was to model its chance of fossilization on 'number of lineages' or diversity, then one would be forced to predict that it would not have much luck. Yet, in fact, the preservation rate is clearly not linked to number of species but to number of individuals. Determining numbers of individuals of a fossil, or indeed extant, organism is fraught with difficulties, and is rarely even attempted. Willis (1922) showed that there was a striking hollow-curve distribution of species abundance, which he attempted to explain by a punctuated model of speciation. If one were to sample such a distribution, then one's chances of selecting a high-abundance species would have some relationship to the number of species sampled: high-abundance species are rare. However, during a radiation, the total abundance – related to carrying capacity – would be the determinant of chances of fossilization, not necessarily the number of species. As a result, estimating how total abundance of a clade might change through the early stages of a diversification must be of more importance than estimating diversity. Given that diversity is usually the only variable known, this would imply in turn having a model that links abundance to diversity during an evolutionary radiation. In part, this link may depend on the balance between sympatric and allopatric speciation mode.

Recent modelling, for example, has suggested that during a diversification, the carrying capacity of a particular environment is first attained rather quickly, followed by instability in the population distribution and break-up into distinct subpopulations and eventual sympatric speciation (Drossel and McKane 2000). Competitive displacement of characters through selection may lead to the introduction of phenotypic and therefore resource exploitation 'no-go' zones, with the surprising theoretical result being that the total number of individuals may actually drop during a radiation (see figs 1 and 2 of Drossel and McKane 2000). The effect of increasing ecological complexity on carrying capacity is of course a famous problem, first mentioned by Darwin, but not yet satisfactorily answered. The initial stages of the radiation will indeed be marked by a rapid increase of population as the newly available resource spectrum is exploited, but this may be on a geologically instantaneous timescale. In any case, the relationship between abundance (proportional to rate of fossilization) and diversity is clearly not at all a straightforward one, and it seems to us that one cannot simply use the latter as a proxy for the former. For these reasons, we remain sceptical of current attempts to model the preservability of the early stages of radiations (Foote *et al.* 1999; Tavaré *et al.* 2002).

Even if the above objections could be taken into account, there remain the particular problems presented by the Cambrian fossil record (see Smith and Peterson 2002 for a concise summary). And here, some attempts to reconcile the fossil record with molecular estimates do not seem to do the job required. Smith and Peterson (2002), following Knoll and Carroll (1999), emphasized that, in general, bilaterian divergence times are being measured by molecular clocks, not metazoan origin itself. They both adopt the model that a tail of stem-group bilaterian diversification in the Precambrian precedes crown-group bilaterian radiation in the Cambrian explosion itself. Yet this formulation conceals an ambiguity between the members of the stem-group of Bilateria and the members of the stem-groups of bilaterian phyla. The importance of this distinction is that we can have some idea about what stem-group members of bilaterian phyla were like on phylogenetic grounds, whereas we have much less idea currently about the stem-group of the Bilateria. The stem-group of the Bilateria is in fact somewhat irrelevant to the molecular problem, because molecular clock estimates are typically of the time of divergence of crown-, not stem-group bilaterians. Even if some stem-group bilaterians were tiny planktonic animals (but for critical discussion of this idea see Budd and Jensen 2000), the upper stem-group would still have been characterized by the progressive appearance of the monophyletic features of crown-group Bilateria and these features must have been assembled in large animals (Budd and Jensen 2000).

Molecular methods have, in general, dated splits between existing phyla such as vertebrates and echinoderms, and as discussed in Budd and Jensen (2000), these sorts of taxa are bound, on phylogenetic grounds, to share important features such as a coelom. Even if caution should be employed in inferring morphological homology from shared developmental systems (e.g. Budd 2001; Erwin and Davidson 2002), there are still classical grounds for thinking that many important metazoan features are homologous. A recent renewed interest in convergence (e.g. Conway Morris 1998a) should not lead to the trap of assuming that similar features must be convergent, a tendency recognized long ago in the memorable words of Derek Ager: 'The fashionable fixation for homeomorphy in many groups brainwashed many of us into thinking: "if they look alike they cannot be related"' (Ager 1993, p. 90). At the very least, the last common ancestor of Bilateria must have possessed muscles and a head, suggesting it was not a tiny planktonic organism. If the Precambrian was really characterized by the slow diversification of the stem-groups of the bilaterian phyla, then *ad hoc* arguments must be employed to explain why animals that can be reconstructed to be of large size and biomechanically complex left neither a trace nor body fossil record. The common argument (e.g. Thomas 1999) that the presence of phyla characterized by small body size that have no fossil record (e.g. kinorhynchs) suggests that similar animals could have 'silently' existed in the Precambrian misses the point: the body plan features of bilaterians such as muscles and the coelom could not have evolved in such an organism (see discussion in Budd and Jensen 2000). Phylogenetic reconstruction of the history of such features, combined with consideration of the biomechanical regime in which they would have been adaptively useful, concisely rules out the possibility of the most ancestral bilaterians being tiny and/or planktonic. The only possibility is that the body plan features of groups of phyla were acquired independently of each other, a feature of some earlier attempts to reconcile molecular and fossil evidence (e.g. Bergström 1989; Davidson *et al*. 1995, and, most recently, Erwin and Davidson 2002). Even so, this can hardly apply to important features of the deuterostome taxa mentioned above.

Further doubt is cast on this model by the increasingly recognized presence of members of the stem-groups of bilaterian phyla in the Cambrian (e.g. Budd 1993; Mooi *et al*. 1994; Conway Morris and Peel 1995; Holmer *et al*. 2002; Williams and Holmer 2002). If they are preserved in the Cambrian, why not any in the Precambrian? One is forced to return to the view that, if there truly is a long Precambrian tail of bilaterian diversification, it was of animals that should have been capable of leaving a trace or body fossil record. The presence of Ediacaran fossils and many other sediments of suitable age and type in the terminal Proterozoic means that one cannot suggest lack of outcrop as a reason for the lack of evidence (Darwin 1985). Similarly, lack of suitable facies (as in Gale *et al*. 2001) cannot be invoked over the long period of time required in the Precambrian to obliterate any benthic fossil record. In particular, the excellent and microscopic preservation in the Duoshantuo Formation and similar sediments, none of which yield convincing bilaterians, is compelling evidence that bilaterians, even tiny ones, were not then widespread. One must therefore fall back on the idea that rarity of the taxa meant that the fossil record simply did not pick them up. The question to be addressed, then, is whether the early stages of the bilaterian (abundance) radiation were truly of this nature – or, indeed, are radiations in general?

The Cambrian explosion as an ecological radiation

Despite some important peculiarities, the events at the beginning of the Cambrian, the 'Cambrian explosion' as defined by Budd *et al*. (2001), should be approached as any other major evolutionary event, that is, from the initial perspective of the neodarwinian synthesis. Undoubtedly, the ecological theory of adaptive radiation (Simpson 1953; Schluter 2000) is the most dominant theory in this field, despite the existence of more heretical 'neutralist' theories of both ecologists and palaeontologists. The ecological theory claims that diversity and phenotypic divergence is largely the result of selective pressures acting within communities. Indeed, the recognition that selection could be divergent in its effect is an important scientific discovery of the last century.

When applied to the Cambrian explosion, the results of this view would be to claim that the numerous important adaptive features that appear in fossils at this time, such as hard parts, are the result of ecological interactions (cf. Bengtson 2002) that have arisen through the sorts of population processes that can be seen ongoing today. Why, then, is the event so problematic? These difficulties can be divided into both methodological problems and peculiarities about the event itself. It is only in recent years that rigorous efforts to understand the phylogenetic patterns displayed by Cambrian organisms have been made (Smith 1988; Briggs and Fortey 1989; Budd 1993, 2002; Conway Morris and Peel 1995), and before this time it was easy to make claims about the nature of the radiation that are in fact contrary to analysis (e.g. that few patterns of relationship could be seen in the Cambrian fauna). These analyses have increasingly shown that the origins of the phyla that have been studied are of a progressive and adaptive nature, more fitting with the 'correlated progression' view (e.g. Kemp 1982) of macroevolution than the more fashionable 'key innovation' one (e.g. Hunter 1998). Naturally enough, the highly fragmentary record, and difficulties of understanding the more problematic taxa, have greatly impeded this effort.

Second, there are problems associated with the event itself. As no one believes today that the pattern presented by the fossil record is entirely artefactual, and it is generally accepted that that there was a genuine ecological radiation affecting plants and animals during this time, then one is entitled to ask what sorts of ecological events were involved. Given the likelihood that the macrobenthic realm was being exploited by highly mobile bilaterians for the first time, with no obvious competitors for resources, the initial expansion of bilaterians is likely to have more clearly approximated to an expansion *in vacuo* than to any other major evolutionary expansion. Such a situation is (on such a large scale) a non-actualistic one, but perhaps could be compared with ecological expansions after mass extinctions (e.g. Conway Morris 1998b). Even then, the situation is not truly comparable. Ecosystems in general seem to remain highly fragile after major extinction events, perhaps partly because of the destruction of the supporting communities that taxa evolved to live in. With the exception of opportunistic pioneer species, the expansion of other species from refugia may be impeded until suitable community structures can be re-established (Erwin 2001). Such considerations may not apply to the earliest stages of bilaterian expansion and, if not, early bilaterian expansion may have been less impeded by the lack of pre-existing community structures to which organisms were adapted.

Another even more important factor would be the lack of biogeographical diversity and invasion resistance. Modern studies are increasingly showing that successful invasion of a new region by a particular species is in general impeded by high biodiversity in the target region (e.g. Stachowicz *et al.* 1999). This resistance would be negligible in the early stages of the bilaterian expansion. The world that the crown-group bilaterians expanded into would be one populated by sponges and cnidarians, together with the stem-group bilaterians (some of which may be presented by the Ediacaran taxa). The essentially sessile nature of all benthic members of these groups would imply a very different ecological setting before and after the bilaterian expansion, and the cnidarian/sponge thickets that would have constituted benthic ecology before this event are unlikely to have offered much resistance to bilaterian expansion. Indeed, intriguing evidence for different niche occupancy of lower metazoans and bilaterians, perhaps a relic of the times before the bilaterian invasion, has recently been presented (Yuan *et al.* 2002). Nor can one argue that the lack of bilaterian community structure would have impeded the early bilaterian pioneers: some of the most productive communities today are monocultures that do not depend on high diversity to support a high biomass.

The important conclusion that one can draw from the above considerations is that bilaterian abundance expansion in its early stages is likely to have been exponential, with both ecological fragility and diversity-dependent invasion resistance being significantly weaker than might normally apply. In particular, all connected or close shelf areas that did not cross climatic zones should have been able to be colonized with high abundance extremely rapidly, certainly faster than the normal resolution of the fossil record. Given a broad tectonic setting of rifting continents (Gubanov 2002; Smith 2000), one might expect that cross-continent transfer would be relatively easy in the early stages of the radiation, with provinciality only becoming established going into the late Early Cambrian. Biogeographical studies in the Cambrian have often attempted to demonstrate 'centre of origin' theories (e.g. Ushatinskaya 1996). However, some authors have suggested that Cambrian patterns of biogeography reveal a long prehistory (in order to have time to generate the distinct patterns seen at first appearance; Fortey *et al.* 1996) suggesting that they are dependent upon ancient patterns of vicariance (e.g. Lieberman 1997, 2002). In particular, Lieberman (2002) suggested that the break-up of Pannotia, around 600–550 Ma, gave rise to the principal vicariant patterns shown by basal trilobites. These arguments are amongst the strongest available for suggesting a deep origin of metazoan clades.

Nevertheless, the choice of trilobites as a study clade raises the problems associated with such an approach in their acutest form. The ancestral trilobites were large, calcareous organisms that, by hypothesis, diversified rapidly and were geographically widespread. They were no doubt capable of leaving distinctive trace fossils; and are among the commonest of Cambrian fossils. Why, then, do trace fossils assignable even to arthropods appear only towards the end of the Nemakit-Daldynian, and body fossils of trilobites themselves only at the base of the Atdabanian, some 25 myr after the base of the Cambrian? Based on the modelling of Tavaré *et al.* (2002), the enormous gap – of perhaps 100 myr or more of implied cryptic trilobite history – would suggest that the fossil record did a much worse job of recording early trilobites than early primates, which have a gap of 'only' approximately 36 myr between their modelled time of origin and the first appearance in the fossil record. Yet it would be widely agreed that trilobites have an excellent fossil record when compared with that of primates. One possible way out of this undoubted dilemma may be to recognize that the continents drifted apart only slowly after Pannotian rifting; and the vicariant events identified by these authors do indeed belong in the Early Cambrian rather than latest Proterozoic, as in Gubanov (2002) and, by implication, Smith (2000).

Testing the Cambrian explosion: some approaches

We have argued above that the Cambrian explosion needs to be considered in its many different guises – especially ecological and biogeographical – if we are to understand its rapidity. Nevertheless, mere argumentation will not resolve the pressing problems presented by mismatches between the fossil record and the molecular clock, where they occur. Can analytical or modelling approaches take us further? We would like to suggest, sketchily, two ways forward: the use of extinction events for estimating diversification speeds, and the use of phylogenetic tests of the stratigraphic record.

Extinction events and estimates of fossilization attainment

The major difficulty that is presented by trying to compare fossil and molecular estimates of divergence times is that there is rarely, if ever, an independent test of either. In the case of fossil evidence, the only information available is the first appearance datum of a clade, combined with the subsequent pattern of diversity. Fundamentally, the best we can say about the clade's pre-fossil history is a probabilistic or likelihood statement, but there are few ways of checking the accuracy of our estimates (one possibility is provided by finds of new fossils that fall outside the known range, and comparing them with the estimate of the true range from the previously existing fossils; Marshall 1990). One under-utilized method in this general context, however, would appear to be provided by mass extinctions. If a clade vanishes from the record at an extinction event, only to reappear some time later, one reason may be a sudden decline in diversity/abundance, followed by a 'lag' period of time during which the numbers of the organism are not high enough to ensure the finding of a fossil record (Wignall and Benton 1999; but see also Twitchett 2000, for a discussion of this model and Gale *et al.* 2000, for the alternative preservation-driven model). If an extinction is extremely severe, then this scenario will approximate to a true origin of the clade and the lag between the extinction and the first appearance in the fossil record will give some measure of the length of time that is required for a diversification to be recorded.

Considerable interest has recently been directed at these 'lag' or 'survival' intervals (see Erwin 2001 for a helpful review; Harries and Little 1999), which suggests that the logistic models applied to such diversifications may underestimate the complexity of the underlying processes. For example, empirical studies sometimes point to prolonged periods of biotic crisis, followed by extremely rapid positive-feedback enhanced recoveries. Although there is some variability, the evidence from these recoveries is that restocking – even with an environmentally disturbed survival period – is unlikely to be more than a few million years, and sometimes less. This also applies to radiation events whose beginning may be accurately fixed by other events, such as the monograptid radiation at the beginning of the Silurian (Koren and Bjerreskov 1999). Although neither of these cases is directly applicable to the Cambrian explosion, they do give some basis for examining the lag times in general that exist between the time of origin of a clade (or the time of its severe restriction) and its subsequent (re)appearance in the fossil record. Both these cases suggest that, contrary to many molecular studies, the lag time is in general no more than a few million years. Clearly, in order to develop this theme further would require careful examination of post-extinction intervals in order to distinguish between the various possibilities that could account for the temporary loss of many taxa within them.

Phylogeny as a test of stratigraphy

Whilst much discussion has recently been generated by the possibility of stratigraphic evidence being employed in phylogenetic reconstruction (e.g. Wills 2001), in this instance,

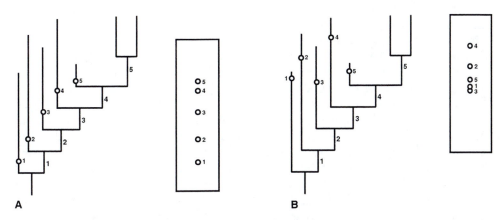

Figure 9.1 A schematic representation of the relationship between order of appearance in the fossil
record and phylogenetic position. The approach adopted is to consider paraphyletic assemblages bounded by particular nodes, and to consider the order of appearance in the fossil
record of the earliest representatives of that phylogenetic interval (cf. Huelsenbeck 1994).
In A, the fossil record is good, and the first representatives of a paraphyletic phylogenetic
interval (marked by the small circles) appear in the same order in the stratigraphic column
(right) as the phylogenetic intervals they represent are arranged in the phylogram (left).
In B, the record, and the gap between the true time of origin of particular paraphyletic
intervals is random but long. As a result, the first appearance of fossil representatives of
each interval is often after the first appearance of a representative of a more derived
paraphyletic interval. The degree of congruence of the fossil record to phylogeny is therefore potentially useful for assessing the length of time between the time of first appearance in the fossil record of a paraphyletic interval and its true time of origin.

the opposite direction of confirmation may also be employed. If bilaterian origins
were genuinely many tens or hundreds of millions of years before the base of the
Cambrian, then the time of appearance of particular clades in the Cambrian record
should be a matter of chance, and not reflect phylogeny. Alternatively, if the order
of appearance does reflect known or postulated phylogeny, then this agreement is strong
– indeed, almost overwhelming – evidence that the true time of origin of the clades
in question was close to the appearance in the record. Naturally enough, even on the
most optimistic reading of the fossil record, there will be a gap between the true appearance of a clade, and its entry in the fossil record. If these gaps are in general small,
then one should expect that they will not perturb the order of appearance in the record,
because the time gaps between successive nodes of a phylogeny will be large relative
to the lags between appearance and true origin (Figure 9.1A). Conversely, if the lag
times are large relative to the temporal gaps between nodes, then one would not expect
the true order of appearance to be reflected in the fossil record (Figure 9.1B). The degree
of correspondence between stratigraphic order of appearance and phylogeny must therefore, in principle, place important constraints on the average lag times, and should
be considered to be a critical test of the hypothesis that true divergence times are greatly
underestimated by the fossil record (for the development of a model essentially
similar to this test, albeit one applied to phylogenetic and not molecular clock studies,
see Huelsenbeck 1994). As with so many issues, therefore, resolution of the molecular clock/fossil record discrepancy could revolve around obtaining and applying the

results from accurate phylogenies. In this particular case, it is important to exclude the timing of fossil origins from phylogenetic reconstruction, because of the unacceptable circularity this could introduce.

This sort of test has largely been ignored, at least for the Cambrian explosion, partly because of the common assumption that all phyla appear more or less at once; that is, there is no temporal resolution of order of appearance. However, this assumption is manifestly untrue. Both trace fossils and body fossils appear progressively through the latest terminal Proterozoic and Early–Middle Cambrian (Budd and Jensen 2000; Budd in press; Jensen in press). The question is whether or not the perceived order is compatible with a reasonable phylogeny or not. Although a satisfactory answer to this question is not yet forthcoming, both because of inadequacies in Cambrian systematics and dating, in broad terms it does seem compatible. This can be seen in two ways: first, by the broad succession of the faunas, and second, by the relative appearance of stem- and crown-group representatives of living clades.

The broad faunal succession of the Cambrian explosion

Recent reassessment of terminal Proterozoic biotas has led to two important results. First, the trace fossil record is much simpler than previously suspected, and progressively diversifies through this period into the Cambrian (e.g. Budd and Jensen 2000; Jensen in press); second, the assignment of Ediacaran taxa to the crown-groups of bilaterian or even cnidarian phyla (e.g. Glaessner 1984) has become increasingly difficult. Whilst the status of nearly all Precambrian presumed metazoan fossils is problematic, none of them can be confidently assigned to the bilaterian crown-group (Budd and Jensen 2000). The earliest mineralizing taxa, such as *Cloudina* (Grant 1990), *Namapoikia* (Wood *et al.* 2002) and *Namacalathus* (Grotzinger *et al.* 2000) all seem in general to be assignable to cnidarian or poriferan grades of organization, and their morphology does not demand placement in the Bilateria. The successive faunas include various tubes that have been assigned to many groups, including the Annelida, but are most reasonably and conservatively thought of as being cnidarian grade (e.g. *Anabarites*; for discussion see Bengtson *et al.* 1990; Kouchinsky and Bengtson 2002; Budd in press). Finally, taxa such as halkieriids and helcionellids, which probably demonstrate stem-group protostome to stem-group mollusc affinities (Conway Morris and Peel 1995; Holmer *et al.* 2002; see discussion in Budd and Jensen 2000), enter the record just above the base of the Cambrian. It should thus be stressed once more that the earliest Cambrian does not in general yield crown-group members of the phyla, which are much more characteristic of the Upper Cambrian and younger sediments (Budd and Jensen 2000; Budd in press).

The overall faunal succession of the Precambrian–Cambrian interval therefore gives the impression of first yielding taxa of poriferan–cnidarian grade (?Ediacarans, plus the first mineralized taxa), followed by taxa that are reasonably assignable to positions deep in the protostome clade, followed by crown-group members of the phyla. All things being equal, one would not expect this succession to occur unless the radiation that generated these taxa was broadly contemporaneous with their first fossils.

This impression of stratigraphic and phylogenetic congruence is reinforced by examining the first appearance of inferred stem- and crown-group taxa within particular clades (Figure 9.2). This diagram should be approached with some caution.

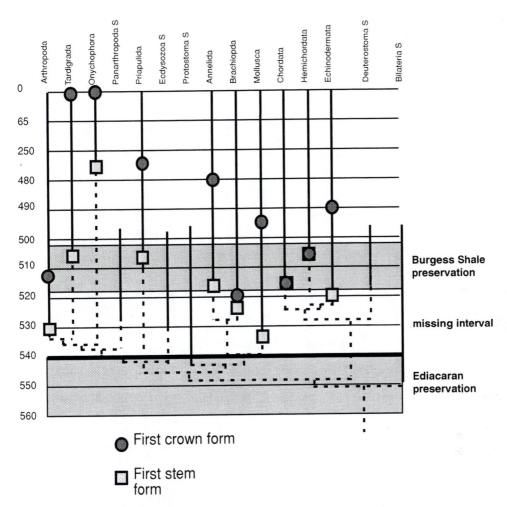

Figure 9.2 Stem- and crown-group appearances of bilaterian phyla. Cambrian timescale as in Budd *et al.* (2001) – note the condensation of the post-Cambrian timescale. For data, see Table 9.1. The terminations of stem lineages are essentially arbitrary. The shaded areas show the coverage of Ediacaran and Burgess Shale exceptional preservation. Note the important 'missing interval' where this model predicts many stem-lineage taxa of the bilaterian phyla should be evolving: a clear test of the model.

The systematics of many Cambrian groups is highly problematic (Budd and Jensen 2000), and the distinction into stem- and crown-group forms is not simple. Many taxa, such as the echinoderms and molluscs, have had wildly differing phylogenetic schemes presented for them, which would greatly alter the time of appearance of the crown-group. For example, crown-group echinoderms might appear in the Middle Cambrian or the Lower Ordovician, according to taste (see discussion in Budd and Jensen 2000). We have here taken a conservative approach, assuming taxa to be in the stem-group unless definitive evidence (i.e. a derived character shared with an extant form) is demonstrable. Although this biases the record towards recovery of

Table 9.1 Data sources for Figure 9.2

Phylum	First stem-group form	First crown-group form	Remarks
Euarthropoda	*Rusophycus*-like trace fossils, Upper Nemakit-Daldynian, Mongolia (Crimes 1987)	Phosphatocopid sp., Comley, UK (Siveter et al. 2001)	Early arthropod trace-fossils are not assignable any higher than the upper stem-group. In addition, recent evidence suggests that trilobites and the clade they fall in may lie in the upper stem-group (Budd 2002), along with many other Cambrian taxa
Tardigrada	Unnamed form, Middle Cambrian, Siberia (Walossek et al. 1995)	*Beorn leggi* (Cooper 1964)	The Cambrian form is clearly a stem-group taxon, as it appears to exhibit at least partial 'short-germband' development
Priapulida	Various species from the Burgess Shale (Conway Morris 1977)	*Priapulites* (Schram 1973)	Taxonomy of all the Cambrian cycloneuralians urgently in need of revision. Position of palaeoscolecids, for example, is uncertain; these may be stem-group cycloneuralians
Annelida	Unnamed taxon from the Sirius Passet fauna (Conway Morris pers. comm.)	?Scolecodonts from the Lower Ordovician	See comments in Budd and Jensen (2000)
Onychophora	?*Helenodora*, (Thompson and Jones 1980)	Extant taxa	No unequivocal characters tying in any of the Cambrian lobopods to the Onychophora have been found; all these taxa may be stem-group forms (Budd 1999)
Brachiopoda	Siberian paterinids, Tommotian (Pelman 1977)	Oboloids, orthids, Atdabanian (Rowell 1977)	Basal brachiopod phylogeny in a state of flux, and new assignments may alter the view presented here (e.g. Williams and Holmer 2002)
Mollusca	*Latouchella*, middle *Purella* Biozone, Nemakit-Daldynian, Siberia (Khomentovsky et al. 1990): if *Purella* itself is molluscan, this dates from the base of the Biozone	Early Ordovician bivalves, cephalopods etc.	Crown-group molluscs probably exist in the Cambrian, but uncertainty about basal molluscan phylogeny is hindering their unequivocal recognition. We do not accept *Kimberella* as a crown- or stem-group mollusc (see text)
Hemichordata	'*Ottoia*' *tenuis* from the Burgess Shale?? (Conway Morris 1986)	Siberian and Baltic pterobranch-like taxa, e.g. *Rhabdotubus* (Bengtson and Urbanek 1986)	Poorly known group, currently under restudy (pers. comm. E. Boulter)
Echinodermata	Helicoplacoids, Atdabanian, Nevada (Durham 1993)	Ordovican crinoids etc.	Problematic taxa such as *Gogia* and *Echmatocrinus* (see discussion in Budd and Jensen 2000) have the potential to alter this view
Chordata	Various taxa from the Chengjiang fauna (e.g. Shu et al. 1996)	As for stem-groups	

early stem-groups, we would argue that this bias is defensible on grounds of parsimony. If it is not known whether or not a taxon possesses a particular derived character, then, all things being equal, it is more parsimonious to assume that it does not possess it than to assume that it does. We hope that dissatisfaction with our results will in some measure spur increased efforts to resolving the systematics of Cambrian fossils.

In our analysis, a remarkable pattern is seen: no crown-group taxa appear before the first stem-group taxa in any of the 10 phyla considered. If stem-group bilaterian radiation occurred deep in the Proterozoic, then one would expect this signal to be lost, and that crown-group taxa would be as likely to appear before stem-group taxa. We regard this pattern as strong evidence that, not only was bilaterian radiation taking place in the latest Proterozoic, as suggested by the broad faunal succession, but that stem-group phyletic radiation was a phenomenon of the earliest Cambrian. Both these pieces of evidence provide critical evidence that the Cambrian explosion was a real event that started off below the crown-group bilaterian level, contrary to the more extreme molecular results. One important factor to note arises from revised timescales of the Cambrian. There is a relatively long period of time between the Ediacaran and Burgess Shale-style preservations, consisting of most of the Nemakit-Daldynian and the Tommotian, where this picture of bilaterian evolution predicts that most phylum stem-groups would be evolving. The test of this model would therefore be to compare any exceptional preservations found from that interval to the later Burgess Shale-type ones: rather than consisting of the same sorts of taxa, the assemblages should be considerably less derived. Some hint of this comes from the many problematic 'small shelly fossils' from this era (e.g. Bengtson *et al.* 1990): it is reasonable to equate 'problematic' with 'stem-group' in many cases!

Discussion

Advocates of the gap between fossil and molecular evidence for the origin of major clades being explicable by the general poverty of the fossil record are sometimes content to rest their case merely on showing – on modelled or other grounds – that such a gap is plausible. Nevertheless, ideally one would want to go beyond this merely theoretical stratum of the problem and ask what is meant to be taking place during this period of time. Is the lack because of low diversity? Or low abundance? Or poor preservability? Or lack of rock? In such a way, it is possible to see that the question of the poverty of the fossil record can be analysed further into its constituent components, each of which can then be critically examined and considered in the context of the particular radiation in question. For the Cambrian explosion, the alleged gap is so large that many possible explanations seem extremely unlikely.

We have argued previously (Budd and Jensen 2000) that the likelihood of the lack of convincing Precambrian crown-group bilaterian fossils cannot be explained by recourse to a step-change in preservation potential (e.g. a rapid increase in body size or development of hard parts). Ancestral bilaterians were of some reasonable size, in order to need muscles and other important organ systems that are almost universally considered to be present in the ur-bilaterian. Furthermore, the total silence of any sort of convincing trace fossil record before about 555–560 Ma argues against the hard part hypothesis, as does the plentiful preservation of non-mineralized, non-bilaterian

organisms throughout much of the Proterozoic, especially in the terminal Proterozoic. Similarly, any possible argument that the lack of bilaterians in the
Proterozoic is because of lack of possible host rocks could not (especially on its own)
explain a total absence for so long. Almost every Cambrian sedimentary rock yields
a few bilaterian fossils of some sort, even the unpromising basal sandstones; no
Proterozoic sedimentary rock before the late terminal Proterozoic has ever yielded
any such thing. If brachiopods and trilobites had been as abundant in the Proterozoic
as they are in the Cambrian, they surely would have been found.

The possible explanation must therefore come down to problems of diversity or
abundance. We have argued here that diversity, although beloved of modellers, is the
wrong metric to use in assessing fossilization potential; and that abundance should
increase extremely rapidly at the base of a radiation, especially with few ecological
impediments to the spread of the new organisms. Furthermore, if the widespread
temporary disappearance of taxa after mass exinctions is owing to restriction of their
abundance, then some generalized metric is available to estimate how long recovery
(considered here as a proxy to time from origination) takes before appearance in the
fossil record. On all of these grounds, it remains extremely hard to imagine that the
time from the origination of crown-group bilaterians to their widespread appearance
in the fossil record was more than a few million years.

If the fossil record is moderately reliable in this regard, then the suggestion must be
that molecular estimates of the origins of bilaterians are largely inaccurate. Potential
sources of inaccuracy have been widely discussed in the literature (e.g. Bromham in
press). More recent ones include the suggestion that molecular clock estimates have
an upwards bias (Bromham et al. 2000; Rodríguez-Trelles et al. 2002) because the
bounds of the molecular estimates of divergence times are asymmetrical (they are rigidly
bound to be non-negative, but non-rigidly bounded at their upper boundary).

In many cases, dissatisfaction with attempts to prove that clock-like conditions
pertain has led to efforts to relax the clock assumption, perhaps by modelling rates
with more complex distributions than the Poisson distribution of the clock model.
Perhaps the most insightful investigation into the complexities of relaxing the clock
assumption is that of Gillespie (1991), who builds up an 'episodic clock' model of
molecular evolution. Noting that the average rate of nucleotide substitution on a site
basis is approximately once per billion years, with protein structure changing on a
timescale of tens of millions of years, he suggests that this sort of timescale is far too
long to account for changes that take place during speciation, especially if the
latter is driven by environmental change on the scale of a few thousand years. The
inevitable consequence must be that at least some molecular evolution takes place in
highly concentrated bursts, separated by long periods of quiescence. Although some
tests of this possibility have not revealed any sudden bursts of molecular evolution
(e.g. Bromham et al. 2000), Aris-Brosou and Ziheng Yang (2000), using a Bayesian
approach on 18S rRNA data, found just that: a burst of rapid evolution during a
bilaterian radiation dated at around 560 Ma. If this pattern is confirmed, then the
apparent conflict between fossil and molecular dates would of course have been resolved;
but more important, the discovery of a relationship between rates of molecular evolution and periods of high morphological evolution would have profound consequences
on how the two are related. Finally, we would like to stress once more that all of
the approaches outlined above rely absolutely on good phylogenies being available,

an approach that would be surely endorsed by many of the protagonists in the Cambrian explosion debate.

Summary

The origins of the bilaterians is a classic problem in the fossil/molecular clock dating field. Here, we have argued that:

(1)   The known fossil record has not been misunderstood, and that there are no convincing bilaterian candidates known from the fossil record until just before the beginning of the Cambrian (c. 543 Ma), even though there are plentiful sediments older than this that should reveal them (including cases of exceptional preservation).

(2)   Phylogenetic and functional grounds cast doubt on the stem-groups of bilaterian phyla being totally planktonic or too small to leave any sort of body or trace fossil.

(3)   Questions of rock and facies preservation only apply to short-term gaps in the record, and become increasingly less viable as the length of time to be explained away increases (the rock record is less complete at the small scale than at the large scale).

(4)   Models of diversity increase at the base of radiations fail to take into account the fact that rate of preservation is related to abundance, rather than to diversity, and that abundance is likely to increase extremely rapidly at the base of a radiation.

(5)   The particular tectonic setting of the Early Cambrian would imply easy shelf-to-shelf transfer of marine taxa between continents, and the lack of competitive exclusion and previous community structure would both assist in rapid spread of taxa. Furthermore, the changes in substrate (Dornbos and Bottjer 2000; Yuan et al. 2002) and development of ecological linkages (Butterfield 1997) would both produce positive feedback results. All of these points give strong grounds for doubting that there was an extensive period of pre-fossil bilaterian history.

(6)   Available tests, in terms of broad faunal succession, stem- and crown-group order of appearance, comparison with recovery after extinctions and with modelling of taxa with poor preservational potential (primates), all strongly suggest that the lag times for the best-preserved taxa in the Cambrian are unlikely to be more than a few million years.

(7)   Early molecular estimates of the time of origin of the bilaterians (e.g. Wray et al. 1996) have been subsequently subjected to considerable discussion and analysis. Current models, including a recent one based on fashionable Bayesian methodology, that do not rely on a strict clock assumption for the rate of molecular change have largely indicated younger dates for the origin of the Bilateria; although this does not necessarily imply that these models are more accurate than previous ones.

Acknowledgements

G.E.B. acknowledges support from the Swedish Scientific Research Council (VR), and S.J. acknowledges support from the National Science Foundation (grant EAR-0074021 to M.L. Droser). We would like to thank Mary L. Droser, James G. Gehling, Philip Donoghue, and Mats Björklund for valuable discussion, and Andrew B. Smith and an anonymous reviewer for constructive and thoughtful criticisms that considerably improved the manuscript.

References

Ager, D.V. (1993) *The Nature of the Stratigraphical Record*, 3rd edn, London: Macmillan.

Aris-Brosou, S. and Ziheng, Y. (2002) 'The effects of models of rate evolution on estimation of divergence dates with special reference to the metazoan 18S rRNA phylogeny', *Systematic Biology*, 51: 703–14.

Barfod, G.H., Albarede, F., Knoll, A.H., Xiao, S.H., Telouk, P., Frei, R. and Baker, J. (2002) 'New Lu–Hf and Pb–Pb age constraints on the earliest animal fossils', *Earth and Planetary Science Letters*, 201: 203–12.

Bengtson, S. (2002) 'Origins and early evolution of predation', *The Paleontological Society, Paper*, 8: 289–317.

Bengtson, S. and Urbanek, A. (1986) '*Rhabdotubus* new genus, a Middle Cambrian rhabdopleurid hemichordate', *Lethaia*, 19: 293–308.

Bengtson, S., Conway Morris, S., Cooper, B.J., Jell, P.A. and Runnegar, B.N. (1990) 'Early Cambrian fossils from South Australia', *Memoirs of the Association of Australasian Palaeontologists*, 9: 1–364.

Bergström, J. (1989) 'The origin of animal phyla and the new phylum Procoelomata', *Lethaia*, 22: 259–69.

Bottjer, D.J., Hagadorn, J.W. and Dorbos, S.Q. (2000) 'The Cambrian substrate revolution', *GSA Today*, 10: 1–7.

Briggs, D.E.G. and Fortey, R.A. (1989) 'The early radiation and relationships of the major arthropod groups', *Science*, 246: 241–3.

Bromham, L.D. (in press) 'What can DNA tell us about the Cambrian Explosion?', *Integrative and Comparative Biology*.

Bromham, L., Penny, D., Rambaut, A. and Hendy, M.D. (2000) 'The power of relative rates tests depends on the data', *Journal of Molecular Evolution*, 50: 296–301.

Budd, G. (1993) 'A Cambrian gilled lobopod from Greenland', *Nature*, 364: 709–11.

—— (1999) 'The morphology and phylogenetic significance of *Kerygmachela kierkegaardi* Budd (Buen Formation, Lower Cambrian, North Greenland)', *Transactions of the Royal Society of Edinburgh: Earth Sciences*, 89: 249–90.

—— (2001) 'Why are arthropods segmented?', *Evolution and Development*, 3: 132–42.

—— (2002) 'A palaeontological solution to the arthropod head problem', *Nature*, 417: 271–5.

—— (in press) 'The Cambrian fossil record and the origin of the phyla', *Integrative and Comparative Biology*.

Budd, G.E. and Jensen, S. (2000) 'A critical reappraisal of the fossil record of the bilaterian phyla', *Biological Reviews*, 75: 253–95.

Budd, G.E., Butterfield, N.J. and Jensen, S. (2001) 'Crustaceans and the Cambrian explosion', *Science*, 294: 2047a.

Butterfield, N.J. (1997) 'Plankton ecology and the Proterozoic–Phanerozoic transition', *Paleobiology*, 23: 247–62.

—— (2000) '*Bangiomorpha pubescens* n. gen., n. sp.: implications for the evolution of sex, multicellularity, and the Mesoproterozoic/Neoproterozoic radiation of eukaryotes', *Paleobiology*, 26: 386–404.

Chen, J.Y., Oliveri, P., Li, C.W., Zhou, G.Q., Gao, F., Hagadorn, J.W., Peterson, K.J. and Davidson, E.H. (2000) 'Precambrian animal diversity: Putative phosphatized embryos from the Doushantuo Formation of China', *Proceedings of the National Academy of Sciences, USA*, 97: 4457–62.

Chen, J.Y., Oliveri, P., Gao, F., Dornbos, S.Q., Li, C.W., Bottjer, D.J. and Davidson, E.H. (2002) 'Precambrian animal life: probable developmental and adult cnidarian forms from southwest China', *Developmental Biology*, 248: 182–96.

Chen, M. and Xiao, Z.Z. (1991) 'Discovery of the macrofossils in the Upper Sinian Doushantuo Formation at Miaho, eastern Yangtze Gorges', *Scientia Geologica Sinica*, 7: 221–31.

Collins, A.G., Lipps, J.H. and Valentine, J.W. (2000) 'Modern mucociliary creeping trails and the bodyplans of Neoproterozoic trace-makers', *Paleobiology*, 26: 47–55.

Conway Morris, S. (1977) 'Fossil priapulid worms', *Special Papers in Palaeontology*, 20: 1–95.

—— (1986) 'The community structure of the Middle Cambrian Phyllopod Bed (Burgess Shale)', *Palaeontology*, 29: 423–67.

—— (1998a) *The Crucible of Creation*, Oxford: Oxford University Press.

—— (1998b) 'The evolution of diversity in ancient ecosystems: a review', *Philosophical Transactions of the Royal Society of London*, B353: 327–45.

Conway Morris, S. and Peel, J.S. (1995) 'Articulated halkieriids from the Lower Cambrian of North Greenland and their role in early protostome evolution', *Philosophical Transactions of the Royal Society, London*, B347: 305–58.

Conway Morris, S., Rasmussen, B., Bengtson, S., Fletcher, I.R. and McNaughton, N.J. (2002) 'Ancient animals or something else entirely?', *Science*, 298: 57–8.

Cooper, K.W. (1964) 'The first fossil tardigrade *Beorn leggi* Cooper, from Cretaceous amber', *Psyche*, 71: 41–8.

Cooper, R.A., Maletz, J., Wang, H.F. and Erdtmann, B.D. (1998) 'Taxonomy and evolution of earliest Ordovician graptoloids', *Norsk Geologisk Tidsskrift*, 78: 3–32.

Copper, P. (1985) 'Fossilized polyps in 430-Myr-old *Favosites* corals', *Nature*, 316: 142–4.

Crimes, T.P. (1987) 'Trace fossils and correlation of late Precambrian and early Cambrian strata', *Geological Magazine*, 124: 97–119.

Darwin, C. (1985) *The Origin of Species*, Harmondsworth, UK: Penguin Classics.

Davidson, E.H., Peterson, K.J. and Cameron, R.A. (1995) 'Origin of bilaterian body plans – evolution of developmental regulatory mechanisms', *Science*, 270: 1319–25.

Dornbos, S.Q. and Bottjer, D.J. (2000) 'Evolutionary paleoecology of the earliest echinoderms: Helicoplacoids and the Cambrian substrate revolution', *Geology*, 28: 839–42.

Droser, M.L., Jensen, S. and Gehling, J.G. (2002) 'Trace fossils and substrates of the terminal Proterozoic–Cambrian transition: Implications for the record of early bilaterians and sediment mixing', *Proceedings of the National Academy of Sciences, USA*, 99: 12572–6.

Drossel, B. and McKane, A. (2000) 'Competitive speciation in quantitative genetic models', *Journal of Theoretical Biology*, 204: 467–78.

Durham, J. (1993) 'Observations on the Early Cambrian helicoplacoid echinoderms', *Journal of Paleontology*, 67: 590–604.

Dzik, J. and Ivantsov, A.Y. (2002) 'Internal anatomy of a new Precambrian dickinsoniid dipleurozoan from northern Russia', *Neues Jahrbuch für Geologie und Paläontologie, Monatshefte*, 7: 385–96.

Erwin, D.H. (2001) 'Lessons from the past: biotic recoveries from mass extinctions', *Proceedings of the National Academy of Sciences, USA*, 98: 5399–403.

Erwin, D.H. and Davidson, E. (2002) 'The last common bilaterian ancestor', *Development*, 129: 3021–32.

Fedonkin, M.A. and Waggoner, B.M. (1997) 'The Late Precambrian fossil *Kimberella* is a mollusc-like bilaterian organism', *Nature*, 388: 868–71.

Foote, M., Hunter, J.P., Janis, C.M. and Sepkoski, J.J. (1999) 'Evolutionary and preservational constraints on origins of biologic groups: divergence times of eutherian mammals', *Science*, 283: 1310–14.

Fortey, R.A., Briggs, D.E.G. and Wills, M.A. (1996) 'The Cambrian evolutionary explosion – decoupling cladogenesis from morphological disparity', *Biological Journal of the Linnean Society*, 57: 13–33.

Gale, A.S., Smith, A.B., Monks, N.E.A., Young, J.A., Howard, A., Wray, D.S. and Huggett, J.M. (2000) 'Marine biodiversity through the Late Cenomanian–Early Turonian: palaeoceanographic controls and sequence stratigraphic biases', *Journal of the Geological Society of London*, 157: 745–57.

Gale, A.S., Smith, A.B. and Monks, N.E.A. (2001) 'Sea-level change and rock-record bias in the Cretaceous: a problem for extinction and biodiversity studies', *Paleobiology*, 27: 241–53.

Gehling, J.G. (1991) 'The case for Ediacaran fossil roots to the metazoan tree', *Memoirs of the Geological Society of India*, 20: 181–223.

Gillespie, J.H. (1991) *The Causes of Molecular Evolution*, Oxford: Oxford University Press.

Glaessner, M. (1984) *The Dawn of Animal Life – a Biohistorical Study*, Cambridge: Cambridge University Press.

Gnilovskaya, M.B., Veis, A.F., Bekker, Yu.R., Olovyanishnikov, V.G. and Rabeen, M.E. (2000) 'Pre-Ediacarian fauna from Timan (Annelidomorphs of the Late Riphean)', *Stratigraphy and Geological Correlation*, 8: 327–52.

Grant, S.W.F. (1990) 'Shell structure and distribution of *Cloudina*, a potential index fossil for the terminal Proterozoic', *American Journal of Science*, 290A: 261–94.

Grazdhankin, D. and Seilacher, A. (2002) 'Underground Vendobionta from Namibia', *Palaeontology*, 45: 57–78.

Grotzinger, J.P., Watters, W.A. and Knoll, A.H. (2000) 'Calcified metazoans in thrombolite-stromatolite reefs of the terminal Proterozoic Nama Group, Namibia', *Paleobiology*, 26: 334–59.

Gubanov, A.P. (2002) 'Early Cambrian palaeogeography and the probable Iberia–Siberia connection', *Tectonophysics*, 352: 153–68.

Harries, P.J. and Little, C.T.S. (1999) 'The early Toarcian (Early Jurassic) and the Cenomanian–Turonian (Late Cretaceous) mass extinctions: similarities and contrasts', *Palaeogeography, Palaeoclimatology, Palaeoecology*, 154: 39–66.

Hoffman, P.F., Kaufman, A.J., Halverson, G.P. and Schrag, D.P. (1998) 'A Neoproterozoic snowball earth', *Science*, 281: 1342–6.

Hofmann, H.J. (1990) 'Precambrian time units and nomenclature – the geon concept', *Geology*, 18: 340–1.

Holmer, L.E., Skovsted, C.B. and Williams, A. (2002) 'A stem group brachiopod from the Lower Cambrian: support for a *Micrina* (halkieriid) ancestry', *Palaeontology*, 45: 875–82.

Huelsenbeck, J.P. (1994) 'Comparing the stratigraphic record to estimates of phylogeny', *Paleobiology*, 20: 470–83.

Huelsenbeck, J.P. and Rannala, B. (1997) 'Maximum likelihood estimation of phylogeny using stratigraphic data', *Paleobiology*, 23: 174–80.

—— (1998) 'Using temporal information in phylogenetics', in J. Weins (ed.) *Phylogenetic Analysis of Morphological Data*, Washington DC: Smithsonian Press, pp. 165–91.

Hunter, J.P. (1998) 'Key innovations and the ecology of macroevolution', *Trends in Ecology and Evolution*, 13: 31–5.

Jägersten, G. (1972) *Evolution of the Metazoan Life Cycle*, London: Academic Press.

Jensen, S. (in press) 'The Proterozoic and earliest Cambrian trace fossil record: patterns, problems and perspectives', *Integrative and Comparative Biology*.

Jensen, S., Gehling, J.G. and Droser, M.L. (1998) 'Ediacara-type fossils in Cambrian sediments', *Nature*, 393: 567–9.

Kemp, T.S. (1982) *Mammal-like Reptiles and the Origin of Mammals*, London: Academic Press.

Khomentovsky, V.V., Val'kov, A.K. and Karlova, G.A. (1990) 'Novye dannye po biostratigrafii perekhodnykh vend-kembrijskikh sloev v bassejne srednegotecheniya r. Aldan' [New data on the biostratigraphy of transitional Vendian–Cambrian strata in the middle reaches of the River Aldan], in V.V. Khomentovsky and A.S. Gibsher (eds) *Pozdnij dokembrij i rannij paleozoj Sibiri*, Novosibirsk: Voprosy regional'noy stratigrafi Institut geologii i geofiziki, Sibirskoe otdelenie, Akademiya nauk SSSR, pp. 3–57.

Kirschvink, J.L. (1992) 'Late Proterozoic low-latitude global glaciation: the snowball Earth', in J.W. Schopf (ed.) *The Proterozoic Biosphere: a Multidisciplinary Study*, Cambridge: Cambridge University Press, pp. 51–2.

Knoll, A.H. and Carroll, S.B. (1999) 'Early animal evolution: emerging views from comparative biology and geology', *Science*, 284: 2129–37.

Koren, T. and Bjerreskov, M. (1999) 'The generative phase and the first radiation event in the Early Silurian monograptid history', *Palaeogeography, Palaeoclimatology, Palaeoecology*, 154: 3–9.

Kouchinsky, A. and Bengtson, S. (2002) 'The tube wall of Cambrian anabaritids', *Acta Palaeontologica Polonica*, 47: 431–44.

Landing, E. (1994) 'Precambrian–Cambrian boundary global stratotype ratified and a new perspective of Cambrian time', *Geology*, 22: 179–82.

Lieberman, B.S. (1997) 'Early Cambrian paleogeography and tectonic history: A biogeographic approach', *Geology*, 25: 1039–42.

—— (2002) 'Phylogenetic analysis of some basal early Cambrian trilobites, the biogeographic origins of the Eutrilobita, and the timing of the Cambrian radiation', *Journal of Palaeontology*, 76: 692–708.

Marshall, C.R. (1990) 'Confidence-intervals on stratigraphic ranges', *Paleobiology*, 16: 1–10.

—— (1997) 'Confidence intervals on stratigraphic ranges with nonrandom distributions of fossil horizons', *Paleobiology*, 23: 165–73.

Martin, M.A., Grazhdankin, D.V., Bowring, S.A., Evans, D.A.D., Fedonkin, M.A. and Kirschvink, J.L. (2000) 'Age of Neoproterozoic bilatarian [*sic*] body and trace fossils, White Sea, Russia: implications for metazoan evolution', *Science*, 288: 841–5.

Mooi, R., David, B. and Marchand, D. (1994) 'Echinoderm skeletal morphologies: classical morphology meets modern phylogenetics', in B. David, A. Guille, J.P. Féral and M. Roux (eds) *Echinoderms Through Time*, Rotterdam: Balkema, pp. 87–95.

Narbonne, G.M. (1998) 'The Ediacara Biota: a terminal Neoproterozoic experiment in the evolution of life', *GSA Today*, 8(2): 1–6.

Pelman, Y.L. (1977) 'Ranne i srednekembriiskjie bezzamkovje brakhiopody Sibirskoy Platformy' [Early and Middle Cambrian inarticulate brachiopods of the Siberian Platform], *Trudy Instituta Geologii geofiziki Sibirskogo Otdelenjia*, 36: 1–168.

Rasmussen, B., Bose, P.K., Sarkar, S., Banerjee, S., Fletcher, I.R. and McNaughton, N.J. (2002a) '1.6 Ga U-Pb zircon ages for the Chorhat Sandstone, lower Vindhayan, India: Possible implications for early evolution of animals', *Geology*, 30: 103–6.

Rasmussen, B., Bengtson, S., Fletcher, I.R. and McNaughton, N.J. (2002b) 'Discoidal impressions and trace-like fossils more than 1200 million years old', *Science*, 296: 1112–15.

Ray, J.S., Martin, M.W., Veizer, J. and Bowring, S.A. (2002) 'U–Pb zircon dating and Sr isotope systematics of the Vindhayan Supergroup, India', *Geology*, 30: 131–4.

Rodríguez-Trelles, F., Tarrío, R. and Ayala, F.J. (2002) 'A methodological bias toward overestimation of molecular evolutionary time scales', *Proceedings of the National Academy of Sciences, USA*, 99: 8812–15.

Rowell, A.J. (1977) 'Early Cambrian brachiopods from the south-western Great Basin of California and Nevada', *Journal of Paleontology*, 51: 68–85.

Runnegar, B. (1995) 'Vendobionta or Metazoa? Developments in understanding the Ediacara "fauna"', *Neues Jahrbuch für Geologie und Paläontologie, Abhandlundgen*, 195: 303–18.

Runnegar, B.N. and Fedonkin, M.A. (1992) 'Proterozoic metazoan body fossils', in J.W. Schopf (ed.) *The Proterozoic Biosphere*, Cambridge: Cambridge University Press, pp. 369–88.

Schluter, D. (2000) *The Ecology of Adaptive Radiation*, Oxford: Oxford University Press.

Schram, F.R. (1973) 'Pseudocoelomates and a nemertine from the Illinois Pennsylvanian', *Journal of Paleontology*, 47: 985–9.

Seilacher, A. (1957) 'An-aktualistiches Wattenmeer', *Paläontologische Zeitschrift*, 31: 198–206.

—— (1999) 'Biomat-related lifestyles in the Precambrian', *Palaios*, 14: 86–93.

Seilacher, D., Bose, P.K. and Pflüger, F. (1998) 'Triploblastic animals more than 1 billion years ago: Trace fossil evidence from India', *Science*, 282: 80–3.

Shu, D.G., Conway Morris, S. and Zhang, X.L. (1996) 'A *Pikaia*-like chordate from the Lower Cambrian of China', *Nature*, 384: 157–8.

Simpson, G.G. (1953) *The Major Features of Evolution*, New York: Columbia University Press.

Siveter, D.J., Williams, M. and Waloszek, D. (2001) 'A phosphatocopid crustacean with appendages from the Lower Cambrian', *Science*, 293: 479–81.

Smith, A.B. (1988) 'Patterns of diversification and extinction in Early Palaeozoic echinoderms', *Palaeontology*, 31: 799–828.

—— (1999) 'Dating the origin of metazoan body plans', *Evolution and Development*, 1: 138–42.

Smith, A.B. and Peterson, K.J. (2002) 'Dating the time of origin of major clades: Molecular clocks and the fossil record', *Annual Review of Earth and Planetary Sciences*, 30: 65–88.

Smith, A.G. (2000) 'Paleomagnetically and tectonically based global maps for Vendian to mid-Ordovician time', in A.Y. Zhuravlev and R. Riding (eds) *The Ecology of the Cambrian Radiation*, New York: Columbia University Press, pp. 11–46.

Stachowicz, J.J., Whitlatch, R.B. and Osman, R.W. (1999) 'Species diversity and invasion resistance in a marine ecosystem', *Science*, 286: 1577–9.

Steiner, M. (1994) 'Die neoproterozoischen Megaalgen Sudchinas' [The Neoproterozoic mega-algae of South China], *Berliner geowissenschaftliche Abhandlungen*, E15: 1–146.

Strauss, D. and Sadler, P.M. (1989) 'Classical confidence limits and Bayesian probability estimates for ends of local taxon ranges', *Mathematical Geology*, 21: 411–27.

Tavaré, S., Marshall, C.R., Will, O., Soligo, C. and Martin, R.D. (2002) 'Using the fossil record to estimate the age of the last common ancestor of extant primates', *Nature*, 416: 726–9.

Thomas, A. (1999) 'Trace fossils and the Cambrian explosion – Reply', *Trends in Ecology and Evolution*, 13: 507.

Thompson, I. and Jones, D.S. (1980) 'A possible onychophoran from the Middle Pennsylvanian Mazon Creek beds of Northern Illinois', *Journal of Paleontology*, 54: 588–96.

Twitchett, R.J. (2000) 'Discussion on Lazarus taxa and fossil abundance at times of biotic crisis', *Journal of the Geological Society of London*, 157: 511–12.

Ushatinskaya, G.T. (1996) 'Brachiopod palaeozoogeography through the Cambrian', in P. Copper and J. Jin (eds) *Brachiopods*, Rotterdam: A.A. Balkema, pp. 275–80.

Vidal, G. and Moczydlowska-Vidal, M. (1997) 'Biodiversity, speciation, and extinction trends of Proterozoic and Cambrian phytoplankton', *Paleobiology*, 23: 230–46.

Walossek, D., Zakharov, A. and Müller, K.J. (1995) ' "Orsten" type phosphatized soft-integument preservation and a new record from the Middle Cambrian Kuonamka Formation in Siberia', *Neues Jahrbuch für Geologie und Paläontologie, Abhandlungen*, 191: 101–18.

Wignall, P.B. and Benton, M.J. (1999) 'Lazarus taxa and fossil abundance at times of biotic crisis', *Journal of the Geological Society of London*, 156: 453–6.

Williams, A. and Holmer, L.E. (2002) 'Shell structure and inferred growth, functions and affinities of the sclerites of the problematic *Micrina*', *Palaeontology*, 45: 845–73.

Willis, J. (1922) *Age and Area*, Cambridge: Cambridge University Press.

Wills, M.A. (2001) 'How good is the fossil record of arthropods? An assessment using the stratigraphic congruence of cladograms', *Geological Journal*, 36: 187–210.

Wood, R.A., Grotzinger, J.P. and Dickson, J.A.D. (2002) 'Proterozoic modular biomineralized metazoan from the Nama Group, Namibia', *Science*, 296: 2383–6.

Wray, G.A., Levinton, J.S. and Shapiro, M. (1996) 'Molecular evidence for deep Precambrian divergences among metazoan phyla', *Science*, 214: 568–73.

Xiao, S.H. (2002) 'Mitotic topologies and mechanics of Neoproterozoic algae and animal embryos', *Paleobiology*, 28: 244–50.

Xiao, S.H., Yuan, X.L. and Knoll, A.H. (2000) 'Eumetazoan fossils in terminal Proterozoic phosphorites?', *Proceedings of the National Academy of Sciences, USA*, 97: 13684–9.

Yuan, X.L., Xiao, S.H., Parsley, R.L., Zhou, C.M., Chen, Z. and Hu, J. (2002) 'Towering sponges in an Early Cambrian Lagerstatte: Disparity between nonbilaterian and bilaterian epifaunal tiers at the Neoproterozoic–Cambrian transition', *Geology*, 30: 363–6.

Zhuravlev, A. Yu. and Riding, R. (eds) (2000) *The ecology of the Cambrian Radiation*, New York: Columbia University Press.

Chapter 10

The origin and early evolution of chordates: molecular clocks and the fossil record

Philip C.J. Donoghue, M. Paul Smith and Ivan J. Sansom

ABSTRACT

Evolutionary biology abounds with theories and scenarios for the origins of the major chordate clades but little attempt has been made to constrain knowledge over the dating of these evolutionary events. The fossil record of early chordates, including stem-gnathostomes and basal crown-gnathostomes, as well as the sister-clade Ambulacraria (Hemichordata plus Echinodermata), is critically re-evaluated. This is achieved through both qualitative and quantitative assessment of the fit of phylogenetic hypotheses to stratigraphic range data, and through assessment of the internal consistency of stratigraphic range data. The results suggest that the fossil record of early chordates is of variable quality; the fossil record of basal chordates appears to be a poor reflection of their evolutionary history, while the fossil record of many stem-gnathostomes, such as conodonts and heterostracans, appears to be very good, albeit poorly understood in places. Thus, palaeontological data provide little constraint on the origin of chordates, craniates, and vertebrates, other than to indicate that these clades were established by 530 Ma. The origin of total-group Gnathostomata has a well-supported fossil estimate of 495 Ma, a date which falls within the error calculations of published molecular clock estimates. The origin of crown-gnathostomes is dated at 457 Ma using the fossil record, with a confidence interval extending to 463 Ma, implying an incomplete record; this lies just outside molecular estimates (e.g. 528 Ma ± 56.4 myr). Finally, the fossil record suggests the divergence of actinopterygians and sarcopterygians at 425 Ma, with a very narrow confidence interval (+ 580 Ka) and falls within molecular estimates (450 Ma ± 35.5 myr). Thus, where internal assessments of palaeontological data imply a good record there is correlation with molecular clock estimates, and where these assessments suggest a poor record there is poor correlation. Where correlation occurs we may assume that our estimates are a good reflection of the true time of divergence of the various clades, and where there is conflict we must assume nothing. We note that even where corroboration between datasets occurs, error bars on divergence times remain too coarse to attempt correlation to evolutionary events in other clades, and extrinsic events in Earth history.

The problem

It is of course anthropocentric bias, but the nature of the evolutionary and environmental events surrounding the origin and early evolution of the phylum Chordata are some of the most extensively researched problems in evolutionary biology. Theories that have sought to account for these events are contingent upon shifts in calibration of the geological timescale, a fossil record that is dynamic both in terms of new discoveries and reinterpretation of the phylogenetic affinities of old finds and, more recently, the introduction of molecular clock estimates for the times of divergence of living clades. It is therefore not surprising that many such hypotheses have fallen purely because events once thought to be coeval are revealed not to be so. But with so many lines of evidence, many of which are independent, there remains the possibility that conflict may give way to consilience, rather than merely to compromise. Recent advances have resulted in a considerable fleshing out of the early fossil record of chordates (Sansom *et al.* 2001; M.P. Smith *et al.* 2001, 2002), the geological timescale is now more finely calibrated than at any time in the past (e.g. Remane 2000), and there is an ever increasing database of molecular sequences for analysing evolutionary relationships and sampling for molecular clock analyses. With these developments, understanding the events surrounding the origin and early evolution of the chordate phylum may now prove more tractable than at any time previously.

The data

The nearest living relatives of the chordates are the echinoderms and hemichordates and, together, these three phyla comprise the Deuterostomia. Living invertebrate chordates are a very depauperate group in comparison with their vertebrate relatives, comprising two or three groups depending upon how the Vertebrata are defined. The most plesiomorphic groups are the tunicates and cephalochordates, and although there has historically been a great deal of prevarication surrounding their interrelationships, the tunicates are now widely recognized as the most basal group of living chordates. The next most inclusive clade, Craniata, includes only the hagfishes in addition to the vertebrates, which are in turn composed of the lampreys plus the Gnathostomata (living jawed vertebrates). Gnathostomes, in turn, are composed of chondrichthyans, and the two most derived groups, the actinopterygians and sarcopterygians, which includes the lineage leading to tetrapods.

These taxonomic groups are defined solely on the basis of living taxa, and so it is possible to provide molecular estimates for the divergence of the various groups without recourse to the fossil record for anything other than internal and/or external calibration. However, the divergence of the various lineages does not equate to the origin of the taxonomic groups, at least not in the sense that most biologists understand these taxa. This is because most of these groups also include fossil taxa, with varying degrees of taxonomic diversity and disparity, which are part of the lineage leading to the crown-group of living taxa, but do not possess the full suite of anatomical characteristics necessary for inclusion within the crown-group (Jefferies 1979). For instance, the extinct osteostracans are a group of jawless vertebrates that share a number of derived characters with gnathostomes that they do not share with lampreys. Hence, osteostracans are resolved as more closely related to gnathostomes than lampreys and represent part of the lineage leading to gnathostomes after its divergence from that last common ancestor shared with lampreys. However, since they lack jaws, osteostracans cannot be considered part of the group 'Gnathostomata' as perceived by most biologists. Current hypotheses of early vertebrate relationships

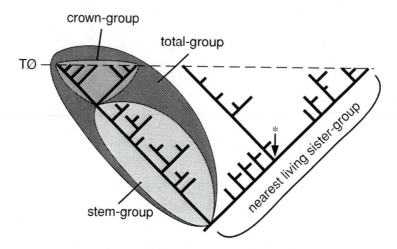

Figure 10.1 The relationships between stem-, crown-, and total-groups. TØ is time zero (present day). * denotes the time of origin of a crown-group and its two constituent total-groups. After Jeffries (1986).

indicate that there are many such groups that are more closely related to gnathostomes than lampreys. Systematists have devised a means of obviating this state of taxonomic purgatory by allying such lineages with their nearest living group to form an inclusive 'total-group' version of the existing taxonomic concept (Figure 10.1). Thus, the extinct members of the gnathostome lineage that fall outside of crown-group Gnathostomata become part of the total-group Gnathostomata, composed of the paraphyletic 'stem'-lineage leading to a 'crown'-group circumscribed by the living representatives of the Gnathostomata, and including all of their descendants, fossil and extant. The difference between the time of origin of the total-group and that of the crown-group (approximately equivalent to the original concept 'Gnathostomata') reflects the time gap between the point of divergence from the lamprey lineage, and the origin of the living clade, palaeontological estimates for the timing of which differ by as much as 60 million years. Both events are approachable through molecular clock theory (because the time of origin of one crown group is also the time of origin of its two constituent total groups (see figure 10.1)), although they are often confused and/or conflated, and it is usually only the origin of the total-group that is calculated owing to the ease of data collection and the reduced reliance upon subsidiary hypotheses of intrarelationships of constituent crown-group taxa.

The fossil record of chordates and their near relatives

Echinoderms

The earliest putative echinoderms are both Neoproterozoic in age. Subsequent to its description, *Tribrachidium* (Glaessner and Wade 1966) has generally been excluded from contention as an echinoderm (e.g. Wills and Sepkoski 1993), but debate concerning the affinities of *Arkarua* (Gehling 1987) continues. Budd and Jensen (2000) have argued that data in support of echinoderm affinity for the latter taxon are tenuous and limited exclusively to the presence of pentameral symmetry. More recent interpretations of echinoderm skeletal homologies following the extraxial–axial theory identify many more echinoderm synapomorphies and symplesiomorphies in the still poorly known anatomy of *Arkarua*. These include a body wall dominated by extraxial rather than axial components, upwardly-oriented perforate extraxial and axial rays, flooring plates that follow the ocular plate rule, and a disc-shaped morphology akin to the edrioasteroids (David and Mooi 1998; Mooi and David 1997, 1998; Mooi 2001; Mooi pers. comm. 2002). The logical extension of this argument is that *Arkarua* is the sister-taxon to all other echinoderms, representing the only member of the echinoderm total-group to lack a stereom skeleton plesiomorphically. However, it should be remembered that all identified homologies are contingent upon the *a priori* assumption of an echinoderm affinity for *Arkarua* and alternative phylogenetic frameworks would lead to a very different interpretation of homologies. Thus, although we will discuss the implications of a Neoproterozoic echinoderm record for the evolutionary history of chordates, this record should not be considered beyond reproach.

Echinoderms are well represented amongst Early Cambrian faunas (see e.g. Smith 1988a,b, 1990), but the precise affinity of these taxa remains the subject of wide-ranging debate. The helicoplacoids are generally considered stem-group echinoderms, slightly more derived than *Arkarua*, but the phylogenetic position of *Camptostroma*, the edrioasteroids, and the carpoids remains contentious, with some authors placing them in stem-echinoderm positions, and others resolving them as members of the echinoderm crown-group; see Smith (1984, 1988a,b, 1990), Sumrall (1997) and David and Mooi (1999; Mooi and David 1998; David *et al.* 2000) for the different arguments, and Mooi (2001) for a compilation of trees reflecting the different hypotheses.

Hemichordates

The early fossil record of hemichordates is limited in large part to the pterobranchs and extends back to the Middle Cambrian (Bengtson and Urbanek 1986; Durman and Sennikov 1993); the record of enteropneusts does not extend beyond the Jurassic (Arduini *et al.* 1981). These data are widely accepted, but the earliest possible record is of *Yunnanozoon* from the Lower Cambrian Chengjiang Lagerstätten of China. The original and most valid interpretation of this organism to date is as a metazoan of unknown affinity (Hou *et al.* 1991). *Yunnanozoon* has subsequently been described both as a chordate (Chen *et al.* 1995; Dzik 1995) and as a hemichordate (Shu *et al.* 1996a), whilst the suspiciously similar *Haikouella* has also been described as a craniate (Chen *et al.* 1999). The mélange of characters exhibited by *Yunnanozoon* (e.g. Dzik 1995) may indicate a more appropriate placement in the deuterostome stem-group.

Chordates

Fossil representatives have been claimed for all living groups of invertebrate chordates and jawless vertebrates as far back as the Early Cambrian. Putative fossil tunicates include *Cheungkongella* (Shu *et al.* 2001a), *Palaeobotryllus* (Müller 1977), and *Peltocystis* (Jefferies *et al.* 1996). Putative acraniate chordates include *Lagynocystis* (Jefferies 1973), *Pikaia* (Conway Morris 1979), *Yunnanozoon* (Chen *et al.* 1995;

Dzik 1995), and *Cathaymyrus* (Shu *et al.* 1996b). Possible fossil representatives or close relatives of the living jawless vertebrates include the Cambrian taxa *Haikouella* (Chen *et al.* 1999; but see above), *Myllokunmingia* and *Haikouichthys* (Shu *et al.* 1999), and the Carboniferous taxa *Gilpichthys* and *Pipiscius* (Bardack and Richardson 1977), *Mayomyzon* (Bardack and Zangerl 1968, 1971), *Myxinikela* (Bardack 1991, 1998), and *Hardistiella* (Janvier and Lund 1983; Lund and Janvier 1986), as well as a number of mitrates such as *Mitrocystites* (Jefferies 1967) and *Placocystites* (Jefferies and Lewis 1978).

In addition, there are a wide variety of fossil jawless vertebrates characterized by an extensively developed dermal 'armour' and historically grouped together as the 'ostracoderms'. These include the anaspids, galeaspids, heterostracans, osteostracans, and thelodonts (see Janvier 1996b for an introduction to these various groups). Amongst the jawed vertebrates, there are also a number of large groups that have no living representatives, principally including the placoderms and acanthodians. There is also a swathe of basal chondrichthyans, actinopterygians, and sarcopterygians that belie the apparent disparity of their living relatives.

The phylogenetic relationships of living and extinct chordates and their near relatives

To compare palaeontological and molecular estimates for the time of divergence of the various chordate clades it is first necessary to resolve the phylogenetic relationships of the living and fossil groups of chordates and their near relatives; palaeontological estimates can then be provided through calibration of the resulting phylogeny to the stratigraphic occurrence of the various groups within the geological timescale.

The calcichordate–stylophoran problem

No discussion of early chordate evolution would be complete without a consideration of the 'calcichordates'. Jefferies (1967 *et seq*) identifies an extinct group of calcite-plated invertebrates, otherwise interpreted as basal echinoderms (Stylophora; e.g. Ubaghs 1968; Paul and Smith 1984), as paraphyletic suites of lineages that interleave the stems of extant echinoderms, cephalochordates, tunicates, and vertebrates. This theory has been criticized on many grounds. Amongst the most substantive of these, independent phylogenetic analyses resolve tunicates as basal chordates (Garcia-Fernàndez and Holland 1994) rather than as the sister-group to the vertebrates, which is a requirement of the 'calcichordate' hypothesis (Jefferies 1986). Furthermore, independent phylogenetic analyses (Peterson 1995) recognize that the cornute and mitrate 'calcichordates' share a number of potential homologies that may only be rejected by weighting other characters that are deemed on the basis of the calcichordate theory to be of greater phylogenetic significance (Ruta 1999). This appears to preclude not only the calcichordate theory, but also Gee's compromise hypothesis that the 'calcichordates' are a paraphyletic ensemble of basal deuterostomes, some of which are more closely related to one or more phyla, than are others (Gee 2001; although it does not preclude the possibility that they are basal deuterostomes). Thus, the stylophorans are not germane to understanding the timing of chordate diversification and we will not discuss them further.

Morphological analysis

The interrelationships of both living and fossil chordates have been the subject of controversy since the origin of systematic classification. Much debate has centred on the relative relationships of the living jawless vertebrates, the hagfishes and lampreys, to living jawed vertebrates, and the implications that this has for the interrelationships of extinct groups of jawless vertebrates and invertebrate chordates. All three possible solutions to the problem of hagfish–lamprey–jawed vertebrate interrelationships have been proposed, but of these, cyclostome monophyly ((hagfish, lamprey) jawed vertebrate) and cyclostome paraphyly (hagfish (lamprey, jawed vertebrate)) have received by far the most attention. Although morphological data were formerly interpreted to support cyclostome monophyly (e.g. Stensiö 1927, 1968; Yalden 1985), the application of phylogenetic systematics to the same dataset led to a revised interpretation of cyclostome paraphyly (Løvtrup 1977; Janvier 1996a, 1981; Hardisty 1982; Forey 1984), a view that is still defended by morphologists (e.g. Janvier 1998; Donoghue et al. 2000; Donoghue and Smith 2001). We will consider the implications of both hypotheses in assessing the completeness of the chordate fossil record.

The hypothesis of relationships that we have adopted to provide palaeontological estimates of divergence times for the various chordate clades is a development of the analysis undertaken by Donoghue et al. (2000), to include the recently discovered groups of invertebrate chordates and basal vertebrates from the Lower Cambrian Chengjiang Lagerstätte. The results of this extended analysis are presented in Figure 10.2 and the codings for additional taxa are included in Appendix 10.1.

Molecular analysis

In contrast to morphological datasets, analyses of molecular datasets universally resolve the living jawless vertebrates as monophyletic. Although phylogenetic analysis of incomplete mitochondrial datasets resolved hagfishes and lampreys as paraphyletic (Suzuki et al. 1995), analysis of the entire mitochondrial genome provides unequivocal support for the monophyly of hagfishes and lampreys (Delarbre et al. 2002). Similarly, small datasets of nuclear DNA have provided support for the paraphyly of the living jawless vertebrates (Suzuki et al. 1995), but larger datasets provide strong support for monophyly (Goodman et al. 1987; Kuraku et al. 1999; Hedges 2001). Analysis of RNA also strongly supports cyclostome monophyly (Stock and Whitt 1992; Mallatt and Sullivan 1998; Mallatt et al. 2001), although analysis of RNA datasets partitioned into small- and large-subunit components provides conflicting support for both hypotheses (Zrzavý et al. 1998).

Resolution of the interrelationships of hagfishes and lampreys is critical to understanding character evolution at the origin of vertebrates and gnathostomes. However, both groups have a comparable fossil record, and there are no known intermediate taxa with a fundamentally earlier or later first appearance in the fossil record than sister and ingroup clades (the reality is quite the opposite). Thus, the difference between the two most likely resolutions of hagfish–lamprey–jawed vertebrate interrelationships is not critical to our understanding of the timing of early chordate diversification or the relationship between the fossil record and molecular clocks.

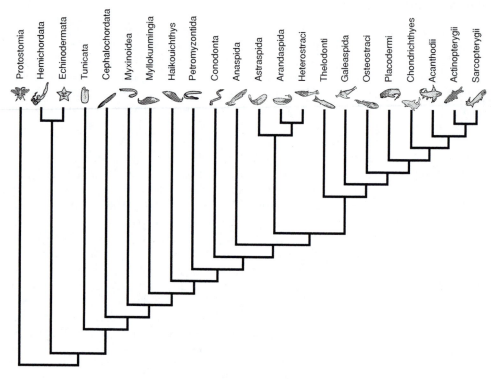

Figure 10.2 The interrelationships of living and extinct groups of chordates and their nearest living relatives.

Stratigraphic analysis

Although stratigraphic range data are often taken at face value in attempts to calibrate molecular clocks and phylogenetic trees, and to provide palaeontological estimates for the divergence times, various techniques exist to provide confidence limits on the first and/or last appearance of taxa based upon the quality of the intervening record. The chief method for this is gap analysis which was originally developed to provide confidence limits on stratigraphic range data in local sections (Marshall 1990), but can be (Marshall 1990) and has been (Bleiweiss 1998) applied to global datasets provided that the distribution of fossil-bearing horizons within the observed range is random. The technique provides a means of determining limits of probability on how far outside the known stratigraphic range of a taxon the true first (and/or last) appearance might occur – this is proportional to the density with which the taxon has been found throughout its known stratigraphic range (Marshall 1990). It follows that the greater the number of horizons from which the taxon has been recorded, the less likely it is that the true range lies far beyond the limits of the known range, and *vice versa*. Gap analysis calculates, at a given level of confidence (e.g. 95 or 99 per cent), an interval within which the true end point (appearance or disappearance) of a stratigraphic range lies (Marshall 1990). At its simplest, the calculation assumes constant fossil recovery potential, but techniques have been developed to

Table 10.1 Confidence intervals calculated on the basis of the internal relationships of the plesions included in the main analysis.

Taxon	Base	Top	n	Sil P > 0.95	Sil P > 0.99	Ord n	Ord P > 0.95	Ord P > 0.99
[a]Myxinoidea	304	0	2	6080	30 400			
[a]Petromyzontida	325	0	2	6500	32 500			
Conodonta	495	418	670	495.345573	495.531871	501	495.312491	495.48115
Arandaspida	477	464	5	491.491653	505.10961			
Astraspis	457	453	36	457.357449	457.5625			
Heterostraci	428	418	110	428.278649	428.431545			
Anaspids	433	418	57	433.824279	434.285667			
Thelodonts	457	418	295	457.399425	457.615699	7	466.065686	466.065686
Galeaspids	438	418	9	447.084309	453.565588			
Osteostracans	433	418	25	434.994205	436.172915			
Placoderms	428	418	11	431.492828	433.848932			
Chondrichthyans	457	418	67	458.810995	459.818422	9	463.359016	463.359016
Acanthodians	446	418	178	446.477934	446.738061	1		
Actinopterygians	425	418	59	425.371054	425.578457			
Sarcopterygians	423	418	1					

'Sil *n*' is the number of records within the interval Cambrian–Silurian, and 'Sil *P* > 0.95' and 'Sil *P* > 0.99' are 95 and 99 per cent confidence intervals on the first appearance of the plesion respectively, based upon the Cambrian–Silurian interval. 'Ord *n*' is the number of records within the interval Cambrian–Ordovician, and 'Ord *P* > 0.95' and 'Ord *P* > 0.99' are 95 and 99 per cent confidence intervals on the first appearance of the plesion, respectively, based upon the Cambrian–Ordovician interval.

[a] The fossil record of hagfishes and lampreys is limited to the Carboniferous and, as a result, the confidence interval calculations are based on their full stratigraphic range, rather than limited to the pre-Devonian as are the other calculations.

incorporate variable recovery potential that may result, for example, from biases in facies preservation arising from changes in relative sea level (Holland 1995, 2000; Marshall 1997; Tavaré *et al.* 2002).

We have calculated 95 and 99 per cent confidence limits for the fossil record of each of the main groups of fossil and living invertebrate chordates and jawless and jawed vertebrates using the combined micro- and macrofossil record. The values are presented in Table 10.1 and are graphically expressed in Figure 10.3.

Assessing congruence between cladograms and stratigraphy

Ghost lineages and their conceptual efficacy

Following the principle that sister taxa are derived from common ancestors and, thus, have an evolutionary history that can be traced back to the point in time at which they diverged from their latest common ancestor, an assessment of the completeness of the fossil record of a taxon can be achieved through comparing the stratigraphic ranges of sister taxa. The inferred range extension of a taxon based on the longevity of its sister taxon is known as a 'ghost lineage' or 'ghost range', a concept introduced by Gauthier *et al.* (1988) and developed by Norell (1992) amongst others. The technique is useful because it provides a means of inferring the existence of unsampled or unsampleable taxa, but it relies upon a number of important and potentially limiting assumptions. First, it must be assumed that the cladogram is a faithful reflection of evolutionary relationships. Second, all the taxa in the cladogram must be

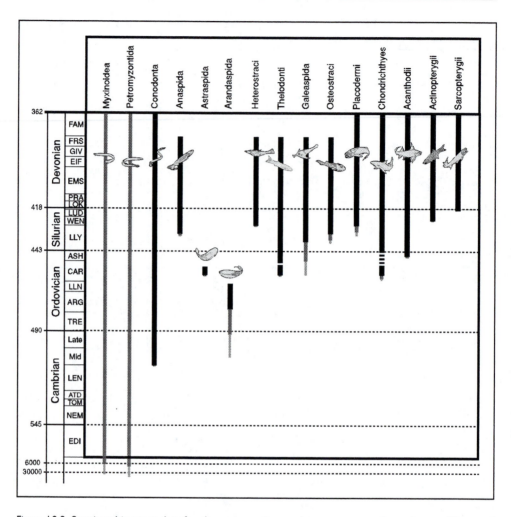

Figure 10.3 Stratigraphic range data for the major extinct and extant groups of vertebrates. Observed
stratigraphic range represented by thick black bars; large gaps within observed range
represented by dashed bars; 95 per cent confidence limit on first appearance represented by medium thickness dark grey bars; 99 per cent confidence limit on first appearance represented by thin light grey bars. These data are also presented in Table 10.1.

monophyletic, since the inclusion of paraphyletic taxa (e.g. ancestors) will lead to an
incorrect inference of a ghost lineage (Wagner 1998; Paul, Chapter 5).

Although many of the nodes in the tree presented in Figure 10.2 are relatively weakly
supported, the overall structure of the tree is well supported. The second assumption
is also justified in that taxa used in the analysis have been scrutinized through character analysis and all exhibit identifiable synapomorphies (e.g. Janvier 1996b). The
results of this analysis indicate that although the fossil record of most groups of
stem-gnathostomes does not begin until the Silurian, all have ghost ranges that
extend a considerable way downwards into the Ordovician (Figure 10.4). This is

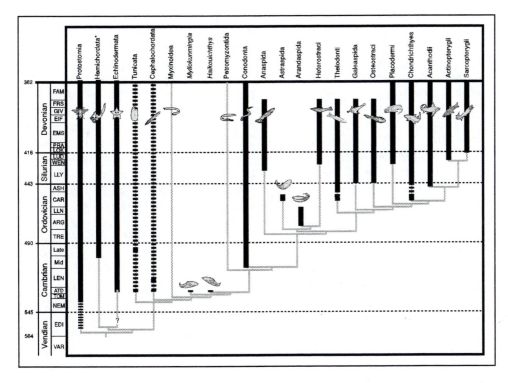

Figure 10.4 Cladogram from Figure 10.2 calibrated to time to reveal inferred ghost lineages; observed stratigraphic range data represented by thick black bars and large gaps represented by thick dashed bars.

surprising given that these organisms have an extensive mineralized component to their skeleton and, thus, might be expected to have a much better fossil record.

Cladogram fit to stratigraphy

Calibrating cladograms to time and inferring ghost lineages provides a useful visual assessment of the completeness of the fossil record. However, the rigour of this technique has been extended through the application of a number of metrics that assess different aspects of the relationship between tree structure and stratigraphic data. These include the Stratigraphic Consistency Index (SCI; Huelsenbeck 1994), which compares the number of stratigraphically consistent cladogram nodes to the total number of cladogram nodes. The Relative Completeness Index (RCI; Benton and Storrs 1994) attempts to measure the overall level of inconsistency in a tree by quantifying the ghost range implied as the difference between the age of origin of branches subtending sister taxa, divided by the observed range length, and expressed as a percentage. A third metric, the Gap Excess Ratio (GER; Wills 1999) combines aspects of both SCI and RCI, expressing the sum of inferred ghost lineages across a tree as a fraction of the total range of possible ghost lineage values based on a common stratigraphic dataset. Other metrics have been devised but they have not been widely applied, either without reason (as in the case of the Manhattan Stratigraphic Measure (Siddall 1998), or because of concerns regarding their efficacy, such as Spearman Rank Correlation (SRC; Gauthier *et al.* 1988; Norell and Novacek 1992).

The SRC has been criticized regarding its appropriateness (Huelsenbeck 1994), its contingency upon temporal spacing (Benton and Storrs 1994; Hitchin and Benton 1996), and the procedural requirement of altering the cladogram before analysis (trees must be pectinate for analysis and so more balanced trees must be pruned *a priori* precluding the full analysis; Huelsenbeck 1994). Neither are the SCI and RCI metrics free of potential artefact. The SCI is handicapped by tree balance such that only fully imbalanced trees can achieve the full theoretical range of values 0.00–1.00; perfectly balanced trees also have theoretical maximum SCI score of 1.00, but the minimum value achievable is 0.50 (Siddall 1996, 1997; Wills 1999). The SCI is also affected by the temporal distribution of first occurrences such that if they are all contemporaneous the SCI will equal 1.00, regardless of tree balance; if no first appearances are contemporaneous, the range of SCI is again contingent upon tree balance such that perfectly balanced trees will have a SCI of 0.50, while fully pectinate trees yield the full range of SCI scores (Wills 1999). The effect of tree balance on the RCI is more complex. A perfect RCI score of 100 per cent is possible only if the first appearance of all taxa is contemporaneous and fully pectinate trees fulfilling these stratigraphic requirements will always achieve this score. However, other topologies may not be able to achieve a perfect RCI score even if the component taxa meet these stratigraphic requirements (Wills 1999). The GER controls for the distribution of range data and is also sensitive to tree balance. However, by randomly reassigning the stratigraphic range data over the tree it is possible to assess whether stratigraphy–cladogram congruence is significantly better than random, while holding the potential biases (stratigraphy, taxon number, tree balance) constant (Wills 1999). Permutation tests can also be applied to SCI calculation and the degree to which these metrics deviate from random provides a measure of confidence in their significance (Wills 1999).

The SCI, RCI, and GER (as well as permutation tests for significance of these indices) were calculated for the overall tree and for the internal record of each of the plesions in the overall tree, using Ghosts 2.3 (Wills 1999) and the results are presented in Table 10.2. Dates for chronostratigraphic boundaries used in the stratigraphy file for the program were obtained from Tucker and McKerrow (1995), Gradstein and Ogg (1996), Saylor *et al.* (1998), Tucker *et al.* (1998), Cooper (1999), Encarnación *et al.* (1999), Knoll (2000), and Remane (2000). Internal relationships of the plesions used in the analysis are presented in Appendix 10.2; the full data matrix, as well as the associated stratigraphy files and the occurrence data on which the cladogram–stratigraphy correlation metrics are based, are available from the senior author upon request.

Results: internal assessment of the quality of the early chordate fossil record

The combined results of the analyses outlined above are presented in Figure 10.5 and imply that, overall, the fossil record of early chordates is much better than has been suggested previously. The SCI analysis indicates that approximately two-thirds of the cladogram nodes are consistent with stratigraphic data, and the RCI and GER

Table 10.2 Cladogram–stratigraphy metrics calculated using Ghosts (Wills 1999) for the plesions included in the main analysis based upon hypotheses of relationships included in the appendix

Taxon	n	SCI	SCIsig	RCI	Gmin	Gmax	MIG	GER	GERsig
Echinodermata	16	0.5	0.95	81.319555	349	1552	705	0.70407	0.975
Chordata	18	0.6875	0.99	94.803759	141	1569	282	0.901261	0.99
Conodonta	37	0.685714	0.99	65.305011	177	2258	637	0.778952	0.99
Anaspida	5	0	0	−33.333333	15	60	60	0	0
Heterostraci	18	0.625	0.99	47.826087	67	965	132	0.927617	0.99
Thelodonti	4	0	0	48.076923	24	54	54	0	0
Galeaspida	7	0.461538	0.876	35.97561	33	346	105	0.769968	0.99
Osteostraci	25	0.5	0.435	63.186813	46	214	134	0.47619	0.94
Placodermi	22	0.55	0.713	61.977186	36	400	200	0.70896	0.993
Chondrichthyes	14	0.5	0.606	72.643375	140	552	415	0.332524	0.736
Acanthodii	8	0.5	0.525	66.924565	61	282	171	0.502262	0.75
Actinopterygii	15	0.230769	0.681	−97.321429	356	777	442	0.795724	0.99
Basal Synapsida[a]	13	0.727273	0.975	80.96	61	462	107	0.885287	0.5
Basal Diapsida[b]	8	0.83	1	40.3	61	241	74	0.927778	0.5

'n' is the number of terminal taxa that the metrics are based upon, 'SCI' is the Stratigraphic Consistency Index (Huelsenbeck 1994), 'SCIsig' is the significance that the SCI value is better than random, 'RCI' is the Relative Completeness Index (Benton and Storrs 1994), 'Gmin' and 'Gmax' are the minimum and maximum possible summation of the temporal ranges of the terminal taxa included in the analysis based upon a rearrangement of the terminals such that they achieve best- and worst-possible fit to stratigraphy. 'MIG' is the Minimum Implied Gap based upon the given topology of relationships and stratigraphic data, 'GER' is the Gap Excess Ratio (Wills 1999) and 'GERsig' is the significance that the RCI and GER values are better than random. The basal synapsid and diapsid metrics were calculated as part of a study by Benton and colleagues, including Benton and Hitchin (1996), and further details can be found at the following url: <http://palaeo.gly.bris.ac.uk/cladestrat/reptiles.html>.

[a] Based on the hypothesis of relationships from Modesto (1995, fig. 19A).
[b] Based on the hypothesis of relationships from de Braga and Reisz (1995, fig. 6).

analyses both indicate that the record is approximately complete. Permutation tests reveal that these values are not significantly worse than random ($P > 0.99$). However, analyses of the stratigraphic data and their correlation to cladograms of the internal relationships of the operational taxa indicate that the quality of the record varies from group to group. For instance, the fossil record of the living jawless vertebrates and invertebrate chordates is so poor that at 95 per cent confidence the first appearance of these groups can only be constrained within an interval that predates the origin of the Earth in some groups (lampreys), and the origin of the universe in others (hagfishes). This is not an altogether surprising result given that these organisms are entirely soft-bodied and the chances of their preservation in the fossil record are very low. However, the same cannot be said for taxa more derived than lampreys, all of which possess a mineralized, and therefore readily fossilizable, component to their anatomy. Again, the fossil record of these groups is of variable quality and the metrics offer conflicting interpretations of the dataset. For instance, confidence limits suggest that our knowledge of the conodont fossil record is very mature; at 95 per cent confidence the first appearance of the group suggests that it lies within a bracket of 346 kyr of the first stratigraphic appearance, and at 99 per cent confidence, within a bracket of 532 kyr of this datum, both of which are beyond the limits of stratigraphic resolution within this interval.

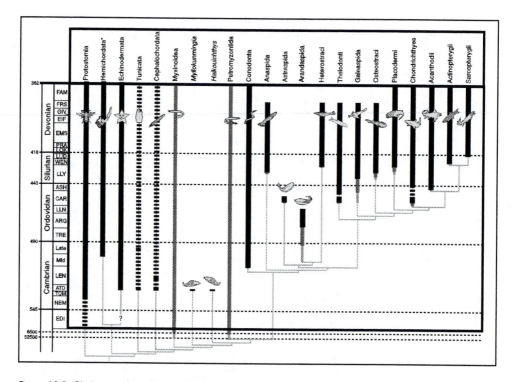

Figure 10.5 Cladogram from Figure 10.3 calibrated to time and including 95 and 99 per cent confidence intervals on stratigraphic range data; medium thickness dark grey lines represent 95 per cent confidence intervals and thin light grey lines represent 99 per cent confidence intervals.

The GER value is less supportive, but nevertheless indicates that the cladogram–stratigraphy correlation is close, within 0.77 of a tree constructed solely on the basis of stratigraphic order of appearance. However, the other cladogram–stratigraphy metrics suggest that the record is only moderately complete; the SCI indicates that only two-thirds of the cladogram nodes are stratigraphically consistent and the RCI suggests that the known record is only two-thirds complete. All values achieve 99 per cent confidence that they are no worse than random. Taking another example of the better groups, assessments of the heterostracan fossil record have also yielded conflicting results; at 95 per cent confidence, the bracket on first appearance is only 278 kyr, and 432 kyr at 99 per cent confidence. This compares well with the GER which indicates that cladogram–stratigraphy correlation is within 0.92 of a tree perfectly concordant to the stratigraphic data. However, the SCI and RCI metrics tell a very different story: less than two-thirds of cladogram nodes are stratigraphically consistent and the heterostracan fossil record represents less than half of the hypothesized evolutionary history of the group.

Although both examples show the same conflicting pattern between metrics, they probably result from two different artefacts in the datasets. Knowledge of the intrarelationships of conodonts is at a relatively immature stage and all existing schemes are based on stratophenetic analysis of the dataset. It is surprising, therefore, that there is not a better tree–stratigraphy correlation. However, in the process of converting the published phylogeny (Sweet 1988, and in Sweet and Donoghue 2001), some 'hard' polytomies have been converted to 'soft' polytomies for the purposes of analysis and, thus, ghost ranges have been artificially extended and statistical scores are artificially lower than might be expected. On the other hand, the intrarelationships of heterostracans have been analysed independently of stratigraphic data (Blieck 1984; Blieck *et al.* 1991; Janvier and Blieck 1993; Janvier 1996b) and although the cladogram–stratigraphy correlation of well-understood groups is good, the SCI and RCI metrics are depressed because many taxa are too poorly known to be included in phylogenetic analyses and have, thus, been placed in a soft polytomy in the most derived position likely, based on character distribution across existing trees. However, given the method of tree construction, it is likely that the short confidence interval and high GER value provide a better assessment of the completeness of the heterostracan fossil record than do the SCI and RCI metrics. In short, the fossil record of both groups is probably very complete but relatively poorly understood, the lack of understanding arising from poor quality data in heterostracans and from poorly resolved relationships in both groups (cf. Benton *et al.* 1999).

Conodonts and heterostracans are important for providing constraints on molecular estimates for the divergence of the living jawless vertebrates and jawed vertebrates, and it is therefore a happy coincidence that they appear to possess a fossil record that exhibits internal consistency. Assessments of the quality of the record of groups that might constrain the divergence of crown-group jawed vertebrates suggest that implied divergence dates may be less reliable. Groups such as the osteostracans, placoderms, and chondrichthyans bracket this diversification event, and potentially provide important upper bounds on divergence timing. However, cladogram–stratigraphy correlation in these groups is poor, generally at a level of 50 per cent for the SCI, at ≤ 0.5 for the RCI (except placoderms which appear to have a fossil record that is internally more consistent), and with a GER ≈ 0.3–0.5; SCI values do not pass a 95 per cent confidence test to determine whether they are not significantly worse than random, although GER and RCI values are generally no worse than random at the same confidence level. Positive correlation between high and low GER and SCI values suggests that poor cladogram–stratigraphy correlation does not arise solely from cladogram inaccuracy, which would normally produce an inverse correlation, although cladogram inaccuracy is possibly an important factor. It is more likely that correlated low GER and SCI values reflect a genuinely poor fossil record and this is corroborated by relatively long confidence intervals on stratigraphic occurrence data in, for example, chondrichthyans, which have a confidence bracket of >1.8 myr. However, this calculation is based on the compilation of Ordovician and Silurian occurrences, only 9 of the 67 of which are Ordovician, and these occurrences are limited to a narrow interval in the Caradoc (Harding Sandstone and its equivalents). Thus, the fossil record of this group appears to be particularly intermittent early on, and a reassessment of confidence limits on first appearance based upon the Ordovician record alone results in a bracket of over six million years. The same situation is true of thelodonts. Inverse correlation between low SCI ($P > 0.95$) and high GER ($P < 0.95$), as in the placoderms and actinopterygians, probably results from a good, but poorly understood, fossil record. The fossil record of placoderms is rich, but attempts to resolve the relationships of the group have thus far proved only variably successful (e.g. Goujet and Young 1995; Goujet 2001). The fossil record of actinopterygians is more gap than record, hence the strongly negative RCI ($P < 0.99$), but the sum of implied ghost ranges is very low compared with the maximum, and very close to the minimum possible by optimizing stratigraphic fit/discordance to the tree (GER 0.79; same P as for RCI).

Problems with assessing the quality of the record

There are two potential problems with regard to this analysis, one relating to the analysis itself, and the second relating to potential artefact in the dataset. First, there is a very poor correlation, absolutely and proportionally, between the confidence intervals on each of the groups, which are derived from internal assessments of the quality of the record within each of the plesions, and the inferred ghost lineages, which are based on analysis at plesion level (compare Figures 10.4 and 10.5). Paul (1998) suggested that this may be an appropriate means of identifying ghost lineages that are an artefact of cladistic methodology, rather than reflecting a true gap in the temporal record of a lineage. We outlined earlier why we think that our analysis is not subject to this kind of artefact (plesions are monophyletic).

The second problem relates to the dataset and has implications for the analysis of confidence intervals and, in turn, their degree of fit to ghost lineages. The calculation of classic confidence intervals assumes that fossil recovery potential is random. Testing this assumption is very difficult when dealing with global compilations of palaeontological data and probably represents the greatest limitation upon the extension of confidence intervals to global datasets. Nevertheless, there is some evidence to suggest that there are two significant biases in the dataset, indicating that the existing dataset is not a random sample of the fossil record. First, the vast majority of known occurrences are from northern Europe, the USA, and South-East Asia, compared with a global fossil collection bias for north-west Europe and North America (e.g. Smith 2001). Although there are numerous fossil records from North America as a whole, the vast majority of taxonomic treatments of North American faunas (especially Arctic Canada) are new taxa, suggesting that although the North American record is being recovered rapidly, it has been sampled only sparsely to date (using the collecting curve analogy we remain on the steep component of the curve). A bias against collecting central Asian faunas appears to be supported by records of spot occurrences in terranes such as Tuva (Afanassieva and Janvier 1985). A virtual absence of 'ostracoderm' faunas, bar thelodonts, from Gondwana after the Ordovician may also suggest a dearth of collecting. However, many basins have been densely sampled, particularly for conodont biostratigraphy, to little avail (the exception to this being the enigmatic pituriaspids; Young 1991). It would appear that the absence of records from this interval does reflect the real absence of most 'ostracoderm' groups in Gondwana during this time (for further discussion see Smith *et al.* 2002). Thus, there is a systematic bias in the sampling of geographical regions, but some gaping holes in the regional distribution of fossil sites result from primary signal rather than an absence of sampling.

Another source of evidence supporting a non-random fossil record stems from the fact that the distribution of many, or even most, groups was facies controlled. Given the differential preservation potential of facies with sea level change, it would be expected that the recovery potential and, thus, the stratigraphic distribution of facies-controlled fossil taxa would be similarly affected (Holland 1995). While the only recourse to removing a geographical collecting bias is systematic sampling of unsampled regions, the effect of a non-random record upon the calculation of confidence intervals on stratigraphic data may be readily overcome, at least in principle. This is achieved through abandoning the uniform recovery potential assumption of classic confidence limits (Paul 1982; Strauss and Sadler 1989; Marshall 1990) and replacing it with a fossil recovery potential function that reflects secular bias resulting from, for example, sea level change (Holland 1995; Marshall 1997). Devising this function can be non-trivial, but in many cases it may be simplified on the basis that it is only change with stratigraphic position that is significant (Marshall 1997).

Our attempts to implement the 'generalized' method of calculating confidence intervals failed on a number of counts. First, the method requires that the stratigraphic position of each fossil occurrence is known with a degree of precision that is not possible with the global dataset of early vertebrates; the stratigraphic position of some occurrences cannot be resolved even to series level. Second, fossil recovery potential functions are incalculable at the taxonomic level at which our analysis has been undertaken. Many of the component lineages (e.g. heterostracans and osteostracans) exhibit an ecological shift through time and phylogeny (Blieck and Janvier 1991; Smith et al. 2002) and so it would have been necessary to divide plesions into much lower taxonomic levels for which fossil recovery potential curves could be produced and implemented. The conflation of these two variables precluded analysis of the entire dataset. As a fallback, and given that it is the time of first appearance of groups that is germane to this study, it was our intention to confine application of the generalized method to the pre-Silurian record alone. This objective is more easily achieved because the secular distribution of vertebrates is much better constrained for the Cambro-Ordovician (mainly because the records are entirely marine), and the ecologies of taxa are less complex than for post-Ordovician vertebrates. However, while it is possible to derive fossil recovery probabilities for each lineage, the calculation of fossil recovery potential functions is precluded by almost total absence of agreement over a eustatic sea level curve for the interval. While we intend to remedy this problem in the near future, it is beyond the scope of the present study. In the interim, we have observed elsewhere (Sansom et al. 2001; Smith et al. 2002) that intracontinental occurrences of Ordovician vertebrates in Laurentia are confined to eustatic highstand episodes. Thus, although it has not proved possible to quantify confidence intervals that consider systematic bias in groups that have their first records in the Ordovician, we may conclude that the base range of the Ordovician groups (bar conodonts) would be revised downwards. To provide constraint on the lower limit of first appearance we note that the absence of records from preceding highstand episodes is significant.

Comparison of molecular and fossil estimates

Origin of chordates, craniates, and vertebrates

Inferences regarding the time of origin of these clades are hampered by the perennial problem of first appearances clustering in the Atdabanian (mid-Early Cambrian). The absence of outgroup representatives of greater age precludes further interpretation

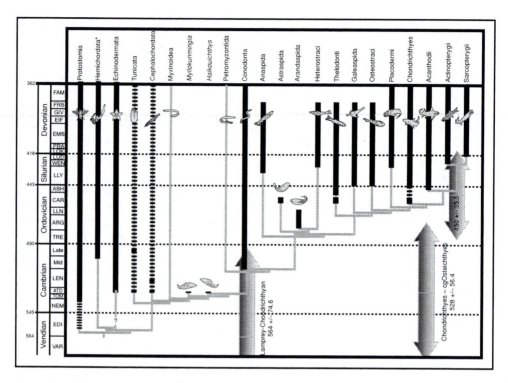

Figure 10.6 Observed stratigraphic range data including inferred ghost lineages and molecular estimates on the divergence of various clades (from Kumar and Hedges 1998).

beyond the conclusion that representatives of these clades are observed, or can be inferred to have been present, at this time (530 Ma). There are two reasons for equivocation over this date. First, putative echinoderm remains are known from the Proterozoic (e.g. *Arkarua* Gehling 1987), possibly providing evidence for a chordate ghost lineage extending back to the Neoproterozoic. Second, the fossil record of these groups is so poor that internal assessments of confidence limits ($P > 0.95$) place a bracket on the evolutionary origin of the cephalochordates, hagfishes and lampreys that is sufficiently broad to encompass any hypothesis that is compatible with their origin within the constraints provided by the origin of the Earth and/or Universe, as well as some that are not. Thus, without recourse to negative evidence, the fossil record is mute with regard to a judicious lower constraint on the timing of origin of chordates, craniates, and vertebrates.

The absence of firm palaeontological data is unfortunate because molecular estimates for the diversification events are strongly discordant with the available evidence (Figure 10.6), although molecular estimates also differ from one another by just as great a degree. The earliest molecular estimate for the divergence of chordates from their sister clade, the Ambulacraria (echinoderms plus hemichordates), is 1001 Ma (Wray *et al.* 1996), while the latest is 590 Ma (Feng *et al.* 1997). The average estimate is 722.75 Ma with a standard deviation of 192 myr ($n = 4$). The time of divergence of craniates and acraniates has been addressed in only two analyses, yielding estimates of 700 Ma (Nikoh *et al.* 1997) and 751 Ma (Hedges 2001). These analyses are non-independent in that the sequences used by Nikoh *et al.* (1997) were also used by Hedges (2001), but yielded results from individual sequences that differ by as much as 100 Ma that can be accounted for by differences in calibration dates and analytical techniques (Hedges 2001). The calibration points used by both analyses are questionable in that they use molecular estimates and, thus, do not provide an adequate or independent test of molecular clock theory. Furthermore, the calibration date for the divergence of arthropods and vertebrates (700 Ma) used by Nikoh *et al.* (1997) is derived from Dayhoff (1978) who provided no form of substantiation for a date that does not accord with any palaeontological data. In addition, the use of molecular calibration points in this analysis appears to render the analytical argumentation circular since the molecular calibration points are applied to molecular sequences that were used in calculating the molecular estimate. Specifically, Hedges (2001) uses calibration points derived from Kumar and Hedges (1998; and Wang *et al.* 1998) that are based, in part, upon aldolase and TPI, which are analysed in Hedges (2001). Given that the only palaeontological calibration point used by Kumar and Hedges (1998) is the bird–mammal divergence date of 310 Ma, Hedges (2001) has, in effect, calibrated multiple times, directly and indirectly, using a single palaeontological datum.

No molecular estimates have been calculated for the divergence of invertebrate craniates and vertebrates. This is because molecular phylogenies consistently resolve hagfishes and lampreys as monophyletic and, hence, there is no distinction between craniates and vertebrates within the field of molecular systematics. However, Hedges (2001) used a subset of his data to provide an estimate for the divergence of hagfishes and lampreys within the context of cyclostome monophyly and arrived at a date of 499 ± 36.8 Ma, using the lamprey–gnathostome molecular estimate of 564 Ma (Kumar and Hedges 1998) as the calibration point. For once, palaeontological data indicate two older spot dates (530 Ma on *Myllokunmigia* and *Haikouichthys*) with a slightly younger and better-constrained date of 495 Ma bracketing the molecular estimates and falling well within the standard error on the molecular estimate.

Overall, in consideration of the origin of chordates and divergence of the lower chordate groups, there is poor correlation both between molecular and palaeontological estimates, and between molecular estimates that are based upon different datasets. Thus, in the absence of any independent constraint that the fossil record may otherwise afford, it would be dangerous to conclude anything other than the fact that chordates, craniates, and vertebrates had diverged by 530 Ma. Furthermore, given the low stochastic value of the calculated error limits on published molecular clock studies, there can be no confidence that the origin of the clade was not significantly before such estimates, based on molecular data alone.

Origin of gnathostomes

Although there is little temporal constraint on the origin of vertebrates afforded by the fossil record, internal assessments of the consistency of the record point to a dramatic increase in the quality of the record after the divergence of the lineages leading to lampreys and living jawed vertebrates (crown-gnathostomes). This coincides with the appearance of mineralized skeletonization within the gnathostome stem-lineage. The fossil record of conodonts exhibits remarkable internal consistency and the first appearance of this basal member of the gnathostome stem-lineage affords an inferred 495 Ma constraint on the latest possible date for divergence, with a confidence bracket of just 346 kyr ($P > 0.95$; and only 532 kyr with $P > 0.99$). The absence of a sister taxon with even an approximately comparable fossil record precludes the possibility that this date can be corroborated through inference of a ghost lineage. The absence of a stepwise geological appearance of successive sister taxa within the sister clade to the Conodonta is problematic; the conodont fossil record implies a significant ghost lineage amongst many of these groups. This signal is in agreement with internal assessments of the quality of the record of these groups (which conclude that it is a poor reflection of the evolutionary history of these groups), the fact that all well-known taxa fall within known clades within the stem-lineage (rather than as individual plesions on the stem-lineage), and knowledge of a large number of vertebrate remains of this age that cannot be assigned to known groups (e.g. Sansom et al. 2001).

Only two molecular clock analyses have addressed the lamprey–gnathostome divergence. Wray et al. (1996), who famously estimated the divergence of Bilateria at 1200 Ma, estimated the divergence of lampreys and gnathostomes at 599 Ma, while Kumar and Hedges (1998) suggested a date of 564 Ma. The entire analysis undertaken by Wray et al. (1996) was reanalysed and robustly criticized by Ayala et al. (1998), who revised upwards all the divergence estimates. This is in accordance with the analysis undertaken by Kumar and Hedges (1998) which, given the enormity of the dataset, must be considered the most robust analysis undertaken to date. Despite the poor internal support for palaeontological dating of the origin of gnathostomes, the average date given by Kumar and Hedges (1998; 559 Ma) corresponds well to the palaeontological estimate, at least when considered within the context of a standard error of ±74.6 myr on the molecular estimate which encompasses an interval extending from the earliest Ordovician to the Neoproterozoic (484.4–633.6 Ma).

Origin of crown-group gnathostomes

Inference of the time of origin of crown-gnathostomes is complicated by equivocation over the affinity of shark-like microremains from the Late Ordovician and early Silurian (for a discussion see Janvier 1998; Sansom et al. 2000, 2001; Smith et al. 2002). These scales are identified as chondrichthyans (total-group Chondrichthyes) on the basis that they possess neck canals and exhibit rigidly patterned areal growth, at present a chondrichthyan synapomorphy and symplesiomorphy, respectively. However, it is not known whether the inclusion of these taxa renders the group paraphyletic or, indeed, whether these characters are exclusive to crown-gnathostomes; only further resolution of the anatomy of these taxa will lead to a resolution of their precise placement within total-group gnathostome systematics. In the meantime, all available evidence suggests that they are representatives of total-group Chondrichthyes and, thus, crown-gnathostomes; by inference they place a lower constraint on the divergence of crown-gnathostomes at 457 Ma, but with a confidence interval extending to 463.4 Ma ($P > 0.95$; accepting the confidence interval based upon the Ordovician record alone).

A molecular estimate for the time of divergence between Chondrichthyes and Osteichthyes, and hence, the time of origin of crown-gnathostomes, was calculated by Kumar and Hedges (1998) at 528 ± 56.4 Ma, encompassing an interval from the mid-Ordovician (471.6 Ma) to late Neoproterozoic (584.4 Ma). The palaeontological estimate derived from a literal reading of the record (457 Ma) narrowly misses the lower bound on the bracket provided by the standard error on the molecular estimate. Confidence limits provided by the Ordovician record alone extend the predicted first appearance closer to the lower bound on the molecular estimate, but the two do not overlap.

Origin of actinopterygians and sarcopterygians

The earliest actinopterygian remains are dated at 425 Ma, and are part of a rich record (thanks, again, to the application of micropalaeontological techniques; e.g. Schultze 1968; Gross 1969; Mårss 1986; Fredhölm 1988a,b). This confers a remarkably short confidence interval (425.58 Ma at $P > 0.95$) on first appearance. The record of pre-Devonian sarcopterygians is much poorer, with only a single known fossil horizon (Zhu and Schultze 1997). With such a poor record, the obvious implication is that the true range extends much further back than present evidence indicates. This is supported by the observation that these occurrences were palaeogeographically remote from each other. However, this earliest sarcopterygian record (423 Ma) is remarkably consistent with the extent of the range of actinopterygians such that there is very little inferred ghost lineage. Thus, we accept 425.58 Ma as a firm lower bound on the origin of crown-osteichthyans and the divergence of two osteichthyan clades.

Kumar and Hedges' (1998) estimate for the divergence of the two extant osteichthyan clades is 450 Ma ± 35.5 myr, encompassing an interval extending from earliest Ordovician (485.5 Ma) to Early Devonian (419.5 Ma). This compares well with the palaeontological estimate, although internal assessments of the quality of the record, particularly of sarcopterygians, suggest that palaeontological data may eventually converge on the mid-range of this molecular estimate.

Discussion

Correspondence between molecular clock estimates for the timing of divergence and palaeontological data indicating the minimum possible date for divergence is very variable. There is a clear correspondence between molecular and palaeontological estimates where there is *a priori* evidence for confidence in the fossil record based upon internal assessments of its quality based upon stratigraphic data alone and the relationship between stratigraphic data and cladogram structure (e.g. the lamprey–gnathostome and actinopterygian–sarcopterygian divergences). Concomitantly, where there is *a priori* evidence for a lack of confidence in the quality of the record on the basis of internal assessments, the molecular and palaeontological estimates are in discord.

Where there is disagreement between palaeontological and molecular estimates it is difficult to reconcile which dataset provides the best approximation of true time of divergence of a particular clade. Palaeontological estimates are limited by their reliance upon negative evidence and although quantitative methods are being developed to assess the plausibility of range extensions in the face of sampled, but barren time intervals (Weiss and Marshall 1999), they are at present limited by the assumptions on which they are based, many of which are extremely controversial. On the other hand, given that it is difficult to reconcile between competing molecular estimates, it is not surprising that it is difficult to arbitrate between palaeontological and molecular estimates. This is partly because, as scientific theories, molecular clock calculations are extremely poorly formulated and, thus, are difficult to test. In many instances, one molecular hypothesis is preferred over another on the basis that it is derived from the greatest dataset, relying upon a law of large numbers approach to molecular clock mechanics (cf. Rodríguez-Trelles et al., Chapter 1), rather than a neutrality theory basis (Zuckerkandl and Pauling 1962, 1965). Thus, they are testable only by other molecular clock calculations, based upon larger, more universal datasets and/or the falsification or augmentation of calibration points. Smith and Peterson (2002) have suggested an explanation for the discrepancy between molecular and palaeontological temporal divergence estimates, arguing that they reflect two quite distinct events, with molecular clocks estimating the time of origin of a clade, and palaeontological estimates recording the diversification of the clade, which they equate to the origin of the total-group and origin of the crown-group – placing undue weight on the evolutionary significance of crown-groups. This follows the widespread assumption that most molecular clock estimates pertain to total-group divergence, but total-groups and crown-groups are hierarchical such that one taxon's total-group is the next more inclusive taxon's crown-group and vice versa. Thus, there is no better correlation between molecular and palaeontological estimates for the origin of crown-groups than total-groups, and the rapprochement fails.

Even when palaeontological and molecular estimates are comparable, molecular clock analyses consistently yield a date that is considerably older than the palaeontological data indicates (except in the instance of the hagfish–lamprey divergence estimate within the context of cyclostome monophyly). Thus, the fossil record of early chordate evolution is either consistently missing the early history of various chordate clades or molecular clock dates consistently overestimate the true time of cladogenesis. To some extent this should be expected. First, because no one argues that the earliest fossil record equates to the origin of a clade; there is a cryptic evolutionary history to all clades, the critical issue is the temporal extent of this period of unrecorded evolutionary history. Second, in a strict interpretation of molecular clock theory, such calculations estimate the time of divergence based on a fossil record comparable with that on which the 'clock' is calibrated, not the true time of origin of a clade and, hence, it has been argued that molecular clock estimates should be conservative. All of the clades in our analysis exhibit stratigraphy–cladogram congruence metrics that are worse than the fossil record of the calibration point on which most molecular clocks are calibrated, the divergence of bird–mammal lineages at 310 Ma (Table 10.2; but see Lee 1999; it should also be noted that although the basal synapsid and diapsid fossil records perform well in the SCI and GER indices, it is likely, given the patchy nature of the record – betrayed by the particularly low RCI for the diapsid lineage – that confidence intervals on the stratigraphic range data for the various plesions would be extensive).

However, there is some circumstantial evidence to suggest that the fossil record of early vertebrates, and total-group gnathostomes in particular, may be more reliable than we would otherwise assume. This stems from the rather surprising degree of correspondence between molecular estimates and palaeontological data, at least in terms of the chronologically consistent ordering of palaeontologically based estimates for the first appearance of successive extant clades. While molecular estimates for the divergence of successive clades have to be chronologically consistent, by definition (they are based upon a direct extrapolation from a hierarchical dataset), the same does not hold true for the fossil data. Indeed, the temporal distribution of fossil remains will only be chronologically consistent if their ordering reflects the hierarchy of evolutionary relationships – which they will do only if the fossil record is preserved with high fidelity. There is also evidence to suggest that molecular clocks may consistently overestimate the date of divergence of clades. This can occur for two non-mutually exclusive reasons. First, constraints on molecular clock estimates are asymmetrical, i.e. they are bound to be non-negative but there are no such constraints at the upper end of the spectrum (Rodríguez-Trelles et al. 2002). Second, overestimation of divergence timing arises from the accumulating inaccuracy associated with extrapolating farther and farther from the calibration date (Springer 1997; Nei et al. 2001) and especially concerns analyses that use single internal palaeontological calibration dates, although it also affects analyses that use multiple external and/or internal calibration dates that are derived from a single palaeontological calibration date. This may be a particular weakness of the analysis by Kumar and Hedges (1998); although encompassing by far the greatest number of sequences in calculating divergence times (658), very few of these were used in calculating the timing of the very oldest divergence events (13 for the origin of jawed vertebrates, 15 for the origin of crown-gnathostomes, 44 for the origin of crown-osteichthyans).

The alternative view, that the fossil record of early chordates is particularly poor, is reflected by the fact that plesiomorphic chordate and deuterostome anatomies have been the subject of debate for over a century and yet the subject remains resolutely intractable. In addition, very few stem-chordates, stem-ambulacrarians, and/or stem-deuterostomes have been identified from the fossil record (e.g. Jefferies et al. 1996; Gee 2001; Shu et al. 2001b) and none (arguably) have escaped critical examination (Ruta 1999; Lacalli 2002). Finally, questions of chordate, craniate, and vertebrate divergence timings cannot be resolved in isolation while debate over the veracity of the Cambrian 'explosion' continues (e.g. Smith 1999; Budd & Jensen, Chapter 9).

The other molecular estimate that exhibits poor correspondence to palaeontological data, the divergence of crown-gnathostomes, predicts a Late Cambrian event and, by inference, a hitherto unrecorded interval of crown-gnathostome evolutionary history that spans the Late Cambrian–Middle Ordovician. Significantly, this interval coincides with recent discoveries of a swathe of new vertebrate taxa, which are assignable to the gnathostome total-group, but no further, based upon the available evidence (Sansom et al. 2001). It is quite possible that these new taxa include further Ordovician representatives of crown-gnathostomes (e.g. Skiichthys Smith and Sansom 1997). The Late Cambrian–Middle Ordovician gap in the record overlaps well with gaps in the records of other groups, for example, the echinoderms (Smith 1988a). This hints at a secular bias in the fossil record as a whole that probably reflects the fact that imperfections in the fossil record are rooted in imperfections in the rock record (cf. Holland 1995, 2000; Smith 2001; A.B. Smith et al. 2001).

Implications for established hypotheses and scenarios

Understanding early chordate evolution using an incomplete fossil record

If nothing else, molecular clocks have provided the stimulus for palaeontologists to look at their datasets anew and provide justification for cherished methodologies. This, in turn, has provided the impetus for the development of old and new methods for assessing the completeness of the fossil record. These internal assessments provide a means of determining degrees of confidence in subsets of a dataset, provide caution-ary limits in reading the evolutionary history of particular clades, and provide predic-tions in our attempts to recover missing components of the record.

The greatest concern of palaeontologists with regard to the mismatch of molecular clocks and the fossil record may be that it indicates not only that the fossil record is substantially incomplete but also, critically, that it is the initial period of the evolu-tionary history of these clades that is missing. In such a case, would not the usefulness of the fossil record, in uncovering the sequences of character change between extant clades, be compromised? The simple answer to this question is no. The chief value of the fossil record is that it reduces error in inferring the sequence of character changes that underlie the establishment of living clades – this has been integral to testing and rejecting models such as, for example, the origin of paired appendages within verte-brates (Coates 1994). Whether or not we have a complete sample of the anatomical designs that have been realized is not relevant; with the fossil record we have a more complete, and continually expanding, understanding of chordate evolution than would be possible using only the living biota. Furthermore, fossils help to prevent the identification of homoplasy as homology in living members of distantly related groups, and identify homologies that might not otherwise be recognized because of the hundreds of millions of years of evolutionary change that has occurred subsequent to the divergence of the clades.

Our understanding of early chordate evolution may well be incomplete but it does not follow that it is incorrect. Further attempts to reconcile the fossil record with the living biota will lead to further refinement not only to the temporal scale of early chordate evolution but also to our understanding of the sequence of character changes that shaped all subsequent events in chordate phylogeny.

Neoproterozoic refugia and the origin of vertebrates

One inevitable development of molecular clock estimates is that attempts are being made to link intrinsic evolutionary change to extrinsic environmental factors. For instance, van Tuinen *et al.* (1998) proposed that the origin of ratites is not just coin-cident with, but inextricably linked to, the separation of Africa and South America during the Early to mid-Cretaceous (see Cracraft 2001 for an excellent analysis). More recently, Hedges (2001; Chapter 2) proposed that the proximity of the molecular clock estimate for the divergence of crown-vertebrates to radiometric dates for the first major Neoproterozoic glaciation (Sturtian; 750–700 Ma) may not be coincidental. It is argued that both the Sturtian and Varanger glaciations (610–570 Ma) would have led to contraction in the topological range of species and, through long-term genetic isolation in small refugia, to considerable speciation. As worthy as this approach may be in demonstrating an integrated approach to the questions of when, where, how, and why vertebrates first evolved, there are two significant problems with regard to this linkage of intrinsic evolutionary and extrinsic environmental factors. First, the nature, timing, and tempo of the Cryogenian period of the Neoproterozoic is utterly unresolved, in terms of the timing, duration, and number of glaciation episodes (Knoll 2000). Second, and more intractably, the standard errors on molecular estimates are currently so vast (and unrealistrically conservative) that they render worthless any attempt to match biotic events to radiometrically dated environmental events.

Evolutionary scenarios based upon palaeontological dating

Although molecular estimates fail to provide the necessary temporal constraint to under-pin attempts to uncover any possible link between intrinsic evolutionary events and extrinsic environmental events, palaeontological data provide no panacea either, at least with regard to the origin and early evolution of vertebrates. It has been recognized for many years that evolutionary history cannot be read directly from the rocks, but many scenarios for the origin of major clades remain current, even though the supporting data have not expanded from those on which they were originally contrived. For instance, Romer's celebrated 'eurypterid influence on vertebrate history' (Romer 1933) is based upon the co-occurrence and vaguely comparable diversity trends of eurypterids and the then earliest skeletonizing vertebrates in the Silurian. Thus, the origin of the skeleton has been attributed to the selection-based effect of predating eurypterids upon early vertebrates. But not only are the earliest known skeletonizing vertebrates now Cambrian in age, and the earliest undisputed 'armoured' vertebrates Ordovician in age, but our phylogenetic tests and internal assessments of the consistency of stratigraphic data both reveal that these lineages probably existed even earlier. Thus, the co-occurrence and evolutionary history of vertebrates and eurypterids is no longer apparent and Romer's evocativer theory must finally be laid to rest.

Similarly, it has been argued that the rise of jawed vertebrates and apparently concomitant demise of skeletonizing jawless vertebrates is the result of competitive displacement (for a summary see Purnell 2001). However, our analyses reveal an extensive cryptic history of early jawed vertebrates that has not been considered in the formulation of the theory, or in attempts to test it. Furthermore, it may not be possible to test such hypotheses adequately on the basis of the currently available dataset.

The bottom line with regard to attempts to link intrinsic and extrinsic events in early vertebrate evolution is that although there are many interesting questions that can be asked, it may not be appropriate to try and answer some of them based upon the available palaeontological dataset, and molecular clock analyses do not at present appear to be even close to capable of overcoming these shortcomings.

Conflict, compromise, or consilience?

Increase in the application of molecular clock theory has led to a considerable period of introspection amongst the palaeontological community, from which two main camps have emerged. There are those who reject molecular clock estimates outright and contend that only the fossil record can provide reliable estimates for the divergence of clades, albeit minimum estimates for the timing of divergence events (e.g. Conway Morris 1997, 2000; Budd and Jensen 2000). Others have capitulated entirely to molecular clock estimates, concluding that use of the fossil record is corrupted by its reliance upon negative evidence (Fortey *et al.* 1996, 1997; Smith 1999; Wills and Fortey 2000; Smith and Peterson 2002). However, neither dataset has a monopoly over the other and, indeed, the two datasets have much mutuality. The inextricable linkage between the fossil record and molecular clock theory is no better exemplified than in the need for palaeontological calibration points in molecular clock analyses, whether they are applied directly or indirectly. Above all, the two databases provide a level of rigour that would not be possible in the absence of one or other dataset, such that molecular clock theory and the fossil record are becoming better understood through reciprocal illumination.

Given the degree of latitude offered by standard error on molecular clock estimates and the lack of internal consistency in the fossil record of early chordates, we are no closer to constraining the times of origin of the chordate, craniate, and vertebrate clades. Indeed, it could be argued that we are even further from providing constrained estimates on the origin of these clades than we were at the outset. Thus, although we understand relatively well what is currently known of early chordate evolution, it appears that what is currently known is by no means all there is to know, and this is particularly the case for the invertebrate chordates, basal vertebrates, and stem-gnathostomes within the Late Cambrian–Middle Ordovician, and lower Silurian intervals. While our knowledge of the invertebrate chordate component of chordate phylogeny will remain contingent upon the chance discovery of fossil remains preserved under exceptional conditions, such a restriction does not obtain for the skeletonizing vertebrates, the remains of which were readily entrained in the fossil record. Targeted examination of previously unsampled environments and palaeogeographical realms will be crucial to resolving the evolutionary history of early vertebrates and stem-gnathostomes in particular. At the same time, development of molecular clock theory, more rigorous composition of molecular clock analyses as scientific hypotheses for testing, and the inclusion of more sequences representative of basal chordates and sister groups are likely to provide better constraints on their time of origin. A more realistic attempt to assess errors on molecular clock estimates is required and this can be developed in hand with more rigorous assessments of the palaeontological data used in calibrating molecular clock analyses. However, unless these errors can be reduced, molecular clock estimates will remain of low practical value; the palaeontological record is imperfect but nevertheless provides the only firm constraint on the timing of clade divergence.

Acknowledgements

Philippe Janvier (Museum National d'Histoire Naturelle, Paris) and Mike Coates (University of Chicago) provided useful reviews of the manuscript. Donoghue was funded through NERC Post Doctoral Research Fellowship GT5/99/ES/2; Smith was funded through NERC Research Grant NER/B/S/2000/00284, and Smith and Sansom were funded through NERC Research Grant GR3/10272.

References

Afanassieva, O.B. and Janvier, P. (1985) '*Tannuaspis*, *Tuvaspis* and *Ilemoraspis*, endemic osteostracan genera from the Silurian and Devonian of Tuva and Khakassia (USSR)', *Geobios*, 18: 493–506.

Arduini, P., Pinna, G. and Teruzzi, G. (1981) '*Megaderaion sinemuriense* n. g. n. sp., a new fossil enteropneust of the Sinemurian', *Atti della Società Italiana di Scienze Naturale e del Museo Civico di Storia Naturale di Milano*, 122: 104–8.

Ayala, F.J., Rzhetsky, A. and Ayala, F.J. (1998) 'Origin of the metazoan phyla: molecular clocks confirm paleontological estimates', *Proceedings of the National Academy of Sciences, USA*, 95: 606–11.

Bardack, D. (1991) 'First fossil hagfish (Myxinoidea): a record from the Pennsylvanian of Illinois', *Science*, 254: 701–3.

—— (1998) 'Relationship of living and fossil hagfishes', in J.M. Jørgensen, J.P. Lomholt, R.E. Weber and H. Malte (eds) *The Biology of Hagfishes*, London: Chapman & Hall, pp. 3–14.

Bardack, D. and Richardson Jr., E.S. (1977) 'New agnathous fishes from the Pennsylvanian of Illinois', *Fieldiana Geology*, 33: 489–510.

Bardack, D. and Zangerl, R. (1968) 'First fossil lamprey: a record from the Pennsylvanian of Illinois', *Science*, 162: 1265–7.

—— (1971) 'Lampreys in the fossil record', in M.W. Hardisty and I.C. Potter (eds) *The Biology of Lampreys*, London: Academic Press, pp. 67–84.

Bengtson, S. and Urbanek, A. (1986) '*Rhabdotubus*, a Middle Cambrian rhabdopleurid hemi-chordate', *Lethaia*, 19: 293–308.

Benton, M.J. and Hitchin, R. (1996) 'Testing the quality of the fossil record by groups and habitats', *Historical Biology*, 12: 111–57.

Benton, M.J. and Storrs, G.W. (1994) 'Testing the quality of the fossil record: paleontological knowledge is improving', *Geology*, 22: 111–14.

Benton, M.J., Hitchin, R. and Wills, M.A. (1999) 'Assessing congruence between cladistic and stratigraphic data', *Systematic Biology*, 48: 581–96.

Bleiweiss, R. (1998) 'Fossil gap analysis supports early Tertiary origin of trophically diverse avian orders', *Geology*, 26: 323–6.

Blieck, A. (1984) 'Les Hétérostracés ptéraspidiformes, agnathes du Silurien–Dévonien du continent Nord-Atlantique et des Blocs Avoisnants: révision systématique, phylogénie, biostratigraphie, biogéographie', *Cahiers de Paléontologie, Centre national de la Recherche scientifique, Paris*, 1–199.

Blieck, A. and Janvier, P. (1991) 'Silurian vertebrates', *Special Papers in Palaeontology*, 44: 345–89.

Blieck, A., Elliot, D.K. and Gagnier, P.-Y. (1991) 'Some questions concerning the phylogenetic relationships of heterostracans, Ordovician to Devonian jawless vertebrates', in M.-M. Chang, Y.-H. Liu and G.-R. Zhang (eds) *Early Vertebrates and Related Problems in Evolutionary Biology*, Beijing: Science Press, pp. 1–17.

Budd, G.E. and Jensen, S. (2000) 'A critical reappraisal of the fossil record of bilaterian phyla', *Biological Reviews*, 74: 253–95.

Chen, J.-Y., Dzik, J., Edgecombe, G.D., Ramsköld, L. and Zhou, G.-Q. (1995) 'A possible Early Cambrian chordate', *Nature*, 377: 720–2.

Chen, J.-Y., Huang, D.-Y. and Li, C.-W. (1999) 'An early Cambrian craniate-like chordate', *Nature*, 402: 518–22.

Coates, M.I. (1994) 'The origin of vertebrate limbs', *Development*, 1994 Supplement: 169–80.

Conway Morris, S. (1979) 'The Burgess Shale (Middle Cambrian) fauna', *Annual Review of Ecology and Systematics*, 10: 327–49.

—— (1997) 'Molecular clocks: defusing the Cambrian explosion?' *Current Biology*, 7: R71–4.

—— (2000) 'Evolution: bringing molecules into the fold', *Cell*, 100: 1–11.

Cooper, R.A. (1999) 'The Ordovician time scale – calibration of graptolite and conodont zones', *Acta Universitatis Carolinae Geologica*, 43: 1–4.

Cracraft, J. (2001) 'Avian evolution, Gondwana biogeography and the Cretaceous–Tertiary mass extinction event', *Proceedings of the Royal Society, London*, B268: 459–69.

David, B. and Mooi, R. (1998) 'Major events in the evolution of echinoderms viewed by the light of embryology', in R. Mooi and M. Telford (eds) *Echinoderms: San Francisco*, Rotterdam: A.A. Balkema, pp. 21–8.

—— (1999) 'Comprendre les échinodermes: la contribution du modèle extraxial–axial', *Bulletin de la Société geologique de France*, 170: 91–101.

David, B., Lefebvre, B., Mooi, R. and Parsley, R. (2000) 'Are homalozoans echinoderms? An answer from the extraxial–axial theory', *Paleobiology*, 26: 529–55.

Dayhoff, M.O. (1978) 'Survey of new data and computer methods of analysis', in M.O. Dayhoff (ed.) *Atlas of Protein Sequence and Structure*, Vol. 5, Supplement 3, Washington DC: National Biochemical Research Foundation, pp. 1–8.

de Braga, M. and Reisz, R.R. (1995) 'A new diapsid reptile from the uppermost Carboniferous (Stephanian) of Kansas', *Palaeontology*, 38: 199–212.

Delarbre, C., Barriel, V., Janvier, P. and Gachelin, G. (2002) 'Complete mitochondrial DNA of the hagfish, *Eptatretus burgeri*: the comparative analysis of mitochondrial DNA sequences strongly supports the cyclostome monophyly', *Molecular Phylogenetics and Evolution*, 22: 184–92.

Donoghue, P.C.J., Forey, P.L. and Aldridge, R.J. (2000) 'Conodont affinity and chordate phylogeny', *Biological Reviews*, 75: 191–251.

Donoghue, P.C.J. and Smith, M.P. (2001) 'The anatomy of *Turinia pagei* (Powrie) and the phylogenetic status of the Thelodonti', *Transactions of the Royal Society of Edinburgh (Earth Sciences)*, 92: 15–37.

Durman, P.N. and Sennikov, N.V. (1993) 'A new rhabdopleurid hemichordate from the Middle Cambrian of Siberia', *Palaeontology*, 36: 283–96.

Dzik, J. (1995) '*Yunnanozoon* and the ancestry of the vertebrates', *Acta Palaeontologica Polonica*, 40: 341–60.

Encarnación, J., Rowell, A.J. and Grunow, A.M. (1999) 'A U–Pb age for the Cambrian Taylor Formation, Antarctica: implications for the Cambrian timescale', *Journal of Geology*, 107: 497–504.

Feng, D.-F., Cho, G. and Doolittle, R.F. (1997) 'Determining divergence times with a protein clock: update and reevaluation', *Proceedings of the National Academy of Sciences, USA*, 94: 13028–33.

Forey, P.L. (1984) 'Yet more reflections on agnathan–gnathostome relationships', *Journal of Vertebrate Paleontology*, 4: 330–43.

Fortey, R.A., Briggs, D.E.G. and Wills, M.A. (1996) 'The Cambrian evolutionary "explosion": decoupling cladogenesis from morphological disparity', *Biological Journal of the Linnean Society*, 57: 13–33.

—— (1997) 'The Cambrian evolutionary "explosion" recalibrated', *BioEssays*, 19: 429–34.

Fredholm, D. (1988a) 'Vertebrate biostratigraphy of the Ludlovian Hemse Beds of Gotland, Sweden', *Geologiska Föreningens i Stockholm Förhandlingar*, 110: 237–53.

—— (1988b) 'Vertebrates in the Ludlovian Hemse Beds of Gotland, Sweden', *Geologiska Föreningens i Stockholm Förhandlingar*, 110: 157–79.

Garcia-Fernàndez, J. and Holland, P.W.H. (1994) 'Archetypal organisation of the amphioxus *Hox* gene cluster', *Nature*, 370: 563–6.

Gauthier, J., Kluge, A.G. and Rowe, T. (1988) 'Amniote phylogeny and the importance of fossils', *Cladistics*, 4: 105–209.

Gee, H. (2001) 'Deuterostome phylogeny: the context for the origin and evolution of chordates', in P.E. Ahlberg (ed.) *Major Events in Early Vertebrate Evolution: Palaeontology, Phylogeny, Genetics and Development*, London: Taylor & Francis, pp. 1–14.

Gehling, J.G. (1987) 'Earliest known echinoderm – a new Ediacaran fossil from the Pound Subgroup of South Australia', *Alcheringa*, 11: 337–45.

Glaessner, M.F. and Wade, M. (1966) 'The Late Precambrian fossils from Ediacara, South Australia', *Palaeontology*, 9: 599–628.

Goodman, M., Miyamoto, M.M. and Czelisniak, J. (1987) 'Pattern and process in vertebrate phylogeny revealed by coevolution of molecules and morphologies', in C. Patterson (ed.) *Molecules and Morphology in Evolution: Conflict or Compromise?*, Cambridge: Cambridge University Press, pp. 141–76.

Goujet, D. (2001) 'Placoderms and basal gnathostome apomorphies', in P.E. Ahlberg (ed.) *Major Events in Early Vertebrate Evolution: Palaeontology, Phylogeny, Genetics and Development*, London: Taylor & Francis, pp. 209–22.

Goujet, D. and Young, G.C. (1995) 'Interrelationships of placoderms revisited', *Geobios*, 19: 89–95.

Gradstein, F.M. and Ogg, J. (1996) 'A Phanerozoic time scale', *Episodes*, 19: 3–5.

Gross, W. (1969) '*Lophosteus superbus* Pander, ein Teleostome aus dem Silur Oesels', *Lethaia*, 2: 15–47.

Hardisty, M.W. (1982) 'Lampreys and hagfishes: analysis of cyclostome relationships', in M.W. Hardisty and I.C. Potter (eds) *The Biology of Lampreys*, London: Academic Press, pp. 165–259.

Hedges, S.B. (2001) 'Molecular evidence for the early history of living vertebrates', in P.E. Ahlberg (ed.) *Major Events in Early Vertebrate Evolution: Palaeontology, Phylogeny, Genetics and Development*, London: Taylor & Francis, pp. 119–34.

Hitchin, R. and Benton, M.J. (1996) 'Congruence between parsimony and stratigraphy: comparisons of three indices', *Paleobiology*, 23: 20–32.

Holland, S.M. (1995) 'The stratigraphic distribution of fossils', *Paleobiology*, 21: 92–109.

—— (2000) 'The quality of the fossil record: a sequence stratigraphic perspective', *Paleobiology*, 26 Supplement: 148–68.

Hou, X., Ramsköld, L. and Bergström, J. (1991) 'Composition and preservation of the Chengjiang fauna – a Lower Cambrian soft-bodied biota', *Zoologica Scripta*, 20: 395–411.

Huelsenbeck, J.P. (1994) 'Comparing the stratigraphic record to estimates of phylogeny', *Paleobiology*, 20: 470–83.

Janvier, P. (1981) 'The phylogeny of the Craniata, with particular reference to the significance of fossil "agnathans"', *Journal of Vertebrate Paleontology*, 1: 121–59.

—— (1996a) 'The dawn of the vertebrates: characters versus common ascent in the rise of current vertebrate phylogenies', *Palaeontology*, 39: 259–87.

—— (1996b) *Early Vertebrates*, Oxford: Oxford University Press.

—— (1998) 'Les vertébrés avant le Silurien', *Geobios*, 30: 931–50.

—— (in press) 'Osteostraci', in H.-P. Schultze (ed.) *Handbook of Palaeoichthyology*.

Janvier, P. and Blieck, A. (1993) 'L. B. Halstead and the heterostracan controversy', *Modern Geology*, 18: 89–105.

Janvier, P. and Lund, R. (1983) '*Hardistiella montanensis* n.gen. et sp. (Petromyzontida) from the Lower Carboniferous of Montana, with remarks on the affinities of lampreys', *Journal of Vertebrate Paleontology*, 2: 407–13.

Jefferies, R.P.S. (1967) 'Some fossil chordates with echinoderm affinities', *Zoological Society of London Symposium*, 20: 163–208.

—— (1973) 'The Ordovician fossil *Lagynocystis pyramidalis* (Barrande) and the ancestry of amphioxus', *Philosophical Transactions of the Royal Society, London*, B265: 409–69.

—— (1979) 'The origin of chordates: a methodological essay', in M.R. House (ed.) *The Origin of Major Invertebrate Groups*, London: Systematics Association, pp. 443–7.

—— (1986) *The Ancestry of the Vertebrates*, London: British Museum (Natural History).

Jefferies, R.P.S. and Lewis, D.N. (1978) 'The English Silurian fossil *Placocystites forbesianus* and the ancestry of the vertebrates', *Philosophical Transactions of the Royal Society, London*, B282: 205–323.

Jefferies, R.P.S., Brown, N.A. and Daley, P.E.J. (1996) 'The early phylogeny of chordates and echinoderms and the origin of chordate left-right asymmetry and bilateral symmetry', *Acta Zoologica (Stockholm)*, 77: 101–22.

Knoll, A.H. (2000) 'Learning to tell Neoproterozoic time', *Precambrian Research*, 100: 3–20.

Kumar, S. and Hedges, S.B. (1998) 'A molecular timescale for vertebrate evolution', *Nature*, 392: 917–20.

Kuraku, S., Hoshiyama, D., Katoh, K., Suga, K. and Miyata, T. (1999) 'Monophyly of lampreys and hagfishes supported by nuclear DNA-coded genes', *Journal of Molecular Evolution*, 49: 729–35.

Lacalli, T.C. (2002) 'Vetulicolians – are they deuterostomes? chordates?' *BioEssays*, 24: 208–11.

Lee, M.S.Y. (1999) 'Molecular clock calibrations and metazoan divergence dates', *Journal of Molecular Evolution*, 49: 385–91.

Long, J.A. (1986) 'New ischnacanthid acanthodians from the Early Devonian of Australia, with comments on acanthodian interrelationships', *Zoological Journal of the Linnean Society* 87: 321–39.

Løvtrup, S. (1977) *The Phylogeny of the Vertebrata*, New York: Wiley.

Lund, R. and Janvier, P. (1986) 'A second lamprey from the Lower Carboniferous (Namurian) of Bear Gulch, Montana (U.S.A.)', *Geobios*, 19: 647–52.

Mallatt, J. and Sullivan, J. (1998) '28S and 18S rDNA sequences support the monophyly of lampreys and hagfishes', *Molecular Biology and Evolution*, 15: 1706–18.

Mallatt, J., Sullivan, J. and Winchell, C.J. (2001) 'The relationship of lampreys to hagfishes: a spectral analysis of ribosomal DNA sequences', in P.E. Ahlberg (ed.) *Major Events in Early Vertebrate Evolution: Palaeontology, Phylogeny, Genetics and Development*, London: Taylor & Francis, pp. 106–18.

Marshall, C.R. (1990) 'Confidence-intervals on stratigraphic ranges', *Paleobiology*, 16: 1–10.

—— (1997) 'Confidence intervals on stratigraphic ranges with nonrandom distributions of fossil horizons', *Paleobiology*, 23: 165–73.

Märss, T. (1986) 'Silurian vertebrates of Estonia and West Latvia', *Fossilia Baltica*, 1: 1–104.

Modesto, S.P. (1995) 'The skull of the herbivorous synapsid *Edaphosaurus boanerges* from the Lower Permian of Texas', *Palaeontology*, 38: 213–39.

Mooi, R. (2001) 'Not all written in stone: interdisciplinary syntheses in echinoderm paleontology', *Canadian Journal of Zoology*, 79: 1209–31.

Mooi, R. and David, B. (1997) 'Skeletal homologies of echinoderms', *Paleontological Society Papers*, 3: 305–35.

—— (1998) 'Evolution within a bizarre phylum: homologies of the first echinoderms', *American Zoologist*, 38: 965–74.

Müller, K.J. (1977) '*Palaeobotryllus* from the Upper Cambrian of Nevada – a probable ascidian', *Lethaia*, 10: 107–18.

Nei, M., Xu, P. and Glazko, G. (2001) 'Estimation of divergence times from multiprotein sequences for a few mammalian species and several distantly related organisms', *Proceedings of the National Academy of Sciences, USA*, 98: 2497–502.

Nikoh, N., Iwabe, N., Kuma, K., Ohno, M., Sugiyama, T., Watanabe, Y., Yasui, K., Zhang, S., Hori, K., Shimura, Y. and Miyata, T. (1997) 'An estimate of divergence time of Parazoa and Eumetazoa and that of Cephalochordata and Vertebrata by Aldolase and Triose Phosphate Isomerase clocks', *Journal of Molecular Evolution*, 45: 97–106.

Norell, M.A. (1992) 'Taxic origin and temporal diversity: the effect of phylogeny', in M.J. Novacek and Q.D. Wheeler (eds) *Extinction and Phylogeny*, New York: Columbia University Press, pp. 89–118.

Norell, M.A. and Novacek, M.J. (1992) 'The fossil record and evolution: comparing cladistic and paleontologic evidence for vertebrate history', *Science*, 255: 1690–3.

Novitskaya, L.I. (1971) *Les amphiaspides (Heterostraci) du Dévonien de la Sibérie*. Cahiers de Paléontologie, Centre national de la Recherche scientifique, Paris, 1–130.

Paul, C.R.C. (1982) 'The adequacy of the fossil record', in K.A. Joysey and A.E. Friday (eds) *Problems of Phylogenetic Recontruction*, London: Academic Press, pp. 75–117.

—— (1998) 'Adequacy, completeness and the fossil record', in S.K. Donovan and C.R.C. Paul (eds) *The Adequacy of the Fossil Record*, Chichester: John Wiley & Sons, pp. 1–22.

Paul, C.R.C. and Smith, A.B. (1984) 'The early radiation and phylogeny of echinoderms', *Biological Reviews*, 59: 443–81.

Peterson, K.J. (1995) 'A phylogenetic test of the calcichordate scenario', *Lethaia*, 28: 25–38.

Purnell, M.A. (2001) 'Scenarios, selection and the ecology of early vertebrates', in P.E. Ahlberg (ed.) *Major Events in Early Vertebrate Evolution: Palaeontology, Phylogeny, Genetics and Development*, London: Taylor & Francis, pp. 187–208.

Remane, J. (2000) 'International stratigraphic chart', International Union of Geological Sciences.

Rodríguez-Trelles, F., Tarrío, R. and Ayala, F.J. (2002) 'A methodological bias toward overestimation of molecular evolutionary time scales', *Proceedings of the National Academy of Sciences, USA*, 99: 8112–15.

Romer, A.S. (1933) 'Eurypterid influence on vertebrate history', *Science* 78: 114–17.

Ruta, M. (1999) 'Brief review of the stylophoran debate', *Evolution & Development*, 1: 123–35.

Sansom, I.J., Aldridge, R.J. and Smith, M.M. (2000) 'A microvertebrate fauna from the Llandovery of South China', *Transactions of the Royal Society of Edinburgh (Earth Sciences)*, 90: 255–72.

Sansom, I.J., Smith, M.M. and Smith, M.P. (2001) 'The Ordovician radiation of vertebrates', in P.E. Ahlberg (ed.) *Major Events in Early Vertebrate Evolution: Palaeontology, Phylogeny, Genetics and Development*, London: Taylor & Francis, pp. 156–71.

Saylor, B.Z., Kaufman, A.J., Grotzinger, J.P. and Urban, F. (1998) 'A composite reference section for terminal Proterozoic strata of southern Namibia', *Journal of Sedimentary Research*, 68: 1223–35.

Schultze, H.-P. (1968) 'Palaeoniscoidea-schuppen aus dem Unterdevon Australiens und Kansas und aus dem Mitteldevon Spitzbergens', *Bulletin of the British Museum (Natural History), Geology*, 16: 343–68.

—— (1992) 'Early Devonian actinopterygians (Osteichthyes, Pisces) from Siberia', in E. Mark-Kurik (ed.) *Fossil Fishes as Living Animals*, Tallinn: Academy of Sciences of Estonia, pp. 233–42.

Shu, D.-G., Zhang, X. and Chen, L. (1996a) 'Reinterpretation of *Yunnanozoon* as the earliest known hemichordate', *Nature*, 380: 428–30.

Shu, D.-G., Conway Morris, S. and Zhang, X.-L. (1996b) 'A *Pikaia*-like chordate from the Lower Cambrian of China', *Nature*, 384: 157–8.

Shu, D.-G., Luo, H.-L., Conway Morris, S., Zhang, X.-L., Hu, S.-X., Chen, L., Han, J., Zhu, M., Li, Y. and Chen, L.-Z. (1999) 'Lower Cambrian vertebrates from south China', *Nature*, 402: 42–6.

Shu, D.-G., Chen, L., Han, J. and Zhang, X.-L. (2001a) 'An early Cambrian tunicate from China', *Nature*, 411: 472–3.

Shu, D.-G., Conway Morris, S., Han, J., Chen, L., Zhang, X.-L., Zhang, Z.-F., Liu, H.-Q., Li, Y. and Liu, J.-N. (2001b) 'Primitive deuterostomes from the Chengjiang Lagerstätte (Lower Cambrian, China)', *Nature*, 414: 419–24.

Siddall, M.E. (1996) 'Stratigraphic consistency and the shape of things', *Systematic Biology*, 45: 111–15.

—— (1997) 'Stratigraphic indices in the balance: a reply to Hitchin and Benton', *Systematic Biology*, 46: 569–73.

—— (1998) 'Stratigraphic fit to phylogenetics: a proposed solution', *Cladistics*, 14: 201–8.

Smith, A.B. (1984) 'Classification of the Echinodermata', *Palaeontology*, 27: 431–59.

—— (1988a) 'Patterns of diversification and extinction in Early Palaeozoic echinoderms', *Palaeontology*, 31: 799–828.

Smith, A.B. (1988b) 'Fossil evidence for the relationships of extant echinoderm classes and their times of divergence', in C.R.C. Paul and A.B. Smith (eds) *Echinoderm Phylogeny and Evolutionary Biology*, Oxford: Clarendon Press, pp. 85–97.

—— (1990) 'Evolutionary diversification of echinoderms during the early Palaeozoic', in P.D. Taylor and G.P. Larwood (eds) *Major Evolutionary Radiations. Systematics Association Special Publication No. 42*, Oxford: Clarendon Press, pp. 265–86.

—— (1999) 'Dating the origin of metazoan body plans', *Evolution & Development*, 1: 138–42.

—— (2001) 'Large-scale heterogeneity of the fossil record: implications for Phanerozoic biodiversity studies', *Philosophical Transactions of the Royal Society, London*, B356: 351–67.

Smith, A.B. and Peterson, K.J. (2002) 'Dating the time of origin of major clades: molecular clocks and the fossil record', *Annual Review of Earth and Planetary Science*, 30: 65–88.

Smith, A.B., Gale, A.S. and Monks, N.E.A. (2001) 'Sea level change and rock bias in the Cretaceons: a problem for extinction and biodiversity studies', *Paleobiology*, 27: 241–53.

Smith, M.M. and Sansom, I.J. (1997) 'Exoskeletal microremains of an Ordovician fish from the Harding Sandstone of Colorado', *Palaeontology*, 40: 645–58.

Smith, M.P., Sansom, I.J. and Cochrane, K.D. (2001) 'The Cambrian origin of vertebrates', in P.E. Ahlberg (ed.) *Major Events in Early Vertebrate Evolution: Palaeontology, Phylogeny, Genetics and Development*, London: Taylor & Francis, pp. 67–84.

Smith, M.P., Donoghue, P.C.J. and Sansom, I.J. (2002) 'The spatial and temporal diversification of Early Palaeozoic vertebrates', in J.A. Crame and A.W. Owen (eds) *Palaeobiogeography and Biodiversity Change: the Ordovician and Mesozoic–Cenozoic Radiations*, Geological Society Special Publication 194: 69–83.

Springer, M.S. (1997) 'Molecular clocks and the timing of the placental and marsupial radiations in relation to the Cretaceous–Tertiary boundary', *Journal of Mammalian Evolution*, 4: 285–302.

Stensiö, E.A. (1927) 'The Downtonian and Devonian vertebrates of Spitsbergen. Part 1. Family Cephalaspidae', *Skrifter om Svalbard og Nordishavet*, 12: 1–391.

—— (1968) 'The cyclostomes, with special reference to the diphyletic origin of the Petromyzontida and Myxinoidea', in T. Ørvig (ed.) *Current Problems in Lower Vertebrate Phylogeny, Nobel Symposium 4*, Stockholm: Almquist & Wiksell, pp. 13–71.

Stock, D.W. and Whitt, G.S. (1992) 'Evidence from 18S ribosomal RNA sequences that lampreys and hagfishes form a natural group', *Science*, 257: 787–9.

Strauss, D. and Sadler, P.M. (1989) 'Classical confidence-intervals and Bayesian probability estimates for ends of local taxon ranges', *Mathematical Geology*, 21: 411–21.

Sumrall, C.D. (1997) 'The role of fossils in the phylogenetic reconstruction of the Echinodermata', *Paleontological Society Papers*, 3: 267–88.

Suzuki, M., Kubokawa, K., Nagasawa, H. and Urano, A. (1995) 'Sequence analysis of vasotocin cDNAs of the lamprey *Lampetra japonica*, and the hagfish, *Eptatretus burgeri*: evolution of cyclostome vasotocin precursors', *Journal of Molecular Endocrinology*, 14: 67–77.

Sweet, W.C. (1988) *The Conodonta: Morphology, Taxonomy, Paleoecology, and Evolutionary History of a Long-extinct Animal Phylum*, Oxford: Clarendon Press.

Sweet, W.C. and Donoghue, P.C.J. (2001) 'Conodonts: past, present and future', *Journal of Paleontology*, 75: 1174–84.

Tavaré, S., Marshall, C.R., Will, O., Soligo, C. and Martin, R.D. (2002) 'Using the fossil record to estimate the age of the last common ancestor of extant primates', *Nature*, 416: 726–9.

Taverne, L. (1997) '*Osorioichthys marginis*, "paleonisciform" from the Fammenian of Belgium, and the phylogeny of the Devonian actinopterygians (Pisces)', *Bulletin de l'Institut Royal des Sciences Naturelles de Belgique*, 67: 57–78.

Tucker, R.D. and McKerrow, W.S. (1995) 'Early Paleozoic chronology: a review in light of new U–Pb zircon ages from Newfoundland and Britain', *Canadian Journal of Earth Sciences*, 32: 368–79.

Tucker, R.D., Bradley, D.C., Ver Straeten, C.A., Harris, A.G., Ebert, J.R. and McCutcheon, S.R. (1998) 'New U–Pb zircon ages and the duration and division of Devonian time', *Earth and Planetary Science Letters*, 158: 175–86.

Ubaghs, G. (1968) 'Stylophora', in R.C. Moore (ed.) *Treatise on Invertebrate Paleontology. Part S. Echinodermata 1(2)*, Boulder and Lawrence KS: Geological Society of America and University of Kansas Press, pp. 496–565.

van Tuinen, M., Sibley, C.G. and Hedges, S.B. (1998) 'Phylogeny and biogeography of ratite birds inferred from DNA sequences of the mitochondrial ribosomal genes', *Molecular Biology and Evolution*, 15: 370–6.

Wagner, P.J. (1998) 'Phylogenetic analysis and the quality of the fossil record', in S.K. Donovan and C.R.C. Paul (eds) *The Adequacy of the Fossil Record*, Chichester: John Wiley & Sons Ltd, pp. 165–87.

Wang, D.Y.C., Kumar, S. and Hedges, S.B. (1998) 'Divergence time estimates for the early history of animal phyla and the origin of plants, animals and fungi', *Proceedings of the Royal Society, London*, B266: 163–71.

Weiss, R.E. and Marshall, C.R. (1999) 'The uncertainty in the true end point of a fossil's stratigraphic range when stratigraphic sections are sampled discretely', *Mathematical Geology*, 31: 435–53.

Wills, M.A. (1999) 'Congruence between phylogeny and stratigraphy: randomization tests and the gap excess ratio', *Systematic Biology*, 48: 559–80.

Wills, M.A. and Fortey, R.A. (2000) 'The shape of life: how much is written in stone?' *BioEssays*, 22: 1142–52.

Wills, M.A. and Sepkoski, J.J. (1993) 'Problematica', in M.J. Benton (ed.) *The Fossil Record 2*, London: Chapman & Hall, pp. 543–54.

Wray, G.A., Levinton, J.S. and Shapiro, L.H. (1996) 'Molecular evidence for deep Precambrian divergences among metazoan phyla', *Science*, 274: 568–73.

Yalden, D.W. (1985) 'Feeding mechanisms as evidence of cyclostome monophyly', *Zoological Journal of the Linnean Society*, 84: 291–300.

Young, G.C. (1991) 'The first armoured agnathan vertebrates from the Devonian of Australia', in M.M. Chang, Y.H. Liu and G.R. Zhang (eds) *Early Vertebrates and Related Problems in Evolutionary Biology*, Beijing: Science Press, pp. 67–85.

Zhu, M. and Schultze, H.-P. (1997) 'The oldest sarcopterygian fish', *Lethaia*, 30: 293–304.

Zrzavý, J., Mihulka, S., Kepka, P., Bezdek, A. and Tietz, D. (1998) 'Phylogeny of the Metazoa based on morphological and 18S ribosomal DNA evidence', *Cladistics*, 14: 249–85.

Zuckerkandl, E. and Pauling, L. (1962) 'Molecular disease, evolution and genic heterogeneity', in M. Kasha and B. Pullman (eds) *Horizons in Biochemistry*, New York: Academic Press, pp. 189–225.

—— (1965) 'Evolutionary divergence and convergence in proteins', in V. Bryson and H.J. Vogel (eds) *Evolving Genes and Proteins*, New York: Academic Press, pp. 97–166.

Appendix 10.1

Codings for taxa augmented to the analysis of Donoghue and Smith (2001):

Myllokunmingia

0??????????????????????????1?10???0???????10010?0???0??????00000000000000000000??0??????
?????????????????000??01

Haikouichthys

0??????????????1??????????10???0???????110?0???0??????00000000000000000000??0??????
?????????????000??01

Appendix 10.2

Tree topologies and occurrence data upon which cladogram–stratigraphy statistics and confidence intervals are based.

Conodonta

((*Proconodontus* ((Fryxellodontidae Pygodontidae)(Cordylodontidae (Ansellidae (Dapsilodontidae Belodellidae)))))(*Teridontus* ((Clavohamulidae (Drepanoistontidae (Acanthodontidae Panderodontidae)))(*Rossodus* (Multioistodontidae (Periodontidae (Rhipidognathidae (Prioniodontidae ((Cyrtoniodontidae ((Chirognathidae Prioniodinidae) Kockelellidae))(Polyplacognathidae (Distomodontidae (Icriodellidae Icriodontidae)))))))))))))); following Sweet & Donoghue (2001), after Sweet (1988).

Heterostracomorpha

(Astraspida (Arandaspididae (Lepidaspididae Tesseraspididae Phialaspidiformes (Corvaspids/Tolypelepids ((Cyathaspidida (Ctenaspididae ((Eglonaspididae Hibernaspididae) (Siberiaspididae (Amphiaspididae Olbiaspididae))))) Anchipteraspididae (Protopteraspididae (Pteraspididae (Protaspididae Psammosteidae))))))))); following Novitskaya (1971), Blieck (1984) and Janvier (1996b).

Anaspida

(*Pharyngolepis*,(*Pterygolepis*,(*Rhyncholepis*,*Lasanius*,*Birkenia*))); following Janvier (1996b).

Thelodontii

(Furcacaudiformes (Thelodontids (Loganellidae Phlebolepidae))); following Donoghue & Smith (2001).

Galeaspida

(Dayongaspididae Hanyangaspis Xiushiuaspididae (((((Eugaleaspidae Tridensaspidae) *Nochelaspis*) *Yunnanogaleaspis*) Sinogaleaspidae) ((((*Duyunolepis* *Neoduyunaspis* *Paraduyunaspis*) *Dongfangaspis* Polybranchiaspidae) *Bannhuanaspis*) Huananaspidida))); following Janvier (1996b).

Osteostraci

(*Ateleaspis* (*Aceraspis* (*Hirella* (*Hemiteleaspis* (*Hemicyclaspis* (Escuminaspididae Tannuaspididae (Cephalaspididae (Mimetaspididae Pattenaspididae)) (Zenaspidida ((*Tauraspis* (*Hapilaspis* Benneviaspididae (Hoelaspididae Boreaspididae))) ((*Procephalaspis* (*Auchenaspis* (*Witaaspis* (*Thyestes* Tremataspididae)))) (*Kiaeraspis* (Axinaspididae Acrotomaspididae))))))))))))); following Janvier (1996b; in press)

Chondrichthyes

(Cladoselachidae (Eugeneodontida Petalodontida) (Inopterygia ((Helodontidae (Cochliodontidae (Echinochimeridae Chimeridae))) (Symmoriidae Stethacanthidae)) (Xenacanthiformes (*Ctenacanthus* (Hybodontiformes Neoselachii))))); following Janvier (1996b).

Placodermi

((((Actinolepida (Phyllolepida Wuttagoonaspida)) (Phlyctaenii (*Gemeundaspis* (Holonematidae ((Homostiidae Buchanosteidae) ((Brachydeiroidea (Coccosteidae Camuropiscidae)) (Dinichthyidae Aspinothoraci))))))) (Ptycytodontida Petalichthyida))

((Yunnanolepididae (Sinolepidae ((Bothriolepidae Microbrachiidae) (Pterichthyoidea Asterolepidae)))) Rhenanida)); following Goujet & Young (1995) and Janvier (1996b).

Acanthodii
(Ischancanthidae (((Climatiidae Gyracanthidae) (Diplacanthidae Culmacanthidae)) (Mesacanthidae (Cheiracanthidae Acanthodidae)))); following Long (1986).

Actinopterygii
(*Lophosteus* (*Andreolepis* (*Naxilepis* (*Orvikuina* (*Ligulalepis Dialipina* (*Cheirolepis* (*Polypterus* (*Osorioichthys* (*Howqualepis* (*Mimia* (*Moythomasia* (*Tegeolepis* (*Stegotrachelus Kentuckia*))))))))))))); following Schultze (1992) and Taverne (1997).

Chapter 11

Bones, molecules, and crown-tetrapod origins

Marcello Ruta and Michael I. Coates

ABSTRACT

The timing of major events in the evolutionary history of early tetrapods is discussed in the light of a new cladistic analysis. The phylogenetic implications of this are compared with those of the most widely discussed, recent hypotheses of basal tetrapod interrelationships. Regardless of the sequence of cladogenetic events and positions of various Early Carboniferous taxa, these fossil-based analyses imply that the tetrapod crown-group had originated by the mid- to late Viséan. However, such estimates of the lissamphibian–amniote divergence fall short of the date implied by molecular studies. Uneven rates of molecular substitutions might be held responsible for the mismatch between molecular and morphological approaches, but the patchy quality of the fossil record also plays an important role. Morphology-based estimates of evolutionary chronology are highly sensitive to new fossil discoveries, the interpretation and dating of such material, and the impact on tree topologies. Furthermore, the earliest and most primitive taxa are almost always known from very few fossil localities, with the result that these are likely to exert a disproportionate influence. Fossils and molecules should be treated as complementary approaches, rather than as conflicting and irreconcilable methods.

Introduction

Modern tetrapods have a long evolutionary history dating back to the Late Devonian. Their origins are rooted into a diverse, paraphyletic assemblage of lobe-finned bony fishes known as the 'osteolepiforms' (Cloutier and Ahlberg 1996; Janvier 1996; Ahlberg and Johanson 1998; Jeffery 2001; Johanson and Ahlberg 2001; Zhu and Schultze 2001). The monophyletic status of the Tetrapoda and that of its major constituent clades – lissamphibians and amniotes – is supported by a large number of morphological characters and by a wide range of molecular data (e.g. Duellman and Trueb 1986; Panchen and Smithson 1987, 1988; Duellman 1988; Gauthier *et al.* 1988a,b; Milner 1988, 1993; Hedges *et al.* 1990; Carroll 1991; Trueb and Cloutier 1991; Hedges and Maxson 1993; Hay *et al.* 1995; Laurin and Reisz 1997, 1999; Feller and Hedges 1998; Laurin 1998a–c; Hedges and Poling 1999; Pough *et al.* 2000). The early evolutionary history of the lissamphibian and amniote crown-groups has been the subject of detailed scrutiny (e.g. Laurin 1991; Laurin and Reisz 1995; Báez and Basso 1996; Lee 1995, 1997a,b; Rieppel and deBraga 1996; deBraga and Rieppel

1997; Gao and Shubin 2001), but the phylogenetic placement of some groups is not agreed upon, as exemplified by current debates about the position of turtles relative to other amniotes (e.g. Reisz and Laurin 1991; Lee 1993, 1995, 1996, 1997a,b; Laurin and Reisz 1995; Rieppel and deBraga 1996; deBraga and Rieppel 1997; Platz and Conlon 1997; Zardoya and Meyer 1998; Hedges and Poling 1999; Rieppel and Reisz 1999; Rieppel 2000).

The last twenty years have witnessed a revived interest in early tetrapod inter-relationships. New discoveries and a refinement of phylogenetic techniques have broadened our understanding of the anatomy and intrinsic relationships of several groups. Research in this field has had a significant impact on the shaping of the tetrapod stem-group (Lebedev and Coates 1995; Coates 1996; Ahlberg and Johanson 1998; Johanson and Ahlberg 2001) and has led to the recognition of a previously unsuspected diversity of Mississippian taxa (e.g. Clack 1994, 1998a–d, 2001, 2002; Milner and Sequeira 1994; Rolfe *et al.* 1994; Smithson 1994; Smithson *et al.* 1994; Lombard and Bolt 1995; Clack and Finney 1997; Paton *et al.* 1999; Bolt and Lombard 2000; Clack and Carroll 2000). However, the interrelationships of the vast majority of Palaeozoic groups are still intensely debated. Lack of congruent results in the most widely discussed, recently published phylogenetic analyses is astonishing. Traditional views on the taxonomic memberships of the lissamphibian and amniote stem-groups (Bolt 1969, 1977, 1979, 1991; Heaton 1980; Panchen and Smithson 1987, 1988; Milner 1988, 1993; Trueb and Cloutier 1991; Carroll 1995; Coates 1996; Lee and Spencer 1997; Sumida 1997; Clack 1998a–d; Paton *et al.* 1999) have been challenged repeatedly, notably in a series of recent papers by Laurin and Reisz (1997, 1999), Laurin (1998a–c) and Laurin *et al.* (2000a,b). As a result, no consensus has emerged on the position of several groups relative to the lissamphibian–amniote phylogenetic split.

The ancestry of lissamphibians, as well as the status and mutual relationships of the three modern lissamphibian orders, are particularly controversial topics (see discussions in Carroll and Currie 1975; Duellmann and Trueb 1986; Duellmann 1988; Bolt 1991; Milner 1988, 1993, 2000; Feller and Hedges 1998; Laurin 1998a–c; Carroll 2000, 2001; Anderson 2001). Several authors have suggested that some or all of the lissamphibian orders are related to dissorophoids, a group of Permo-Carboniferous and Lower Triassic temnospondyls (e.g. Bolt 1969, 1977, 1979, 1991; Lombard and Bolt 1979; Bolt and Lombard 1985; Milner 1988, 1990, 1993, 2000; Trueb and Cloutier 1991; Boy and Sues 2000; Holmes 2000; Rocek and Rage 2000a,b; Yates and Warren 2000; Gardner 2001). However, much discussion centres on the identity of the immediate sister taxon to frogs, salamanders, and caecilians (also known as gymnophionans). The temnospondyl theory of lissamphibian origin has been revived recently by Carroll (2001) and Carroll and Bolt (2001). These authors hypothesize that the ancestry of frogs and salamanders is rooted into two distinct families of dissorophoids, the amphibamids and branchiosaurids, respectively. Caecilians, however, are thought to be related to tuditanomorph microsaurs (one of the most diverse groups of lepospondyls). In particular, the Lower Permian genus *Rhynckonkos* has been regarded as the most derived stem-group gymnophionan (Carroll and Currie 1975; Carroll and Gaskill 1978; Milner 1993; Carroll 2000, 2001). In Laurin's (1998a–c) and Laurin and Reisz's (1997, 1999) analyses, temnospondyls are a plesion on the tetrapod stem-group (see also below), whereas lepospondyls form a paraphyletic array of stem-lissamphibians. Within this paraphyletic array, lysorophids (long-bodied, Pennsylvanian to Lower Permian tetrapods characterized by a broad orbitotemporal fenestration; Wellstead 1991) are considered to be the nearest Palaeozoic relatives of crown-lissamphibians.

The evolutionary implications of alternative hypotheses of early tetrapod relationships will be considered elsewhere together with a new, comprehensive cladistic analysis recently completed by the authors (Ruta *et al.* 2003). A summary of the results of this analysis (Figure 11.1) and a review of the chronology of major events in the evolutionary history of early tetrapods are presented here. We explore the implications of conflicting phylogenetic hypotheses on estimates of the time of divergence between lissamphibians and amniotes, and compare morphology-based 'time trees' (for the use of this term, equivalent to Smith's, 1994, X-trees, see Hedges 2001) with those deriving from recent molecular analyses (e.g. Feller and Hedges 1998; Kumar and Hedges 1998; Hedges 2001). Several questions are addressed in this chapter:

(1) Do different morphology-based cladistic analyses of primitive tetrapods imply different chronological estimates of the separation between lissamphibians and amniotes, or the origin of the lissamphibian and amniote crown-groups?
(2) Are palaeontological and molecular time trees in serious conflict with each other, and what is the source of this conflict?
(3) What is the bearing of fossils on time tree reconstruction, especially when integrated with the results of molecular analyses?

Materials and methods

Which consensus for early tetrapods?

In a series of seminal papers, Smithson (1985), Panchen and Smithson (1987, 1988), Milner (1993), and Ahlberg and Milner (1994) discussed the pattern of character disribution in the apical part of the tetrapod stem-group and in the basal portion of the crown-group. A common feature of these studies is the separation of most Palaeozoic tetrapods into two distinct lineages ultimately leading to lissamphibians and amniotes. Several subsequent analyses (Carroll 1995; Lebedev and Coates 1995; Coates 1996; Clack 1998b,d; Paton *et al.* 1999) have supported the basal dichotomy between these two clades (see Laurin 1998a, Laurin and Reisz 1999, Clack 2000, and Clack and Carroll 2000, for a historical perspective on the classification of early tetrapods). Panchen and Smithson's (1988) scheme of relationships is the most eloquent example of a balanced cladogram (*sensu* Smith 1994): major tetrapod clades are equally distributed on the 'batrachomorph' and 'reptiliomorph' branches of the crown-group (equivalent to the lissamphibian and amniote stem-groups, respectively; see Coates 1996). According to Panchen and Smithson (1988), the evolutionary separation between lissamphibians and amniotes is a Late Devonian event, since the Famennian *Ichthyostega* appears as the least derived plesion on the lissamphibian stem-group. Other stem-lissamphibian plesions include, in crownward order, nectrideans (Bossy and Milner 1998), colosteids (Smithson 1982; Hook 1983; Godfrey 1989), microsaurs (Carroll and Gaskill 1978), and temnospondyls (Milner 1988, 1990, 1993). In Panchen and Smithson's (1988) scheme, baphetids (Beaumont 1977; Beaumont and

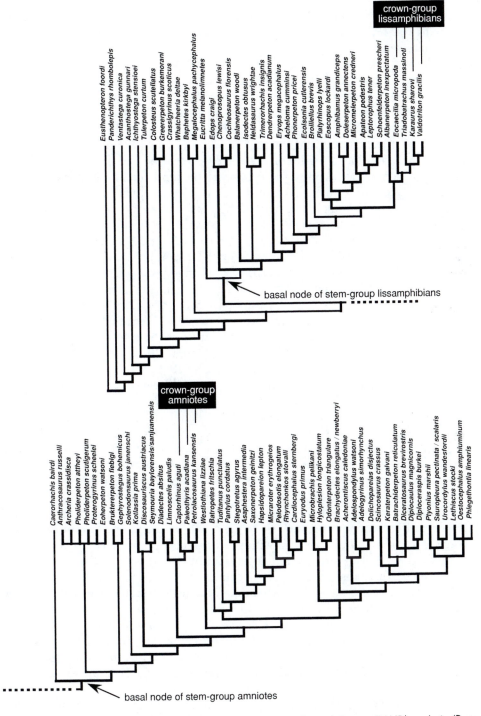

Figure 11.1 Strict consensus of 60 most parsimonious trees derived from a PAUP* analysis (Ruta *et al.* 2003). The stem-tetrapod and lissamphibian portions of the trees are in the upper half of the figure, the amniote portion in the lower half.

Smithson 1998; Milner and Lindsay 1998), anthracosauroids (Smithson 1985, 2000), seymouriamorphs (Laurin 2000), and diadectomorphs (Romer 1946; Heaton 1980; Berman *et al.* 1992, 1998; Lombard and Sumida 1992) are progressively more derived stem-amniotes.

A series of key discoveries have been instrumental in redefining our concept of the most primitive tetrapods as well as in our understanding of the pattern of morphological change at the 'fish'–tetrapod transition (e.g. Coates and Clack 1990, 1991; Coates 1996; Jarvik 1996; Clack 1998b). As a result, the stem-tetrapod affinities of most Devonian taxa, including *Acanthostega* and *Ichthyostega*, are now universally accepted (but see Lebedev and Coates 1995, and Coates 1996, for a discussion of the possible stem-amniote affinities of *Tulerpeton*). Regardless of the phylogenetic placement of Devonian taxa, comparisons between the most recent published phylogenies reveal a drastic shift from dichotomously branching to pectinate tree topologies, implying an increase in the number of stem-group branching events. The studies of Ahlberg and Milner (1994), Carroll (1995), Lebedev and Coates (1995), Coates (1996), Clack (1998b,d), and Paton *et al.* (1999) support Panchen and Smithson's (1988) conclusions with regard to the basal dichotomy of Palaeozoic groups. These analyses tackle such diverse problems as the broad pattern of relationships between major tetrapod groups (Carroll 1995), the reconstruction of the sequence of anatomical changes in taxa spanning the 'fish'–tetrapod transition (Lebedev and Coates 1995; Coates 1996), and the placement of various problematic Mississippian tetrapods (e.g. *Crassigyrinus*, *Whatcheeria*, *Eucritta*) known to display a mixture of characters otherwise considered to be unique to separate clades (Clack 1998b,d, 2000, 2001, 2002; Paton *et al.* 1999).

Laurin and Reisz's (1997, 1999) and Laurin's (1988a–c) analyses have cast doubt on the deep separation of Palaeozoic tetrapods between lissamphibian-related and amniote-related taxa. Their cladograms suggest that several early tetrapods, such as *Crassigyrinus*, *Tulerpeton*, *Whatcheeria*, and baphetids, are equally closely related to lissamphibians and amniotes. These results challenge long-recognized patterns of character change and distribution near the base of the tetrapod crown-clade. In particular, traditional groups such as temnospondyls, embolomeres, gephyrostegids, and seymouriamorphs are regarded as discrete radiations preceding the lissamphibian–amniote phylogenetic split. The fossil membership of Laurin and Reisz's (1997, 1999) and Laurin's (1988a–c) crown-group is smaller than in previous works. Importantly, lissamphibians now sit at the crownward end of a paraphyletic assemblage of lepospondyls, in contrast with previous suggestions that the latter may form a highly diverse clade of stem-amniotes (Carroll 1995; but see also Carroll 2001). Anderson's (2001) analysis agrees with Laurin and Reisz's (1997, 1999) and Laurin's (1988a–c) conclusions that lepospondyls are stem-lissamphibians (although only *Eocaecilia* is used in Anderson's work), and that seymouriamorphs, embolomeres, and temnospondyls (represented, respectively, by *Seymouria*, *Proterogyrinus*, and a clade consisting of *Balanerpeton* and *Dendrerpeton*) are progressively less derived stem-tetrapod plesions. The diadectomorph *Limnoscelis* identifies the stem-amniote branch of Anderson's (2001) cladogram (Berman 2000; Clack and Carroll 2000, and references therein).

Very few early tetrapod groups have survived the intense phylogenetic reshuffling of recent analyses. Among those that have, diadectomorphs appear repeatedly as the nearest relatives of crown-amniotes; likewise, the stem-tetrapod affinities of colosteids and most Devonian forms have been retrieved consistently by different authors, despite differences in taxon sample size and the use of contrasting character ordering, weighting, and coding regimes (see also Ruta *et al.*, in press). These data suggest (although not conclusively) that some regions of the tetrapod tree are better corroborated and more stable than others (Panchen and Smithson 1987, 1988; Sumida and Lombard 1991; Berman *et al.* 1992; Lombard and Sumida 1992; Sumida *et al.* 1992; Laurin and Reisz 1997, 1999; Lee and Spencer 1997; Sumida 1997; Berman *et al.* 1998; Laurin 1998a–c; Paton *et al.* 1999; Berman 2000; Clack 2001).

Methodological note

The strict consensus topologies deriving from the most widely discussed published datasets including *Caerorhachis* are considered here (Figures 11.2–11.7). The strict consensus trees resulting from our new analysis (Figures 11.8–11.9) and from experiments of character removal (Figure 11.10) are also illustrated. As in Ruta *et al.*'s (2001) paper, Lebedev and Coates' (1995) and Clack's (1998b,d) analyses have been omitted, since they are superseded by Coates' (1996) and Paton *et al.*'s (1999) works, respectively. Strict consensus trees are plotted on a stratigraphical scale resolved down to stage level (geological timescale based on Briggs and Crowther 2001, and references therein). For simplicity, stages are drawn to the same length, and not proportional to their actual duration, although dates in millions of years before present (Ma) are appended, where possible, to stage names. In addition, the known ranges of major early tetrapods groups are used (Benton 1993), instead of specific occurrences of individual species. The use of whole ranges permits rapid and easy comparisons between tree shapes, and circumvents the problem of comparing time trees built on different taxon samples for each group. Internodes within monophyletic groups are represented by vertical bars of fixed, arbitrary length (except where ghost ranges are present; Smith 1994). This length represents merely a graphical expedient and does not imply an equal time for the origin of adjacent nodes. It has, however, the inconvenient effect of generating chronologically 'deep' origin events for some groups, depending upon the number of internodes and the placement of the stratigraphically oldest members of a group. Since the actual time occurring between adjacent nodes is unknown, the age of a node leading to two sister taxa is conservatively taken to coincide with the age of the older taxon.

Where species or genera are used as Operational Taxonomic Units (OTUs), it is possible to identify the point of divergence between sister groups, even if whole stratigraphical ranges are employed. For example, in Anderson's (2001) tree, the stem-caecilian *Eocaecilia micropoda* is the sister taxon to brachystelechid microsaurs. Therefore, the divergence of caecilians can be graphically plotted *within* the stratigraphical range of microsaurs instead of at the base of such a range (Figure 11.7). Paraphyletic groups pose problems when whole ranges are used. A possible way around this consists of splitting the ranges of large groups into the smaller ranges in which their component subgroups occurred. For simplicity, however, only total ranges are employed here, whereas paraphyletic groups are denoted by names in inverted commas.

With regards to Coates' (1996) analysis, Ruta *et al.* (2001) pointed out that introduction of corrected scores for digit number and coronoid fangs in some taxa

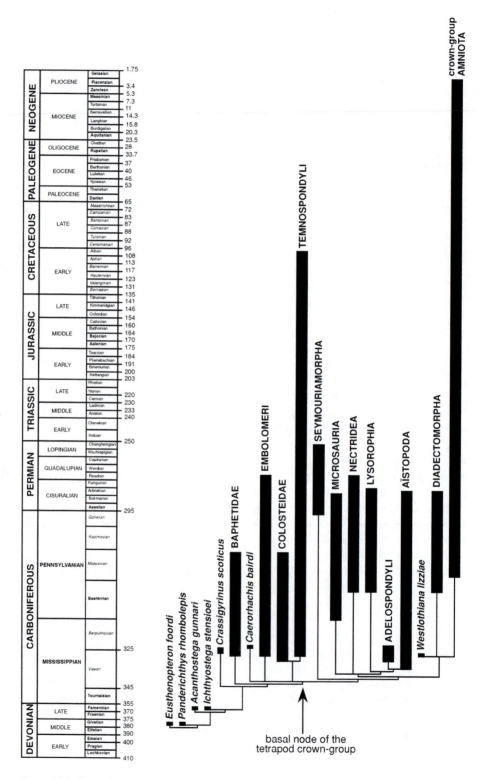

Figure 11.2 Carroll's (1995) analysis plotted on a timescale.

causes loss of phylogenetic resolution among the majority of post-panderichthyid Devonian tetrapods (*Tulerpeton* appears as a basal stem-amniote; Figure 11.3). It can be shown, however, that such poor resolution is due exclusively to the unstable position of *Metaxygnathus* and *Ventastega*. The branching sequence of remaining Devonian taxa is the same as that retrieved by Coates (1996).

Ahlberg and Clack's (1998) analysis (Figure 11.4) poses additional difficulties, because several traditional groups (notably anthracosaurs and temnospondyls) appear as polyphyletic, overlapping arrays of taxa (comments in Ruta *et al.* 2001). However, an expanded version of their dataset (see Appendix 11.1 for a list of the new characters added and their description) retrieves traditional groups after a reweighted run. The strict consensus of the resulting five equally parsimonious trees is discussed here (Figure 11.11), instead of Ahlberg and Clack's original consensus cladogram. A single origin for anthracosaurs and temnospondyls is obtained after analysing their expanded matrix. Lower jaw data can be shown to carry phylogenetic signal in derived portions of the tetrapod tree as well as in the crownward portion of the stem-group. However, the degree to which this signal matches that yielded by other characters is, at present, difficult to evaluate. The impact of lower jaw characters on tree topology must await exhaustive treatment of additional data (e.g. Bolt and Lombard 2001) coded for a larger number of taxa.

Definition and content of Tetrapoda

Any fossil taxon that can be shown, based on a formal character analysis, to be phylogenetically more closely related to extant lissamphibians *and* amniotes than to any other extant monophyletic group is, by definition, a stem-group tetrapod. If a fossil taxon is more closely related to *either* lissamphibians *or* amniotes, it is a crown-group tetrapod (Hennig 1966; Jefferies 1979; Craske and Jefferies 1989). Justification in support of a total-group (or stem-based) clade Tetrapoda will be provided elsewhere (Ruta *et al.* 2003; but see discussions in Ahlberg and Clack 1998, Laurin *et al.* 2000a,b, and Anderson 2001 for alternative nomenclatural solutions).

Briefly, we do not advocate an apomorphy-based definition of the Tetrapoda that excludes the 'fish-like' portion of the tetrapod stem-group (Ahlberg and Clack 1998), nor do we restrict the name Tetrapoda to the crown-clade (Laurin 1998a). Instead, we favour an operational definition (Coates *et al.* 2000), whereby all taxa that belong in the total-group of the extant clade Tetrapoda, but which are not members of the crown-group, are simply referred to as stem-group tetrapods (see also Budd 2001; Jeffery 2001).

Fossil evidence for the origin of crown-tetrapods

With few exceptions, published analyses postulate that the origin of the tetrapod crown-group had occurred by the mid- to late Viséan (e.g. Paton *et al.* 1999). A Late Devonian divergence between lissamphibians and amniotes was first proposed by Panchen and Smithson (1987, 1988), who interpreted *Ichthyostega* as a basal stem-group lissamphibian (see also above). Lebedev and Coates (1995) and Coates (1996) also suggested that the origin of the tetrapod crown-group was a Late Devonian event, but in this case, the hypothesized divergence time was based upon their interpretation of the

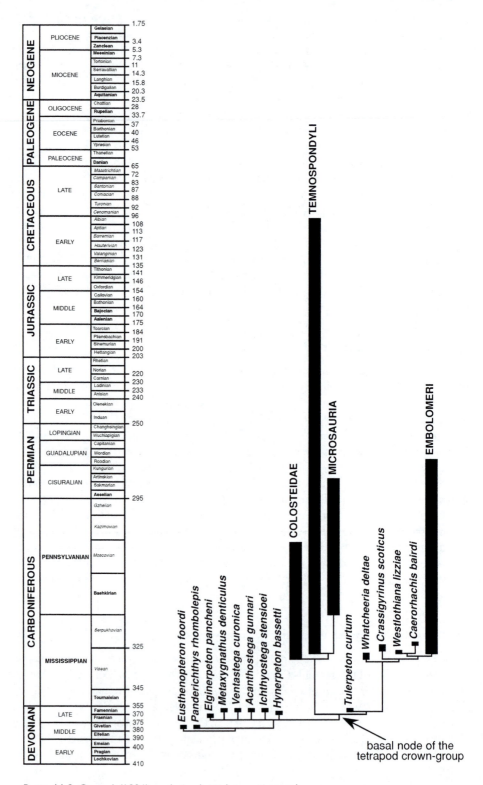

Figure 11.3 Coates' (1996) analysis plotted on a timescale.

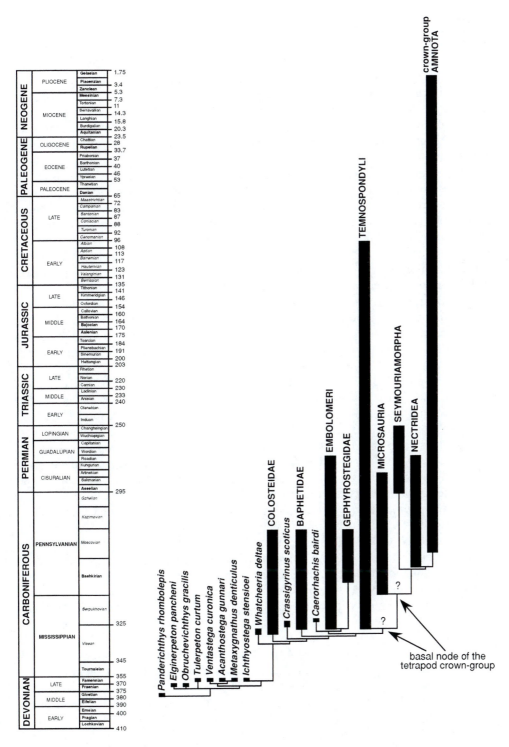

Figure 11.4 Ahlberg and Clack's (1998) analysis plotted on a timescale. The two arrows point to the positions of the basal node of the tetrapod crown-group based upon the derivation of lissamphibians from a lepospondyl or a temnospondyl ancestor.

Famennian *Tulerpeton curtum* (Lebedev 1984) as a stem-group amniote (see also Clack and Carroll 2000). *Tulerpeton* has been neglected in most recent analyses, despite the fact that it is known from well preserved, although incomplete, postcranial material (see Lebedev and Clack 1993 and Ahlberg and Clack 1998 for a discussion of cranial and lower jaw elements attributed to this taxon). However, subsequent studies (e.g. Ahlberg and Clack 1998; Clack 2002; Ruta *et al.* 2003) concur in assigning *Tulerpeton*, as well as all other Devonian taxa, to the tetrapod stem-group. Nevertheless, we acknowledge that hypotheses of a Late Devonian separation between lissamphibians and amniotes are consistent with some recent molecular analyses (e.g. Kumar and Hedges 1998; Hedges 2001).

Mississippian tetrapods are rare. Incomplete remains from the mid-Tournaisian site of Horton Bluff, Nova Scotia (Clack and Carroll 2000), are the oldest documented examples, but these specimens cannot be diagnosed unambiguously as 'batrachomorph' or 'reptiliomorph'. Some isolated humeri appear to be morphologically intermediate between those of *Tulerpeton* (Lebedev and Coates 1995) and the 'anthracosauroid' *Eoherpeton* (Smithson 1985; Clack and Carroll 2000), while others are more similar to colosteid humeri (Godfrey 1989). Additional specimens include femora as well as endochondral and dermal shoulder girdle elements. Recently discovered, mid-Viséan remains from central Queensland represent the only record of Carboniferous tetrapods from East Gondwana (Thulborn *et al.* 1996). Although fragmentary, this fauna is thought to include the earliest known representatives of colosteids and 'anthracosauroids' (*fide* Thulborn *et al.* 1996; Clack and Carroll 2000).

The next oldest Mississippian record is represented by a *Whatcheeria*-like animal from the late Tournaisian of Scotland (Clack and Finney 1997). Like *Whatcheeria* (Lombard and Bolt 1995; Bolt and Lombard 2000), the new tetrapod reveals an array of 'reptiliomorph', 'batrachomorph', and primitive features. The manual character analysis of Lombard and Bolt (1995) and the computer-assisted analyses of Coates (1996), Clack (1998b,d), and Paton *et al.* (1999) concur in assigning *Whatcheeria* to the basal portion of the 'reptiliomorph' branch of the tetrapod tree. Certain recent, comprehensive analyses (Laurin and Reisz 1997, 1999; Laurin 1998a–c; Anderson 2001) have ignored *Whatcheeria*. Other studies, including Ahlberg and Clack's (1998) and our own (Ruta *et al.*, 2003), suggest that *Whatcheeria* is a stem-tetrapod. The exceptional preservation and abundant material of *Whatcheeria* provide an important data source for comparative anatomical and phylogenetic studies of early tetrapods. The sequence of branching events in the crownward part of the tetrapod stem-group is the subject of much current debate and may ultimately lead to a re-assessment of the polarity of several characters. A detailed study of *Whatcheeria* and the new *Whatcheeria*-like animal from the Scottish Tournaisian will certainly prove to be crucial in this respect.

Casineria kiddi, a 340 million years old, incomplete skeleton from Gullane, Cheese Bay, Scotland (lower part of late Viséan), is the next animal to be considered. Regarded as the earliest undisputed amniote, it is the oldest tetrapod showing a pentadactyl forelimb, and predates the uppermost Viséan fauna from the Scottish site of East Kirkton (see below). Its relatively low, squared off neural spines are reminiscent of those of such primitive 'anthracosauroids' as *Silvanerpeton* and *Eldeceeon* (Clack 1994; Smithson 1994), whereas its long, curved ribs, separate scapular and coracoid ossifications and proportions of the manus are similar to those of certain embolomeres and various basal crown-group amniotes. However, different combinations of these features have also been observed in other taxa, such as certain microsaurs and *Whatcheeria* (Carroll and Gaskill 1978; Lombard and Bolt 1995; Bolt and Lombard 2000). Although Paton *et al.*'s (1999) cladistic analysis identified *Casineria* as a basal amniote, it is noteworthy that it failed to resolve its position relative to such diverse taxa as *Westlothiana*, *Captorhinus*, *Petrolacosaurus*, and *Paleothyris* (Ruta *et al.* 2003).

Several other taxa with possible 'reptiliomorph' affinities have been included in our analysis. *Westlothiana lizziae* from East Kirkton is usually regarded as one of the most primitive stem-group amniotes (Smithson 1989; Smithson and Rolfe 1990; Smithson *et al.* 1994). However, Laurin and Reisz (1999) placed this taxon as the closest outgroup to the tetrapod crown-clade. Conversely, our analysis strengthens Smithson *et al.*'s (1994) conclusions and offers a novel perspective for interpreting the puzzling mixture of 'lepospondyl' as well as basal amniote features in *Westlothiana* (see also the analysis in Anderson 2001). *Caerorhachis bairdi*, probably from the lowermost Serpukhovian of Scotland, was originally described as a basal temnospondyl (Holmes and Carroll 1977), but has been reinterpreted as a basal stem-amniote by Ruta *et al.* 2003 (see also discussion in Milner and Sequeira 1994; Coates 1996).

The late Viséan *Crassigyrinus scoticus* has been the subject of controversy ever since its discovery. Panchen (1985) and Panchen and Smithson (1990) redescribed its cranial and postcranial anatomy. Panchen and Smithson (1988) placed it on the 'reptiliomorph' branch of their tetrapod cladogram, either as sister taxon to 'anthracosauroids', or as sister taxon to a clade encompassing 'anthracosauroids' and seymouriamorph as sister group to diadectomorphs plus crown-group amniotes. Further preparation of the material resulted in a reassessment of the morphology of the palate (Clack 1996), snout, and skull roof (Clack 2000) leading to the recognition of an array of plesiomorphic features. Some recent analyses (Coates 1996; Clack 1998b,d; Paton *et al.* 1999), place *Crassigyrinus* as a basal embolomere, whereas Laurin and Reisz (1997, 1999), Ahlberg and Clack (1998) and Laurin (1998a–c) identify it as a crownward stem-tetrapod. The latter conclusion is also supported by Ruta *et al.* (2003).

The nature of the conflict

From the account above, it is clear that the base of the tetrapod crown-group has uncertain boundaries. In fact, only the amniote affinities of *Casineria* (see also below) remain uncontroversial. The existence of incongruent tree topologies is due to several causes that are not mutually exclusive. The use of incomplete or poorly preserved taxa is likely to result in multiple, equally parsimonious solutions. While implicitly assumed in all fossil-based studies, the influence of such taxa on cladogram topology remains largely unexplored (but see Wilkinson 1995; Anderson 2001; Kearney 2002). However, as demonstrated by Coates (1996), incomplete taxa (e.g. *Hynerpeton*; Daeschler *et al.* 1994) do not necessarily behave as 'rogue' OTUs. Sometimes, the presence of just one unambiguous synapomorphy is sufficient to stabilize the affinities of fragmentary material.

Another potential source of character conflict is the fact that various taxa sharing features with two or more different groups deliver confounding signals. In simple cases, variations in the taxon sample are likely to affect the outcome of an analysis through 'attraction' of such 'chimaera'-like taxa. However, the effects of taxon and/or character deletions/inclusions are not predictable. In those cases in which an optimal 'balance' of taxa and characters is achieved, the position of key fossils may remain unresolved. Clack's (2001) analysis provides an excellent example of this taxon/ character interplay. Specifically, a clade consisting of *Eucritta* and baphetids forms a trichotomy with temnospondyls and a diverse group including *Crassigyrinus*, *Whatcheeria*, gephyrostegids, and embolomeres (but see also Clack 1998a) in the two equally parsimonious trees discussed by Clack (2001).

Several groups of early tetrapods are so specialized that they provide little or no indication as to their possible ancestry or sister group. Carroll (2001) has emphasized this observation repeatedly, identifying the apparent excess of apomorphies and widespread homoplasy as responsible for obscuring relationships among basal crown-group tetrapods. However, while homoplasy might be widespread, we think it unlikely that the current tetrapod database contains insufficient phylogenetic signal. Thus, a quick inspection of published analyses reveals that the structure of several matrices is not random. A comparison between two of the most comprehensive datasets – Carroll's (1995) and Laurin and Reisz's (1999) – serves to illustrate this point. Despite the use of different taxon and character samples, Carroll's (1995) and Laurin and Reisz's (1999) cladograms are mostly congruent. Crown-lissamphibians are placed among lepospondyls in Laurin and Reisz's study, but are excluded from Carroll's analysis. If lissamphibians are not taken into account, the sequences of branching events in Carroll's (1995) and Laurin and Reisz's (1999) tree topologies are remarkably similar. Minor differences concern the mutual relationships of the lepospondyl orders, the position of *Westlothiana* (grafted to a diadectomorph–amniote clade in Carroll's analysis, but sister taxon to a diadectomorph–amniote–lepospondyl clade in Laurin and Reisz's) and the pattern of sister group relationships between baphetids, colosteids, and temnospondyls (all three groups branch from adjacent nodes in both analyses). It is also noteworthy that (excluding Laurin and Reisz's location of lepospondyls) the branching sequence in the basal stretch of the putative stem-amniote groups (e.g. embolomeres, gephyrostegids, seymouriamorphs, *Westlothiana*) resembles that proposed by several earlier authors (e.g. Lombard and Sumida 1992; Smithson *et al.* 1994; Lee and Spencer 1997; Sumida 1997).

Results

A new analysis for early tetrapods

Recent advances in our knowledge of early tetrapod anatomy have contributed to an expanded and refined database (Trueb and Cloutier 1991; Coates 1996; Laurin and Reisz 1997, 1999; Ahlberg and Clack 1998; Clack 1998b; Laurin 1998a–c; Lombard and Bolt 1999; Paton *et al.* 1999; Bolt and Chatterjee 2000; Schoch and Milner 2000; Yates and Warren 2000; Bolt and Lombard 2001). In our analysis, we have sought to use the maximum practical range of taxon exemplars, consistent with methodological arguments arising from a series of recent studies (Nixon and Davis 1991; Anderson 2001; Prendini 2001; Salisbury and Kim 2001; Ruta *et al.* 2003).

The new data matrix encompasses 90 tetrapod species coded for 213 cranial and 94 postcranial characters. The results support the hypothesis of a deep evolutionary split between stem-lissamphibians and stem-amniotes. Further major features of these results are summarized as follows (Figure 11.1):

(1) The post-panderichthyid part of the tetrapod stem-group includes, in crownward order, *Ventastega curonica*, *Acanthostega gunnari*, *Ichthyostega stensioei*, *Tulerpeton curtum*, Colosteidae, *Crassigyrinus scoticus*, *Whatcheeria deltae*, and Baphetidae.
(2) *Caerorhachis bairdi*, embolomeres, gephyrostegids, *Solenodonsaurus janenschi*, seymouriamorphs, a clade consisting of *Westlothiana lizziae* plus lepospondyls, and diadectomorphs are progressively more crownward stem-amniotes.
(3) Within lepospondyls, microsaurs are paraphyletic relative to lysorophids, adelospondyls (including *Acherontiscus*), and a clade encompassing nectrideans plus aïstopods.
(4) *Eucritta melanolimnetes* is basal to temnospondyls, which form a paraphyletic array of taxa relative to crown-lissamphibians.
(5) Albanerpetontids and a diverse dissorophoid clade consisting of branchiosaurids, micromelerpetontids, and amphibamids are successively more outlying sister groups of crown-lissamphibians.
(6) Caecilians are the sister group to a salientian–caudate clade.

The tetrapod crown-group is bracketed at its base by *Eucritta* and *Caerorhachis*, a pair of Scottish taxa noted for their mixture of features otherwise considered to be characteristic of such different groups as temnospondyls, baphetids, and 'anthracosauroids' (Clack 1998b, 2001; Ruta et al. 2001). A comprehensive treatment of the characters and results of the new analysis is presented elsewhere (Ruta et al. 2003). PAUP* 4.0b10 (Swofford 1998; see Ruta et al. 2003 for details of the search settings used) finds 60 shortest trees at 1303 steps. If *Casineria* (Paton et al. 1999) and *Silvanerpeton* (Clack 1994) are included in the analysis, then a strict consensus of the resultant 120 equally parsimonious trees shows considerable loss of resolution in the basal part of the amniote stem-group. The polytomy subtends *Casineria*, *Silvanerpeton*, embolomeres, gephyrostegids, *Solenodonsaurus*, *Discosauriscus*, *Kotlassia*, and *Seymouria*. However, an agreement subtree shows that *Silvanerpeton* branches from the amniote stem between *Caerorhachis* and embolomeres (see also Clack 1994), but that *Casineria* is a 'rogue' taxon. Despite its uncertain placement, *Casineria* emerges, consistently, as a stem-amniote, in partial agreement with Paton et al.'s (1999) conclusions.

The new analysis supports traditional views on the amniote affinities of 'anthracosaurs', seymouriamorphs, and diadectomorphs (Panchen and Smithson 1987, 1988; Lombard and Sumida 1992; Lee and Spencer 1997; Sumida 1997), and identifies temnospondyls as a paraphyletic grade group on the lissamphibian stem (Milner 1988, 1990, 1993, 2000). The general results resemble most closely those obtained by Carroll (1995), especially with regards to the monophyly of lepospondyls and their placement on the amniote stem. We are currently evaluating the nature of the lepospondyl groups and the degree of support (morphological as well as statistical) assigned to various nodes within this assemblage (see also Ruta et al. 2003). Thus, while the position of microsaurs on the amniote stem-group is also retrieved in experiments of taxon and/or character deletion and reweighting, the placement of remaining lepospondyls can be affected drastically. For instance, when post-cranial data are omitted from the analysis, the relationships of remaining lepospondyls change significantly: they are relocated as stem-group tetrapods, as the sister group to colosteids. Similar results are obtained if nectrideans and lysorophids are excluded from the dataset. In this case, aïstopods are paired with adelospondyls and, together, they form the sister group to colosteids. The evolutionary implications of these results have yet to be explored in depth. Carroll (1999) has suggested that similarities between lepospondyls and primitive amniotes (especially in the configuration of the vertebrae) represent convergent features related to precocious ossification attained at a small body size. However, the stem-amniote position of microsaurs is not affected by deletion of postcranial characters (Figure 11.10). It is possible that lepospondyl monophyly in the original analysis results from the cumulative effect of implied reversals and optimizations of missing entries related to cranial and postcranial features. Further work in this area is needed.

The results match those of certain previous studies, especially with regards to the position of lissamphibians and the branching pattern in the basal part of the amniote stem. This is unsurprising, because the matrix includes, so far as possible, the majority of characters used in previous analyses (details in Ruta *et al.* 2003), as well as further data from smaller morphological sets (e.g. Trueb and Cloutier 1991). As an additional test of the performance of character subsets, we excluded lower jaw data. Removal of these has no major effect on the overall tree topology. The latter matches the results retrieved in the original analysis, except that crown-lissamphibians are more deeply nested in the derived portion of the temnospondyl tree, whereas most tuditanomorphs are collapsed in a large polytomy. We conclude that cranial and postcranial characters are not in conflict with lower jaw data (but see discussion in Ahlberg and Clack 1998).

Elsewhere (Coates *et al.* 2000), we pondered a few of the biological implications of taxon rearrangements in Laurin's (1998a–c) preferred tree topology, in which lysorophids are the hypothesized closest relatives to frogs, salamanders, and caecilians. We concur with Carroll (2001) and Carroll and Bolt (2001) that hardly any feature of crown-lissamphians can be identified as a convincing synapomorphy shared uniquely by lysorophids with each of the three lissamphibian orders. However, Laurin *et al.* (2000b) correctly point out that grafting lissamphibians to temnospondyls is a much worse fit for their data than the topology retrieved from earlier analyses (e.g. Laurin 1998a–c). Prompted by Laurin *et al.*'s (2000b) suggestion that additional phylogenetic analyses should be performed to test the origin of lissamphibians, we have added characters that have been proposed previously as putative shared features of temnospondyls and lissamphibians (e.g. Bolt 1969, 1977, 1979, 1991; Milner 1988, 1990, 1993, 2000; Trueb and Cloutier 1991; Gardner 2001; Ruta *et al.*, in press). Our analysis favours dissorophoids as the closest relatives of lissamphibians among the vast array of Palaeozoic tetrapods.

A recent study by Yeh (2002) on the effect of miniaturization on the skeleton of frogs has shown that, although paedomorphosis is responsible for the loss of several skull bones in miniaturized vertebrates, there is no simple correlation between such losses and small size. However, several bones that ossify late during development, such as quadratojugals, columellae, and palatines, are also those that are lost most frequently. In most anurans, such bones are usually post-metamorphic. Therefore, their loss is plausibly linked to paedomorphosis. In addition, miniaturization may affect members of the same clade in profoundly different ways. Interestingly, the medial skull elements of miniaturized frogs (e.g. parasphenoid) are transversely expanded, whereas the lateral elements (e.g. pterygoids) are laterally compressed. Certain bones are shortened in comparison with their homologues in non-miniaturized frogs (e.g. maxilla, quadratojugal, vomer). Several of these features are also recorded in certain dissorophoids. Striking similarities between the ontogenetic changes in the skull of various modern lissamphibians and those of amphibamids and branchiosaurids add strength to the temnospondyl hypothesis of lissamphibian ancestry (Milner 1988, 1990, 1993, 2000; Schoch 1992, 1995, 1998; Boy and Sues 2000; Carroll 2001). The list of 'absence' features that link lysorophids to lissamphibians in Laurin and Reisz's (1997, 1999) and Laurin's (1998a–c) analyses calls for a cautious treatment of character losses and characters associated with small size. As noted by Milner (1988), examples of convergence among fossil and extant amphibians are widespread. Therefore, the assessment of their relationships cannot rely upon comparisons between very few representatives of Palaeozoic and Recent groups or upon selection of a limited number of putative shared derived similarities. Instead, efforts should be directed towards the recognition of the group in which the internal relationships best reflect the most coherent, inter-nested set of lissamphibian synapomorphies. We argue that temnospondyls show a coherent nested set of this type.

Crown-tetrapod origin and the apex of the tetrapod stem-group

The following analyses were considered: Carroll 1995 (Figure 11.2); Coates 1996 (Figure 11.3); Ahlberg and Clack 1998 (Figure 11.4; see also Figure 11.11 in Appendix 11.1); Laurin and Reisz 1999 (Figure 11.5); Paton et al. 1999 (Figure 11.6); Anderson 2001 (Figure 11.7). For each analysis, the inferred minimum age for the lissamphibian-amniote phylogenetic separation is bracketed between 325 and 345 Ma (mid- to late Viséan), in agreement with the conclusions of several previous works (e.g. Clack 1998b,d, 2001, 2002; Paton et al. 1999; see also comments in Coates et al. 2000). Importantly, divergence time estimates are not affected by the relative positions of unstable/rogue taxa (e.g. baphetids, Caerorhachis, Crassigyrinus, Eucritta, Whatcheeria, and various lepospondyl groups) or by the degree of tree balance. For example, comparisons between Laurin and Reisz's (1999) analysis (Figure 11.5) and ours (Figure 11.8) reveal a decrease in stem-tetrapod groups, a decrease in stem-lissamphibian groups, and an increase in stem-amniote groups. Both analyses, however, place aïstopods within the tetrapod crown-group (as stem-lissamphibians or stem-amniotes, respectively). These findings necessarily imply a mid-Viséan age as a minimum hypothesis for the date of the lissamphibian-amniote separation (Figure 11.9). This is largely based on the mid-Viséan occurrence of the earliest known aïstopod, Lethiscus (Wellstead 1982).

Stratigraphical data can provide no more than the best approximation of the lissamphibian–amniote divergence time, based on the available sample of fossil material. The absence of an adequate Tournaisian tetrapod record (Coates and Clack 1995; Lebedev and Coates 1995; Coates 1996; Clack and Finney 1997; Paton et al. 1999; Clack and Carroll 2000; Clack 2002), relative to that from flanking stages, affects

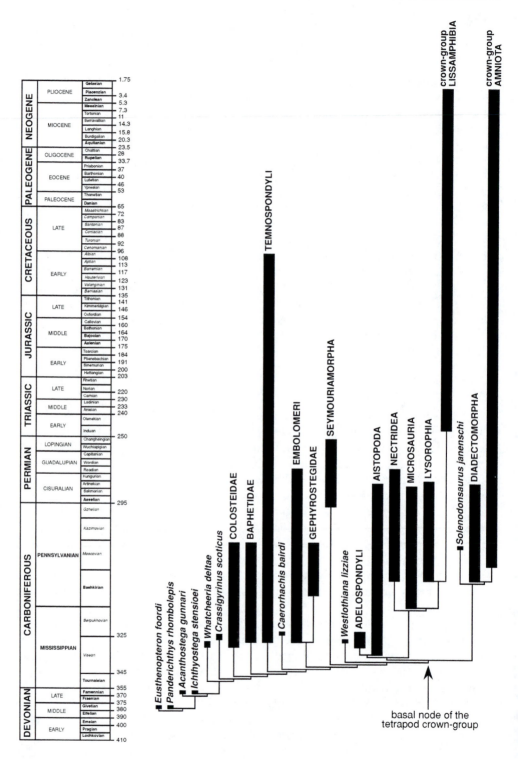

Figure 11.5 Laurin and Reisz's (1999) analysis plotted on a timescale.

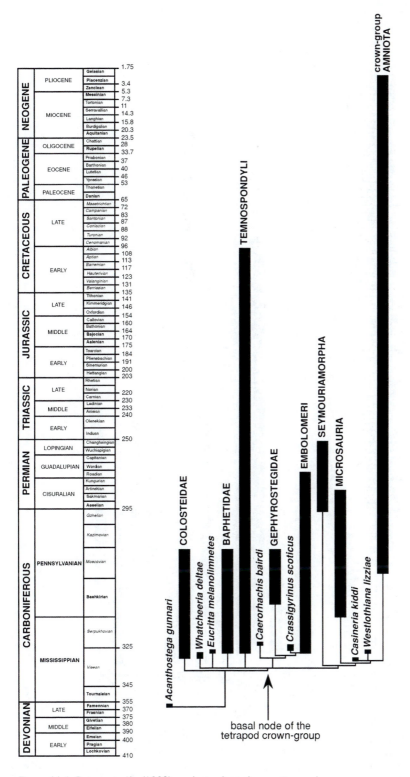

Figure 11.6 Paton *et al.*'s (1999) analysis plotted on a timescale.

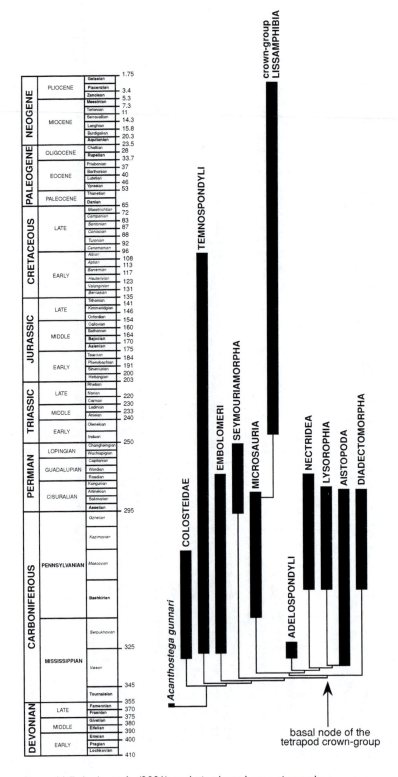

Figure 11.7 Anderson's (2001) analysis plotted on a timescale.

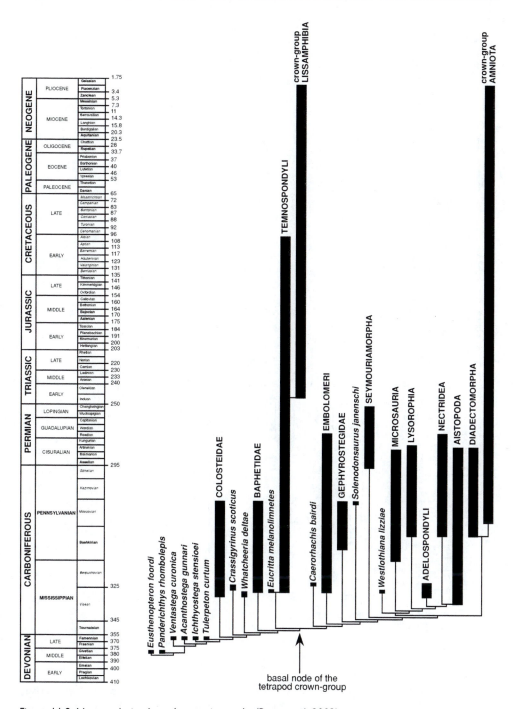

Figure 11.8 New analysis plotted on a timescale (Ruta *et al.* 2003).

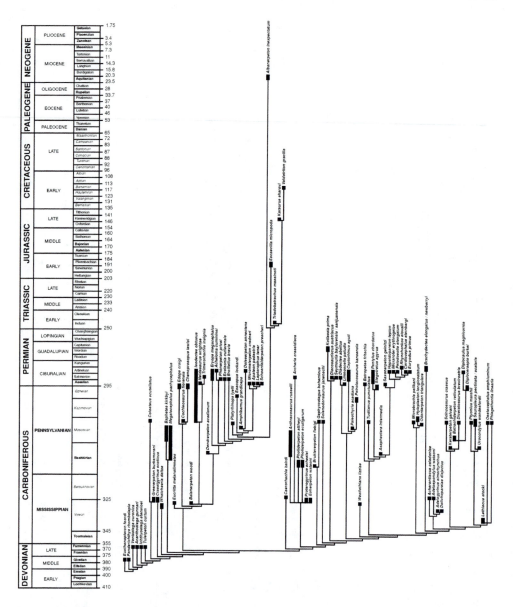

Figure 11.9 Stratigraphical plot of strict consensus of 60 most parsimonious trees at 1303 steps (CI = 0.2357; RI = 0.6744; RC = 0.1641) derived from analysis of cranial and postcranial characters and resolved down to species level (Ruta *et al.* 2003).

theories of divergence times, insofar as cladogenetic events can only be plotted within the Viséan and later, or within the Famennian and earlier periods. Therefore, the apparent consensus between widely conflicting tree topologies about the time of origin of the Tetrapoda is only significant because all recent analyses fail to place this event before the Devonian–Carboniferous boundary. However, the observed 'time signal' is not exclusively under stratigraphical control, since alternative phylogenies based on novel character and/or taxon combinations could move the crown-tetrapod origin event to either side of the Tournasian gap. The quality of the signal is nevertheless compromised by the patchiness of the contributing data. Consequently, character deletion experiments (see Ruta *et al.* 2003, and discussion above) are likely to have only minimal effects. Thus, while postcranial character removal increases the number of stem-tetrapod taxa (Figure 11.10), this causes only a small change in the minimum estimate of the crown-group divergence time, from mid- to late Viséan.

Crown-lissamphibian origin

There is general agreement on the taxonomic composition of the basal portion of the lissamphibian crown-group (Báez and Basso 1996; Gao and Shubin 2001). The Early Triassic stem-salientian *Triadobatrachus massinoti* from Madagascar is the earliest undisputed crown-lissamphibian, and predates the basal members of the caudate and caecilian orders – *Karaurus sharovi* and *Eocaecilia micropoda* from the Late and Early Jurassic, respectively (Ivakhnenko 1978; Milner 1988, 1993, 2000; Rage and Rocek 1989; Jenkins and Walsh 1993; Carroll 2000; Rocek and Rage 2000b). The Middle Jurassic karaurid caudate *Kokartus honorarius* (Nessov 1988; Nessov *et al.* 1996) is older than *Karaurus*, but is usually regarded as a paedomorphic relative of the latter. Problematic taxa such as *Triassurus sixtelae* Ivakhnenko, 1978, variously interpreted as a Triassic stem-caudate or as a temnospondyl larva (review in Milner 2000), are too poorly known. Likewise, the systematic affinities of various Jurassic 'salamander-like' taxa (e.g. batrachosauroidids and scapherpetontids) remain uncertain (Milner 2000).

The total analysis implies the existence of a mid-Pennsylvanian to Early Triassic ghost lineage connecting albanerpetontids plus crown-lissamphibians with a dissorophoid assemblage consisting of the amphibamid, micromelerpetontid, and branchiosaurid families. The duration of this lineage is disconcertingly longer than that postulated by previous studies (Permian to Early Triassic; e.g. Trueb and Cloutier 1991; Milner 1993) and found also in the cranial analysis, wherein the Early Permian *Broiliellus* is the immediate sister taxon to albanerpetontids and crown-lissamphibians (Figure 11.10). Taken at face value, these results suggest the existence of as yet unknown Permo-Carboniferous taxa into which lissamphibian ancestry is rooted (but see discussion in Ruta *et al.* 2003). No crown-lissamphibian has been recorded in the Late Permian. We point out, however, that our analysis does not consider all known dissorophoids (reviewed in Milner 1990), most of which require revision. A resolution of the sister group relationships between the three lissamphibian orders and one or more specific dissorophoid taxa must await a thorough phylogenetic analysis of crownward temnospondyls. Interestingly, Shishkin (1998) discussed a relict and possibly neotenous dissorophoid (*Tungussogyrinus*) from the Permian–Triassic boundary in Siberia. However, the affinities of this fossil are uncertain.

Crown-amniote origin

The earliest known, undisputed crown-amniotes date back to the mid-Pennsylvanian (Carroll 1988, 1991; Benton 1991, 1993, 2001; Carroll and Currie 1991; Hopson

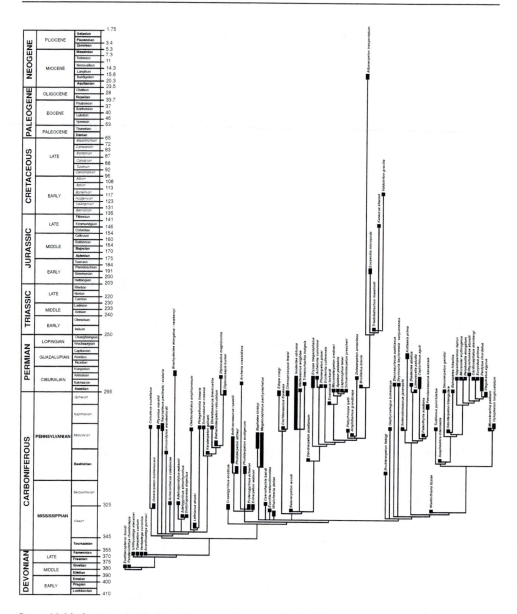

Figure 11.10 Stratigraphical plot of strict consensus of 100 440 most parsimonious trees at 945 steps (CI = 0.2447; RI = 0.6835; RC = 0.17) derived from analysis of cranial characters and resolved down to species level (Ruta *et al.* 2003).

1991). Both stem-diapsids and primitive synapsids are represented in the Moscovian (families Protorothyrididae and Ophiacodontidae, respectively) and in the Kasimovian (families Petrolacosauridae and Edaphosauridae, respectively). The divergence between mammals and sauropsids, placed at about 310 Ma in the Pennsylvanian, has been widely used to calibrate molecular clocks (Kumar and Hedges 1998; Hedges 2001; Van Tuinen *et al.* 2001). The timing of this event is not affected by the controversial locations of turtles and of various Permo-Carboniferous and Triassic forms in current amniote phylogenies (e.g. Laurin 1991; Laurin and Reisz 1995; Lee 1995, 1997a, b; Rieppel and deBraga 1996; deBraga and Rieppel 1997; Rieppel and Reisz 1999; Rieppel 2000).

The sequence of branching events in the basal part of the amniote stem-group is not agreed upon. Our analysis reflects established views on the position of such groups as seymouriamorphs and diadectomorphs, but differs from many previous analyses in the relatively basal position of *Solenodonsaurus* (Lee and Spencer 1997; Laurin and Reisz 1999). Virtually no fossils have been proposed as immediate sister groups to amniotes, crownward of diadectomorphs (Sumida and Lombard 1991; Berman *et al.* 1992; Lombard and Sumida 1992; Sumida *et al.* 1992; Laurin and Reisz 1997, 1999; Lee and Spencer 1997; Sumida 1997; Berman *et al.* 1998; Laurin 1998a–c; Berman 2000; Ruta *et al.* 2003).

Dating phylogenetic events

Comparison between morphological and molecular analyses

According to Kumar and Hedges (1998) and Hedges (2001), lissamphibians and amniotes diverged at around 360 Ma ± 14.7 myr in the Famennian (Late Devonian; see also Panchen and Smithson 1987, 1988; Lebedev and Coates 1995; Coates 1996). The upper boundary of this time interval falls within the upper part of the Tournaisian, whereas the lower boundary coincides with the basal part of the Frasnian. In all cases, the lissamphibian–amniote divergence is postulated to have occurred earlier than available fossil evidence suggests. The mismatch between molecular and morphological data cannot be explained easily (see Smith 1999, for a comparable example involving metazoan divergence dates, and other chapters in this volume). However, we note that the mid- to late Viséan separation between lissamphibians and amniotes inferred from morphological analyses falls slightly short of the upper boundary of Kumar and Hedges' (1998) and Hedges' (2001) time interval. As mentioned above, some tetrapod humeri from Tournaisian sediments in Nova Scotia (Clack and Carroll 2000) resemble in their general proportions those of certain Viséan 'reptiliomorphs', notably *Eoherpeton* (Smithson 1985), although the evidence is not compelling. Furthermore, the same sediments have yielded putative colosteid-like humeri, suggesting the occurrence of deeper branching events for at least some tetrapod groups. In addition, the presence of the stem-amniote *Casineria* in the middle part of the late Viséan shows that 'reptiliomorph' diversification was already under way by about 340 Ma (Paton *et al.* 1999).

It is much more difficult to reconcile a Frasnian age for the lissamphibian–amniote phylogenetic split (the lower boundary of Kumar and Hedges' 1998, and Hedges' 2001, time interval) with available fossil data (summary in Coates 2001). All Devonian tetrapods with limbs postdate fish-like stem-tetrapods, such as *Panderichthys* and *Eusthenopteron*. None of them are currently regarded as a member of the crown-group. Also, the basal node of the tetrapod crown-group cannot be rooted into known Devonian taxa (e.g. Panchen and Smithson 1988) without implying an impressive series of convergent character-changes in the most basal portions of the lissamphibian and amniote stem-groups. However, putative tetrapod trackways recorded in Australia,

Ireland, and Scotland (see reviews in Clack 1998a, 2000) could be used to hypothe-size the existence of an as yet unrecorded radiation of limbed tetrapods during the Frasnian–Famennian. The dating of several track-bearing sediments is disputed, but in certain cases, a Middle to Late Devonian age has been postulated.

A better agreement between morphological and molecular time-calibrated trees is evident by comparing minimum estimates of crown-lissamphibian origins (see also above). Thus, both morphological analyses and molecular studies (e.g. Báez and Basso 1996; Feller and Hedges 1998; Gao and Shubin 2001) support an early Mesozoic divergence for crown-lissamphibians. According to Feller and Hedges (1998), the Early Triassic age of *Triadobatrachus* implies that the three orders of lissamphibians originated in the Palaeozoic under the traditional hypothesis of a sister group relationship between salientians and caudates. Indeed, the morphology of *Triado-batrachus* appears almost exactly intermediate between that of more derived frogs and various derived dissorophoids (Milner 1988; Rocek and Rage 2000a,b). Although the gymnophionan–caudate clade [= Procera] proposed by Feller and Hedges (1998) may imply a later evolutionary event for the origin of caecilians and salamanders relative to frogs, this branching sequence is *not* incompatible with the possibility that pre-Jurassic (or even Late Palaeozoic) representatives of caecilians and salamanders may be discovered. Although Feller and Hedges (1998) found morphological support for their Procera, it is at present difficult to propose a suitable candidate for the stem-group membership of this clade (but see McGowan and Evans 1995).

Conflict or compromise?

Agreement between morphology and molecules in reconstructing the timing of major evolutionary events is rare. Discrepancies between different data sources for several taxonomic groups are well documented. In the case of metazoans, birds, and mammals, for instance, molecular analyses indicate that these groups are twice as old as their oldest fossil representatives. Instances of molecular estimates falling short of morphological estimates exist, but are much rarer (e.g. Easteal and Herbert 1997). Several factors have been identified as responsible for the mismatch between molecules and morphology (cf. Cooper and Fortey 1998; Benton 1999; Smith 1999), including the presumed rarity of ancestral forms of major groups (let alone problems with the recognition of ancestors), their preservation potential, and their possible occurrence in places that have not yet been subject to thorough scrutiny. Further-more, failure to distinguish between the origin of the living members of a Recent clade (crown-group diversification) and the date of separation of the latter from its extant sister group (total-group divergence) may lead to biased assessments of origination times (e.g. Easteal 1999). For instance, assuming the accuracy of our new hypothesis of tetrapod relationships (Ruta *et al.* 2003), a time interval of about 30 million years separates the earliest undisputed crown-amniotes from *Casineria*. On the lissamphibian stem, the time interval between the earliest undisputed crown-lissamphibian, *Triadobatrachus*, and the earliest known temnospondyls is about 75 million years (Figure 11.9). Furthermore, as pointed out by van Tuinen *et al.* (2001), fossil-based calibrations of molecular clocks are inevitably sensitive to fossil dating and phylogeny reconstruction (for a comprehensive discussion, see also Wagner 2000). For this reason, they emphasize the importance of introducing confidence limits around such widely used, fossil-based calibration tools as the synapsid–diapsid divergence time (see above).

Another important issue is represented by the erratic behaviour of molecular clocks, a discussion of which was presented by Ayala (1999). Briefly, several factors (e.g. population size, time elapsed between generations, species-specific occurrences of genetic mutations, changes in protein functions, and changes in the adaptation of organisms to their environments) may speed up or slow down molecular clocks (Cooper and Fortey 1998; Benton 1999; Smith 1999). Examination of combined information from a large number of genes has been proposed as an effective tool to reduce drastically the errors introduced by limited sequence data (e.g. Kumar and Hedges 1998; Ayala 1999; Hedges 2001; Stauffer *et al.* 2001). The discussion thus far shows that the most problematic incongruence between molecular and morphological time trees concerns the age of the tetrapod crown-group radiation. This lack of agreement could result from inaccuracies of molecular clock estimates. Smith (1999) has summarized cases in which rates of molecular evolution might change dramatically, both at the start of clade radiation, and in terminal portions of the tree relative to deeper nodes. For example, if genetic changes in a sufficient number of gene families were slowed down at the beginning of the crown-amniote radiation (one of the most widely used calibration points; Feller and Hedges 1998; Kumar and Hedges 1998; Hedges 2001), then molecular data would deliver an excessively early origination date; certainly much older than that estimated from fossils. We note that such a model of varying molecular clock-speed is consistent with the greater agreement between molecular and morphological estimates of crown-lissamphibian origin (since the crown-lissamphibian radiation is far more recent than that of crown-tetrapods).

Sample bias and 'site' effect

Improved molecular methods and techniques (e.g. Hedges 2001), and increased consistency of divergence times, between different gene samples and calibration points (e.g. Stauffer *et al.* 2001), make it appear *a priori* that the mismatch between palaeontological and molecular estimates for divergence times is caused by deficiencies of the fossil record. However, this is strongly disputed in the case of certain groups (notably, birds and mammals; Benton 1999). Sample bias is an important factor when dealing with palaeontological data. Benton and Hitchin (1996) and Benton *et al.* (2000) used cladograms from a wide range of groups to test the quality of the fossil record, which they acknowledge as decreasing dramatically backwards in time. Older fossils are more liable to physical and chemical destruction than younger ones. The former are often more difficult to interpret and to place in a phylogenetic context than the latter. In addition, it is reasonable to assume that taxa that lie phylogenetically close to cladogenetic events are rare.

Benton and Hitchin (1996) and Benton *et al.* (2000) argue that, although the 'completeness' of the fossil record may be lower in the Palaeozoic than in the Cenozoic, its 'adequacy' in recounting major evolutionary events is maintained. Newly discovered taxa are more likely to fit within well-established higher categories, and to redefine only lower ranks (e.g. splitting or clumping genera and species). It follows that differences in fossil dating are only significant at the level of fine chronostratigraphical subdivisions (e.g. stages). The quality of the fossil record is thus interpreted as more or less uniform when families are used as OTUs and the strati-graphical column is scaled to stages. Therefore, it is unlikely that discoveries of new members of well-characterized Palaeozoic tetrapod clades will have any impact upon the branching sequence and chronology of key events in tetrapod history (although they may cast new light on the intrinsic relationships of the groups to which they belong). Nevertheless, certain discoveries are crucial, as in the case of fossils displaying mosaics of features previously considered diagnostic of higher level, distinct clades (Clack 2001; Ruta *et al.* 2001). Moreover, the methodology employed by Benton and Hitchin (1996) and Benton *et al.* (2000) treats phylogenetic reconstruction as independent of sampling order, even though sampling intensity (the probability that a taxon is sampled per given unit time) affects phylogenetic accuracy (Wagner 2000, and references therein).

Large gaps in the early tetrapod record, most notably the Tournaisian, persist. Sample quality from this time interval is thus extremely poor compared with more recent deposits (Benton 1999). In fact, most discoveries of early tetrapods have resulted from pro-longed, concentrated collecting efforts in a limited number of stratigraphical horizons (e.g. Wood *et al.* 1985; Rolfe *et al.* 1994), although fortuitous finds remain an occasional source of important new data (Clack and Finney 1997; Paton *et al.* 1999). Consequently, certain key fossil sites have a disproportionate influence, most partic-ularly East Kirkton in the Scottish late Viséan (Rolfe *et al.* 1994). Key East Kirkton taxa responsible for pegging divergence dates on the tree include the putative stem-lissamphibians *Balanerpeton* and *Eucritta*, and the stem-amniote *Westlothiana* (Figure 11.1). There is nothing unique to East Kirkton and early tetrapod phylogeny in this respect; such site effects are applicable to the vast majority of fossil-based estimates of evolutionary timing.

Phylogenetic reconstructions cannot be regarded as finished works, because the discovery of just one new fossil may overturn previous hypotheses about character distribution and polarity. Therefore, fossil-based estimates of major evolutionary events are not necessarily in conflict with, or challenged by, existing molecular estimates. However, we point out that this is true only if molecular estimates exceed those implied by morphology. If fossil estimates exceed molecular estimates, then it appears to us that a real conflict exists. As suggested by Stauffer *et al.* (2001), one of the best uses for molecular clock time trees is their ability to provide a framework to evaluate (and, possibly, constrain) palaeontological hypotheses of divergence. Therefore, in agreement with Hedges and Maxson (1997), molecular and palaeontological data are best used as complementary approaches to dating phylogenetic events.

Acknowledgements

We thank Drs Philip Donoghue and Paul Smith (University of Birmingham, UK) for inviting us to contribute this paper to the one-day symposium '*Telling the evolutionary time: molecular clocks and the fossil record*' at the Third Biennial Meeting of the Systematics Association, Imperial College, University of London, UK. We are grateful to them and to Dr Per E. Ahlberg (NHM, London, UK) for their editorial comments, stylistic suggestions, and constructive criticism of an earlier draft of this work. We benefited from exchange of ideas with Prof. S. Blair Hedges (Pennsylvania State University, USA). Marcello Ruta acknowledges the financial support provided by the Palaeontological Association. This work is part of a research project funded by BBSRC Advanced Research Fellowship no. 31/AF/13042 awarded to Michael I. Coates.

References

Ahlberg. P.E. and Clack, J.A. (1998) 'Lower jaws, lower tetrapods – a review based on the Devonian genus *Acanthostega*', *Transactions of the Royal Society of Edinburgh: Earth Sciences*, 89: 11–46.

Ahlberg, P.E. and Johanson, Z. (1998) 'Osteolepiforms and the ancestry of tetrapods', *Nature*, 395: 792–4.

Ahlberg, P.E. and Milner, A.R. (1994) 'The origin and early diversification of tetrapods', *Nature*, 368: 507–14.

Anderson, J.S. (2001) 'The phylogenetic trunk: maximal inclusion of taxa with missing data in an analysis of the Lepospondyli (Vertebrata, Tetrapoda)', *Systematic Biology*, 50: 170–93.

Ayala, F.J. (1999) 'Molecular clock mirages', *BioEssays*, 21: 71–5.

Báez, A.M. and Basso, N.G. (1996) 'The earliest known frogs of the Jurassic of South America: review and cladistic appraisal of their relationships', *Münchner Geowissenschaftliche Abhandlungen, Reihe A (Geologie und Paläontologie)*, 30: 131–58.

Beaumont, E.H. (1977) 'Cranial morphology of the Loxommatidae (Amphibia: Labyrinthodontia)', *Philosophical Transactions of the Royal Society, London*, B280: 29–101.

Beaumont, E.H. and Smithson, T.R. (1998) 'The cranial morphology and relationships of the aberrant Carboniferous amphibian *Spathicephalus mirus* Watson', *Zoological Journal of the Linnean Society*, 122: 187–209.

Benton, M.J. (1991) 'Amniote phylogeny', in H.-P. Schultze and L. Trueb (eds) *Origins of the Higher Groups of Tetrapods: Controversy and Consensus*, Ithaca: Cornell University Press, pp. 317–30.

—— (1993) *The Fossil Record 2*, London: Chapman & Hall.

—— (1999) 'Early origins of modern birds and mammals: molecules vs. morphology', *BioEssays*, 21: 1043–51.

—— (2001) *Vertebrate Palaeontology*, London: Chapman and Hall.

Benton, M.J. and Hitchin, R. (1996) 'Testing the quality of the fossil record by groups and by major habitats', *Historical Biology*, 12: 111–57.

Benton, M.J., Wills, M.A. and Hitchin, R. (2000) 'Quality of the fossil record through time', *Nature*, 403: 534–7.

Berman, D.S. (2000) 'Origin and early evolution of the amniote occiput', *Journal of Paleontology*, 74: 938–56.

Berman, D.S., Sumida, S.S. and Lombard, R.E. (1992) 'Reinterpretation of the temporal and occipital regions in *Diadectes* and the relationships of diadectomorphs', *Journal of Paleontology*, 66: 481–99.

Berman, D.S., Sumida, S.S. and Martens, T. (1998) '*Diadectes* (Diadectomorpha: Diadectidae) from the early Permian of central Germany, with description of a new species', *Annals of the Carnegie Museum*, 67: 53–93.

Bolt, J.R. (1969) 'Lissamphibian origins: possible protolissamphibian from the Lower Permian of Oklahoma', *Science*, 166: 888–91.

—— (1977) 'Dissorophoid relationships and ontogeny, and the origin of the Lissamphibia', *Journal of Paleontology*, 51: 235–49.

—— (1979) '*Amphibamus grandiceps* as a juvenile dissorophid: evidence and implications', in M.H. Nitecki (ed.) *Mazon Creek Fossils*, New York: Academic Press, pp. 529–63.

Bolt, J.R. (1991) 'Lissamphibian origins', in H.-P. Schultze and L. Trueb (eds) *Origins of the Higher Groups of Tetrapods: Controversy and Consensus*, Ithaca: Cornell University Press, pp. 194–222.

Bolt, J.R. and Chatterjee, S. (2000) 'A new temnospondyl amphibian from the late Triassic of Texas', *Journal of Paleontology*, 74: 670–83.

Bolt, J.R. and Lombard, R.E. (1985) 'Evolution of the tympanic ear and the origin of frogs', *Biological Journal of the Linnean Society*, 24: 83–99.

—— (2000) 'Palaeobiology of *Whatcheeria deltae*, a primitive Mississippian tetrapod', in H. Heatwole and R.L. Carroll (eds) *Amphibian Biology, 4: Palaeontology*, Chipping Norton: Surrey Beatty & Sons, pp. 1044–52.

—— (2001) 'The mandible of the primitive tetrapod *Greererpeton*, and the early evolution of the tetrapod lower jaw', *Journal of Paleontology*, 75: 1016–42.

Bossy, A.K. and Milner, A.C. (1998) 'Order Nectridea Miall, 1875', in P. Wellnhofer (ed.) *Handbuch der Paläoherpetologie, Teil 1: Lepospondyli*, Munich: Pfeil Verlag, pp. 73–131.

Boy, J.A. and Sues, H.-D. (2000) 'Branchiosaurs: larvae, metamorphosis and heterochrony in temnospondyls and seymouriamorphs', in H. Heatwole and R.L. Carroll (eds) *Amphibian Biology, 4: Palaeontology*, Chipping Norton: Surrey Beatty & Sons, pp. 1150–97.

Briggs, D.E.G. and Crowther, P.R. (eds) (2001) *Palaeobiology II*, London: Blackwell Press.

Budd, G. (2001) 'Climbing life's tree', *Nature*, 412: 487.

Carroll, R.L. (1988) *Vertebrate Paleontology and Evolution*, New York: Freeman.

—— (1991) 'The origin of reptiles', in H.-P. Schultze and L. Trueb (eds) *Origins of the Higher Groups of Tetrapods: Controversy and Consensus*, Ithaca: Cornell University Press, pp. 331–53.

—— (1995) 'Problems of the phylogenetic analysis of Paleozoic choanates', in M. Arsenault, H. Leliévre and P. Janvier (eds) Studies on Early Vertebrates: (VIIth International Symposium, Parc de Miguasha, Quebec), *Bulletin du Museum National d'Histoire Naturelle, Paris (Series 4)*, 17 (C): 389–445.

—— (1999) 'Homology among divergent Paleozoic tetrapod clades', in G.R. Bock and G. Cardew (eds) *Homology,* Chichester: Wiley, pp. 47–64.

—— (2000) '*Eocaecilia* and the origin of caecilians', in H. Heatwole and R.L. Carroll (eds) *Amphibian Biology, 4: Palaeontology*, Chipping Norton: Surrey Beatty & Sons, pp. 1402–11.

—— (2001) 'The origin and early radiation of terrestrial vertebrates', *Journal of Paleontology*, 75: 1202–13.

Carroll, R.L. and Bolt, J.R. (2001) 'The Paleozoic divergence of frogs and salamanders', *Journal of Vertebrate Paleontology*, 21 (3 – supplement): 38A.

Carroll, R.L. and Currie, P.J. (1975) 'Microsaurs as possible apodan ancestors', *Zoological Journal of the Linnean Society*, 57: 229–47.

—— (1991) 'The early radiation of diapsid reptiles', in H.-P. Schultze and L. Trueb (eds) *Origins of the Higher Groups of Tetrapods: Controversy and Consensus*, Ithaca: Cornell University Press, pp. 354–424.

Carroll, R.L. and Gaskill, P. (1978) 'The Order Microsauria', *Memoirs of the American Philosophical Society*, 126: 1–211.

Clack, J.A. (1994) '*Silvanerpeton miripedes*, a new anthracosauroid from the Viséan of East Kirkton, West Lothian, Scotland', *Transactions of the Royal Society of Edinburgh: Earth Sciences*, 84: 369–76.

—— (1996) 'The palate of *Crassigyrinus scoticus*, a primitive tetrapod from the Lower Carboniferous of Scotland', in A.R. Milner (ed.) *Studies on Carboniferous and Permian Vertebrates, Special Papers in Palaeontology*, 52: 55–64.

—— (1998a) 'Devonian tetrapod trackways and trackmakers; a review of the fossils and footprints', *Palaeogeography, Palaeoclimatology and Palaeoecology*, 130: 227–50.

—— (1998b) 'A new Early Carboniferous tetrapod with a mèlange of crown-group characters', *Nature*, 394: 66–9.

—— (1998c) 'The neurocranium of *Acanthostega gunnari* and the evolution of the otic region in tetrapods', *Zoological Journal of the Linnean Society*, 122: 61–97.

—— (1998d) 'The Scottish Carboniferous tetrapod *Crassigyrinus scoticus* (Lydekker) – cranial anatomy and relationships', *Transactions of the Royal Society of Edinburgh: Earth Sciences*, 88: 127–42.

—— (2000) 'The origin of tetrapods', in H. Heatwole and R.L. Carroll (eds) *Amphibian Biology, 4: Palaeontology*, Chipping Norton: Surrey Beatty & Sons, pp. 979–1029.

—— (2001) '*Eucritta melanolimnetes* from the Early Carboniferous of Scotland, a stem tetrapod showing a mosaic of characteristics', *Transactions of the Royal Society of Edinburgh: Earth Sciences*, 92: 75–95.

—— (2002) 'An early tetrapod from "Romer's Gap"', *Nature*, 418: 72–6.

Clack, J.A. and Carroll, R.L. (2000) 'Early Carboniferous tetrapods', in H. Heatwole and R.L. Carroll (eds) *Amphibian Biology, 4: Palaeontology*, Chipping Norton: Surrey Beatty & Sons, pp. 1030–43.

Clack, J.A. and Finney, S.M. (1997) 'An articulated tetrapod specimen from the Tournaisian of western Scotland', *Journal of Vertebrate Paleontology*, 17 (3 – supplement): 38A.

Cloutier, R. and Ahlberg, P.E. (1996) 'Morphology, characters, and the inter-relationships of basal sarcopterygians', in M.L.J. Stiassny, L.R. Parenti and G.D. Johnson (eds) *Inter-relationships of Fishes*, London: Academic Press, pp. 325–37.

Coates, M.I. (1996) 'The Devonian tetrapod *Acanthostega gunnari* Jarvik: postcranial anatomy, basal tetrapod inter-relationships and patterns of skeletal evolution', *Transactions of the Royal Society of Edinburgh: Earth Sciences*, 87: 363–421.

—— (2001) 'Origin of tetrapods', in D.E.G. Briggs and P.R. Crowther (eds) *Palaeobiology II*, London: Blackwell Press, pp. 74–9.

Coates, M.I. and Clack, J.A. (1990) 'Polydactyly in the earliest known tetrapod limbs', *Nature*, 347: 66–9.

—— (1991) 'Fish-like gills and breathing in the earliest known tetrapod', *Nature*, 352: 234–6.

—— (1995) 'Romer's Gap – tetrapod origins and terrestriality', in M. Arsenault, H. Lelièvre and P. Janvier (eds) *Studies on Early Vertebrates: (VIIth International Symposium, Parc de Miguasha, Quebec)*, *Bulletin du Museum National d'Histoire Naturelle, Paris (Series 4)*, 17 (C): 373–88.

Coates, M.I., Ruta, M. and Milner, A.R. (2000) 'Early tetrapod evolution', *Trends in Ecology and Evolution*, 15: 327–8.

Cooper, A. and Fortey, R.A. (1998) 'Evolutionary explosions and the phylogenetic fuse', *Trends in Ecology and Evolution*, 13: 151–6.

Craske, A.J. and Jefferies, R.P.S. (1989) 'A new mitrate from the Upper Ordovician of Norway, and a new approach to subdividing a plesion', *Palaeontology*, 32: 69–99.

Daeschler, E.B., Shubin, N., Thomson, K.S. and Amaral, W.W. (1994) 'A Devonian tetrapod from North America', *Science*, 265: 639–42.

deBraga, M. and Rieppel, O. (1997) 'Reptile phylogeny and the inter-relationships of turtles', *Zoological Journal of the Linnean Society*, 120: 281–354.

Duellman, W.E. (1988) 'Evolutionary relationships of the Amphibia', in B. Fritzsch, M.J. Ryan, W. Wilczynski, T.E. Hetherington and W. Walkowiak (eds) *The Evolution of the Amphibian Auditory System*, New York: Wiley, pp. 13–34.

Duellman, W.E. and Trueb, L. (1986) *Biology of Amphibians*, New York: McGraw-Hill.

Easteal, S. (1999) 'Molecular evidence for the early divergence of placental mammals', *BioEssays*, 21: 1052–8.

Easteal, S and Herbert, G. (1997) 'Molecular evidence from the nuclear genome for the time frame of human evolution', *Journal of Molecular Evolution*, 44: 121–32.

Feller, A.E. and Hedges, S.B. (1998) 'Molecular evidence for the early history of living amphibians', *Molecular Phylogeny and Evolution*, 9: 509–16.

Gao, K.-Q. and Shubin, N.H. (2001) 'Late Jurassic salamanders from northern China', *Nature*, 410: 574–7.

Gardner, J.D. (2001) 'Monophyly and affinities of albanerpetontid amphibians (Temnospondyli; Lissamphibia)', *Zoological Journal of the Linnean Society*, 131: 309–52.

Gauthier, J.A., Kluge, A.G. and Rowe, T. (1988a) 'The early evolution of the Amniota', in M.J. Benton (ed.) *The Phylogeny and Classification of the Tetrapods, 1: Amphibians, Reptiles, Birds*, Oxford: Clarendon Press, pp. 103–55.

—— (1988b) 'Amniote phylogeny and the importance of fossils', *Cladistics*, 4: 105–209.

Godfrey, S.J. (1989) 'The postcranial skeletal anatomy of the Carboniferous tetrapod *Greererpeton burkemorani*', *Philosophical Transactions of the Royal Society, London*, B323: 75–133.

Hay, J.M., Ruvinsky, I., Hedges, S.B. and Maxson, L.R. (1995) 'Phylogenetic relationships of amphibian families inferred from DNA sequences of mitochondrial 12S and 16S ribosomal RNA genes', *Molecular Biology and Evolution*, 12: 928–37.

Heaton, M.J. (1980) 'The Cotylosauria: a reconsideration of a group of archaic tetrapods', in A.L. Panchen (ed.) *The Terrestrial Environment and the Origin of Land Vertebrates*, London: Academic Press, pp. 497–551.

Hedges, S.B. (2001) 'Molecular evidence for the early history of living vertebrates', in P.E. Ahlberg (ed.) *Major Events in Early Vertebrate Evolution: Palaeontology, Phylogeny, Genetics and Development*, London: Taylor & Francis, pp. 119–34.

Hedges, S.B. and Maxson, L.R. (1993) 'A molecular perspective on Lissamphibian phylogeny', *Herpetological Monographs*, 7: 27–42.

—— (1997) 'Complementary uses of molecules and morphology', *Molecular Phylogenetics and Evolution*, 8: 445.

Hedges, S.B. and Poling, L.L. (1999) 'A molecular phylogeny of reptiles', *Science*, 283: 998–1001.

Hedges, S.B., Moberg, K.D. and Maxson, L.R. (1990) 'Tetrapod phylogeny inferred from 18S and 28S ribosomal RNA sequences and a review of the evidence for amniote phylogeny', *Molecular Biology and Evolution*, 7: 607–33.

Hennig, W. (1966) *Phylogenetic Systematics*, Urbana: University of Illinois Press.

Holmes, R.B. (2000) 'Palaeozoic temnospondyls', in H. Heatwole and R.L. Carroll (eds) *Amphibian Biology, 4: Palaeontology*, Chipping Norton: Surrey Beatty & Sons, pp. 1081–120.

Holmes, R.B. and Carroll, R.L. (1977) 'A temnospondyl amphibian from the Mississippian of Scotland', *Bulletin of the Museum of Comparative Zoology, Harvard University*, 147: 489–511.

Hook, R.W. (1983) '*Colosteus scutellatus* (Newberry), a primitive temnospondyl amphibian from the Middle Pennsylvanian of Linton, Ohio', *American Museum Novitates*, 2770: 1–41.

Hopson, J.A. (1991) 'Systematics of the nonmammalian Synapsida and implications for patterns of evolution in synapsids', in H.-P. Schultze and L. Trueb (eds) *Origins of the Higher Groups of Tetrapods: Controversy and Consensus*, Ithaca: Cornell University Press, pp. 635–93.

Ivakhnenko, M.F. (1978) 'Urodelans from the Triassic and Jurassic of Soviet Central Asia', *Paleontological Journal*, 1978: 84–9.

Janvier, P. (1996) *Early Vertebrates*, Oxford: Clarendon Press.

Jarvik, E. (1996) 'The Devonian tetrapod *Ichthyostega*', *Fossils and Strata*, 40: 1–206.

Jefferies, R.P.S. (1979) 'The origin of chordates – a methodological essay', in M.R. House (ed.) *The Origin of Major Invertebrate Groups, Systematics Association Special Volume* 12: 443–77.

Jeffery, J.E. (2001) 'Pectoral fins of rhizodontids and the evolution of pectoral appendages in the tetrapod stem-group', *Biological Journal of the Linnean Society*, 74: 217–36.

Jenkins, F.A., Jr. and Walsh, D. (1993) 'An Early Jurassic caecilian with limbs', *Nature*, 365: 246–50.

Johanson, Z. and Ahlberg, P.E. (2001) 'Devonian rhizodontids and tristichopterids (Sarcopterygii; Tetrapodomorpha) from East Gondwana', *Transactions of the Royal Society of Edinburgh: Earth Sciences*, 92: 43–74.

Kearney, M. (2002) 'Fragmentary taxa, missing data, and ambiguity: mistaken assumptions and conclusions', *Systematic Biology*, 51: 369–81.

Kumar, S. and Hedges, S.B. (1998) 'A molecular timescale for vertebrate evolution', *Nature*, 392: 917–20.

Laurin, M. (1991) 'The osteology of a Lower Permian eosuchian from Texas and a review of diapsid phylogeny', *Zoological Journal of the Linnean Society*, 101: 59–95.

—— (1998a) 'The importance of global parsimony and historical bias in understanding tetrapod evolution. Part I. Systematics, middle ear evolution, and jaw suspension', *Annals des Sciences Naturelles, Zoologie*, 19: 1–42.

—— (1998b) 'The importance of global parsimony and historical bias in understanding tetrapod evolution. Part II. Vertebral centrum, costal ventilation, and paedomorphosis', *Annals des Sciences Naturelles, Zoologie*, 19: 99–114.

—— (1998c) 'A reevaluation of the origin of pentadactyly', *Evolution*, 52: 1476–82.

—— (2000) 'Seymouriamorphs', in H. Heatwole and R.L. Carroll (eds) *Amphibian Biology, 4: Palaeontology*, Chipping Norton: Surrey Beatty & Sons, pp. 1064–80.

Laurin, M. and Reisz, R.R. (1995) 'A reevaluation of early amniote phylogeny', *Zoological Journal of the Linnean Society*, 113: 165–223.

—— (1997) 'A new perspective on tetrapod phylogeny', in S.S. Sumida and K.L.M. Martin (eds) *Amniote Origins: Completing the Transition to Land*, London: Academic Press, pp. 9–59.

—— (1999) 'A new study of *Solenodonsaurus janenschi*, and a reconsideration of amniote origins and stegocephalian evolution', *Canadian Journal of Earth Sciences*, 36: 1239–55.

Laurin, M., Girondot, M. and de Ricql's, A. (2000a) 'Early tetrapod evolution', *Trends in Ecology and Evolution*, 15: 118–23.

—— (2000b) 'Reply', *Trends in Ecology and Evolution*, 15: 328.

Lebedev, O.A. (1984) 'First discovery of a Devonian tetrapod vertebrate in USSR', Doklady Akademii Nauk SSSR, 278: 1470–3.

Lebedev, O.A. and Clack, J.A. (1993) 'Upper Devonian tetrapods from Andreyevka, Tula region, Russia', *Palaeontology*, 36: 721–34.

Lebedev, O.A. and Coates, M.I. (1995) 'The postcranial skeleton of the Devonian tetrapod *Tulerpeton curtum* Lebedev', *Zoological Journal of the Linnean Society*, 114: 307–48.

Lee, M.S.Y. (1993) 'The origin of the turtle body plan: bridging a famous morphological gap', *Science*, 261: 1716–20.

—— (1995) 'Historical burden in systematics and the inter-relationships of "Parareptiles"', *Biological Reviews*, 70: 459–547.

—— (1996) 'Correlated progression and the origin of turtles', *Nature*, 379: 812–15.

—— (1997a) 'Pareiasaur phylogeny and the origin of turtles', *Zoological Journal of the Linnean Society*, 120: 197–280.

—— (1997b) 'Reptile relationships turn turtle', *Nature*, 389: 245.

Lee, M.S.Y. and Spencer, P.S. (1997) 'Crown-clades, key characters and taxonomic stability: when is an amniote not an amniote?', in S.S. Sumida and K.L.M. Martin (eds) *Amniote Origins: Completing the Transition to Land*, London: Academic Press, pp. 61–84.

Lombard, R.E. and Bolt, J.R. (1979) 'Evolution of the tetrapod ear: an analysis and reinterpretation', *Biological Journal of the Linnean Society*, 11: 19–76.

—— (1995) 'A new primitive tetrapod *Whatcheeria deltae* from the Lower Carboniferous of Iowa', *Palaeontology*, 38: 471–94.

—— (1999) 'A microsaur from the Mississippian of Illinois and a standard format for morphological characters', *Journal of Paleontology*, 73: 908–23.

Lombard, R.E. and Sumida, S.S. (1992) 'Recent progress in understanding early tetrapods', *American Zoologist*, 32: 609–22.

McGowan, G. and Evans, S.E. (1995) 'Albanerpetontid amphibians from the Early Cretaceous of Spain', *Nature*, 373: 143–5.

Milner, A.C. and Lindsay, W. (1998) 'Postcranial remains of *Baphetes* and their bearing on the relationships of the Baphetidae (= Loxommatidae)', *Zoological Journal of the Linnean Society*, 122: 211–35.

Milner, A.R. (1988) 'The relationships and origin of living amphibians', in M.J. Benton (ed.) *The Phylogeny and Classification of the Tetrapods, 1: Amphibians, Reptiles, Birds*, Oxford: Clarendon Press, pp. 59–102.

—— (1990) 'The radiations of temnospondyl amphibians', in P.D. Taylor and G.P. Larwood (eds) *Major Evolutionary Radiations*, Oxford: Clarendon Press, pp. 321–49.

—— (1993) 'The Paleozoic relatives of lissamphibians', *Herpetological Monographs*, 7: 8–27.

—— (2000) 'Mesozoic and Tertiary Caudata and Albanerpetontidae', in H. Heatwole and R.L. Carroll (eds) *Amphibian Biology, 4: Palaeontology*, Chipping Norton: Surrey Beatty & Sons, pp. 1412–44.

Milner, A.R. and Sequeira, S.E.K. (1994) 'The temnospondyl amphibians from the Viséan of East Kirkton, West Lothian, Scotland', *Transactions of the Royal Society of Edinburgh: Earth Sciences*, 84: 331–61.

Nessov, L.A. (1988) 'Late Mesozoic amphibians and lizards of Soviet Middle Asia', *Acta Zoologica Cracoviensia*, 31: 475–86.

Nessov, L.A., Fedorov, P.V., Potanov, D.O. and Golovyeva, L.S. (1996) 'The structure of the skulls of caudate amphibians collected from the Jurassic of Kirgizstan and the Cretaceous of Uzbekistan', *Vestnik Sankt-Petersburgiskogo Universiteta, Geologia*, 7: 3–11.

Nixon, K.C. and Davis, J.I. (1991) 'Polymorphic taxa, missing values and cladistic analysis', *Cladistics*, 7: 233–41.

Panchen, A.L. (1985) 'On the amphibian *Crassigyrinus scoticus* Watson from the Carboniferous of Scotland', *Philosophical Transactions of the Royal Society, London*, B309: 505–68.

Panchen, A.L. and Smithson, T.R. (1987) 'Character diagnosis, fossils, and the origin of tetrapods', *Biological Reviews*, 62: 341–438.

—— (1988) 'The relationships of the earliest tetrapods', in M.J. Benton (ed.) *The Phylogeny and Classification of the Tetrapods, 1: Amphibians, Reptiles, Birds*, Oxford: Clarendon Press, pp. 1–32.

—— (1990) 'The pelvic girdle and hind limb of *Crassigyrinus scoticus* (Lydekker) from the Scottish Carboniferous and the origin of the tetrapod pelvic skeleton', *Transactions of the Royal Society of Edinburgh: Earth Sciences*, 81: 31–44.

Paton, R.L., Smithson, T.R. and Clack, J.A. (1999) 'An amniote-like skeleton from the Early Carboniferous of Scotland', *Nature*, 398: 508–13.

Platz, J.E. and Conlon, J.M. (1997) 'and turn back again', *Nature*, 389: 246.

Pough, F.H., Andrews, R.M., Cadle, J.E., Crump, M.L., Savitzky, A.H. and Wells, K.D. (2000) *Herpetology*, 2nd edn, Upper Saddle River: Prentice Hall.

Prendini, L. (2001) 'Species or supraspecific taxa as terminals in cladistic analysis? Groundplans versus exemplars revisited', *Systematic Biology*, 50: 290–300.

Rage, J.-C. and Rocek, Z. (1989) 'Redescription of *Triadobatrachus massinoti* (Piveteau, 1936) an anuran amphibian from the early Triassic', *Palaeontographica Abteilung A*, 206: 1–16.

Reisz, R.R. and Laurin, M. (1991) '*Owenetta* and the origin of turtles', *Nature*, 349: 324–6.

Rieppel, O. (2000) 'Turtles as diapsid reptiles', *Zoological Scripta*, 29: 199–212.

Rieppel, O. and deBraga, M. (1996) 'Turtles as diapsid reptiles', *Nature*, 384: 453–5.

Rieppel, O. and Reisz, R.R. (1999) 'The origin and early evolution of turtles', *Annual Review of Ecology and Systematics*, 30: 1–22.

Rocek, Z. and Rage, J.-C. (2000a) 'Anatomical transformations in the transition from temno-spondyl to proanuran stages', in H. Heatwole and R.L. Carroll (eds) *Amphibian Biology, 4: Palaeontology*, Chipping Norton: Surrey Beatty & Sons, pp. 1274–82.

—— (2000b) 'Proanuran stages (*Triadobatrachus, Czatkobatrachus*)', in H. Heatwole and R.L. Carroll (eds) *Amphibian Biology, 4: Palaeontology*, Chipping Norton: Surrey Beatty & Sons, pp. 1283–94.

Rolfe, W.D.I., Clarkson, E.N.K. and Panchen, A.L. (eds) (1994) Volcanism and Early Terrestrial Biotas, *Transactions of the Royal Society of Edinburgh: Earth Sciences*, 84: 175–464.

Romer, A.S. (1946) 'The primitive reptile *Limnoscelis* restudied', *American Journal of Science*, 244: 149–88.

Ruta, M., Milner, A.R. and Coates, M.I. (2001) 'The tetrapod *Caerorhachis bairdi* Holmes and Carroll from the Lower Carboniferous of Scotland', *Transactions of the Royal Society of Edinburgh: Earth Sciences*, 92: 229–61.

Ruta, M., Coates, M.I. and Quicke, D.L.J. (2003) 'Early tetrapod relationships revisited', *Biological Reviews*, 78: 251–345.

Salisbury, B.A. and Kim, J. (2001) 'Ancestral state estimation and taxon sampling density', *Systematic Biology*, 50: 557–64.

Schoch, R.R. (1992) 'Comparative ontogeny of Early Permian branchiosaurid amphibians from Southwestern Germany', *Palaeontographica Abteilung A*, 222: 43–83.

—— (1995) 'Heterochrony in the development of the amphibian head', in K.J. McNamara (ed.) *Evolutionary Change and Heterochrony*, New York: John Wiley & Sons, pp. 107–24.

—— (1998) 'Homology of cranial ossifications in urodeles', *Neues Jahrbuch für Geologie und Paläontologie, Monatshefte*, 1998: 1–25.

Schoch, R.R. and Milner, A.R. (2000) *Handbuch der Paläoherpetologie: Teil 3B, Stereospondyli*, Munich: Pfeil Verlag.

Shishkin, M.A. (1998) '*Tungussogyrinus*, a relict neotenic dissorophoid (Amphibia, Temnospondyli) from the Permo-Triassic of Siberia', *Palentologicheskij Zhurnal*, 5: 521–31.

Smirnov, S. (1986) 'The evolution of the urodele sound-conducting apparatus', in Z. Rocek (ed.) *Studies in Herpetology, Proceedings of the European Herpetological Meeting*, Prague: Charles University, pp. 55–8.

Smith, A.B. (1994) *Systematics and the Fossil Record*, London: Blackwell Science.

—— (1999) 'Dating the origin of metazoan body plans', *Evolution and Development*, 1: 138–42.

Smithson, T.R. (1982) 'The cranial morphology of *Greererpeton burkemorani* Romer (Amphibia: Temnospondyli)', *Zoological Journal of the Linnean Society*, 76: 29–90.

—— (1985) 'The morphology and relationships of the Carboniferous amphibian *Eoherpeton watsoni* Panchen', *Zoological Journal of the Linnean Society*, 85: 317–410.

—— (1989) 'The earliest known reptile', *Nature*, 314: 676–8.

—— (1994) '*Eldeceeon rolfei*, a new reptiliomorph from the Viséan of East Kirkton, West Lothian, Scotland', *Transactions of the Royal Society of Edinburgh: Earth Sciences*, 84: 377–82.

—— (2000) 'Anthracosaurs', in H. Heatwole and R.L. Carroll (eds) *Amphibian Biology, 4: Palaeontology*, Chipping Norton: Surrey Beatty & Sons, pp. 1053–64.

Smithson, T.R. and Rolfe, W.D.I. (1990) '*Westlothiana* gen. nov.: naming the earliest known reptile', *Scottish Journal of Geology*, 26: 137–8.

Smithson, T.R., Carroll, R.L., Panchen, A.L. and Andrews, S.M. (1994) '*Westlothiana lizziae* from the Viséan of East Kirkton, West Lothian, Scotland, and the amniote stem', *Transactions of the Royal Society of Edinburgh: Earth Sciences*, 84: 383–412.

Stauffer, S.L., Walker, A., Ryder, O.A., Lyons-Weiler, M. and Hedges, S.B. (2001) 'Human and ape molecular clocks and constraints on paleontological hypotheses', *Journal of Heredity*, 92: 469–74.

Sumida, S.S. (1997) 'Locomotor features of taxa spanning the origin of amniotes', in S.S. Sumida and K.L.M. Martin (eds) *Amniote Origins: Completing the Transition to Land*, London: Academic Press, pp. 353–98.

Sumida, S.S. and Lombard, R.E. (1991) 'The atlas-axis complex in the Late Paleozoic diadectomorph amphibian *Diadectes* and the characteristics of the atlas-axis complex across the amphibian to amniote transition', *Journal of Paleontology*, 65: 973–83.

Sumida, S.S., Lombard, R.E. and Berman, D.S. (1992) 'Morphology of the atlas-axis complex of the late Palaeozoic tetrapod suborders Diadectomorpha and Seymouriamorpha', *Philosophical Transactions of the Royal Society, London*, B336: 259–73.

Swofford, D.L. (1998) *PAUP*: Phylogenetic analysis using parsimony (*and other methods)*, Version 4.0b4a, Sunderland, Massachusetts: Sinauer Associates.

Thulborn, R.A., Warren, A.A., Turner, S. and Hamley, T. (1996) 'Carboniferous tetrapods from Australia', *Nature*, 381: 777–80.

Trueb, L. and Cloutier, R. (1991) 'A phylogenetic investigation of the inter- and intrarelationships of the Lissamphibia (Amphibia: Temnospondyli)', in H.-P. Schultze and L. Trueb (eds) *Origins of the Higher Groups of Tetrapods: Controversy and Consensus*, Ithaca: Cornell University Press, pp. 223–313.

van Tuinen, M., Porder, S. and Hadly, E.A. (2001) 'Putting confidence limits around molecular and fossil divergence dates', *Journal of Vertebrate Paleontology*, 21 (3 – supplement): 110A.

Wagner, P.J. (2000) 'Phylogenetic analyses and the fossil record: tests and inferences, hypotheses and models', in D.H. Erwin and S.L. Wing (eds) *Deep Time: Paleobiology's Perspective*, Lawrence: The Paleontological Society, pp. 341–71.

Wellstead, C.F. (1982) 'A Lower Carboniferous aïstopod amphibian from Scotland', *Palaeontology*, 25: 193–208.

—— (1991) 'Taxonomic revision of the Lysorophia, Permo-Carboniferous lepospondyl amphibians', *Bulletin of the American Museum of Natural History*, 209: 1–90.

Wilkinson, M. (1995) 'Coping with abundant missing entries in phylogenetic inference using parsimony', *Systematic Biology*, 44: 501–14.

Wood, S.P., Panchen, A.L. and Smithson, T.R. (1985) 'A terrestrial fauna from the Scottish Lower Carboniferous', *Nature*, 314: 355–6.

Yates, A.M. and Warren, A.A. (2000) 'The phylogeny of the "higher" temnospondyls (Vertebrata: Choanata) and its implications for the monophyly and origins of the Stereospondyli', *Zoological Journal of the Linnean Society*, 128: 77–121.

Yeh, J. (2002) 'The effect of miniaturized body size on skeletal morphology in frogs', *Evolution*, 56: 628–41.

Zardoya, R. and Meyer, A. (1998) 'Complete mitochondrial genome suggests diapsid affinities of turtles', *Proceedings of the National Academy of Sciences, USA*, 95: 14226–31.

Zhu, M. and Schultze, H.-P. (2001) 'Inter-relationships of basal osteichthyans', in P.E. Ahlberg (ed.) *Major Events in Early Vertebrate Evolution: Palaeontology, Phylogeny, Genetics and Development*, London: Taylor & Francis, pp. 289–314.

Appendix 11.1

Ahlberg and Clack's (1998) matrix of lower jaw characters is reproduced below. For convenience, characters are divided into groups of five separated by a space. For a description of characters 1–50 (upper row), the reader is referred to their publication. Eleven new characters (51–61; lower row), typed in bold, are added. *Caerorhachis* (Holmes and Carroll 1977; Ruta *et al.* 2001) is included. As in the original analysis, question marks denote missing or inapplicable characters. Character 48 (postsplenial pit line) was changed from 1 (absent) to ? (inapplicable) in *Diploceraspis*, *Sauropleura*, *Eocaptorhinus*, and *Ophiacodon* (Ruta *et al.* 2001). The new characters are as follows:

51. Rearmost extension of mesial lamina of splenial closer to anterior margin of adductor fossa (0) than to anterior end of lower jaw (1).

52. Lateral exposure of dentary smaller (0) or greater (1) than lateral exposure of angular.

53. Absence (0) or presence (1) of at least one Meckelian foramen comparable in length with the adductor fossa.

54. Absence (0) or presence (1) of small posterior Meckelian foramen between prearticular and angular.

55. Absence (0) or presence (1) of small posterior Meckelian foramen between prearticular, postsplenial, and angular.

56. Absence (0) or presence (1) of intermediate Meckelian foramen between prearticular and postsplenial.

57. Absence (0) or presence (1) of condition: maximum depth of mesial lamina of splenial comparable with maximum depth of prearticular when both are measured at the level of the mid-length of the adductor fossa.

58. Absence (0) or presence (1) of retroarticular process.

59. Absence (0) or presence (1) of condition: posterior coronoid exposed in lateral view.

60. Absence (0) or presence (1) of condition: mesial lamina of angular deeper than prearticular when both are measured at the level of the anterior margin of the adductor fossa.

61. Absence (0) or presence (1) of condition: mesial margin of posterior coronoid shorter than that of mid-coronoid and up to about two-thirds as long as the latter.

Panderichthys
00000 00000 00000 0000? 00000 00000 00000 00000 00000 00000 000?0 0000? 0

Elginerpeton
??00? 0?000 01111 00100 01010 00111 11000 ?1??0 01??0 0001? ?00?0 0000? ?

Obruchevichthys
????? 0?00? 0?0?? 00101 01??? ?0111 11??? ????0 ??0?0 1??1? ????0 0???? ?

Ichthyostega
00001 00001 11111 ??001 01000 00011 ??000 01000 1?000 00010 100?? ?000? 0

Ventastega
00001 00001 11001 11001 01000 00111 10000 01000 01000 00010 ?00?? ?000? 0

Metaxygnathus
10000 00001 ?1010 11001 0101? 00?11 11000 00000 01000 10?10 000?? ?000? ?

Acanthostega
10000 00000 11111 11001 01101 00010 11000 01000 01000 10110 000?? ?000? 0

Whatcheeria
?0?0? ?0001 1111? 00001 01?01 00001 100?? ?10?? ????0 2011? ?00?? ??00? ?

Tulerpeton
?00?? ??000 ?1?1? ?0?01 0???? ????? ????? ????? ????0 30?1? ????? ????? ?

Crassigyrinus
00100 00131 11111 10011 01101 00?01 01000 1??00 0?100 30110 1000? ?0000 0

Greererpeton
00111 00121 11011 00001 01101 10001 01000 ?0000 1?100 20110 1000? ?0000 0

Megalocephalus
00101 10031 11111 00011 01102 00001 01011 10010 1?100 30111 10000 00000 0

Pholiderpeton
00101 11131 11111 00001 01101 10001 00010 ?1110 1?110 30111 10101 01011 0

Cochleosaurus
00101 11131 11111 00001 01101 1100? ??011 20010 1?111 ??111 000?? ?00?0 ?

Phonerpeton
00101 11131 11111 00001 01101 0100? ??011 ?0111 1?111 ??111 00000 00010 0

Eoherpeton
00101 1?111 11111 00001 ?1101 0???? ??010 ?01?0 0?111 ??111 10001 10000 1

Proterogyrinus
001?1 ?0131 11111 00001 01101 1???? ??0?0 ?11?? 1?111 ??111 1010? ?1001 0

Gephyrostegus
00101 11111 11??1 00001 01101 1100? ??000 201?0 1?111 ??111 1010? ?1010 0

Balanerpeton
00101 0?131 11111 00001 01102 0100? ??011 ?0001 1?101 ??111 00000 00000 0

Platyrhinops
0?101 01131 11111 00001 01101 0100? ??011 ?0??1 1?101 ??111 ??0?? ?00?0 ?

Microbrachis
01101 01131 11111 00001 01101 0100? ??011 ?0000 1?110 31111 10010 00000 0

Discosauriscus
01101 11131 11111 00001 11101 0100? ??011 ?01?0 1?110 31111 11010 00010 0

Eocaptorhinus
0112? ?1031 ????1 00001 11101 0100? ??1?? ?0000 1?111 ???11 1100? ?10?1 ?

Diploceraspis
0112? ?00?1 11111 00001 11101 0100? ??1?? ?0000 1?111 ???01 1001? ?0100 ?

Sauropleura
011?? ????? ????? 00001 11101 11??? ??1?? ?0000 01111 ???11 1010? ?0101 ?

Ophiacodon
00111 ?1031 1?1?1 00001 11101 0100? ??1?? ?0100 1?111 ????1 1100? ?100? ?

Caerorhachis
01101 01131 11111 ???11 01101 00001 01010 ?0010 00111 ???01 1?001 10000 1

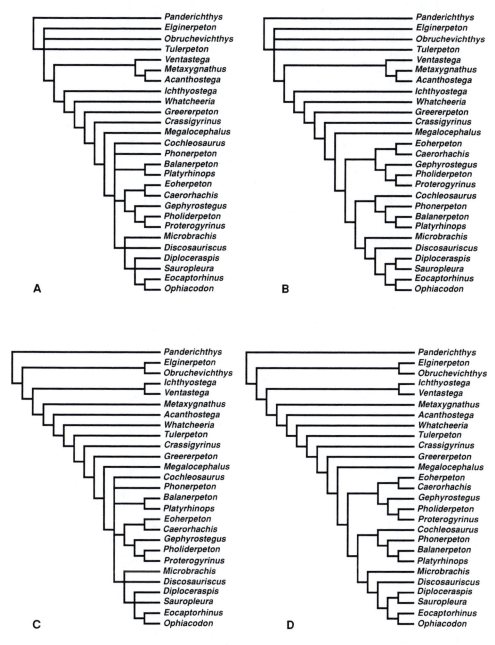

Figure 11.11 A strict consensus of 25 fundamental trees derived from the analysis of an expanded version of Ahlberg and Clack's (1998) dataset with all characters unweighted and unordered; B, strict consensus of five fundamental trees obtained after reweighting characters by their rescaled consistency indices values (best fit); C, strict consensus of five fundamental trees obtained when characters 36 and 46 of Ahlberg and Clack (1998) were ordered, but leaving all characters unweighted; D, single tree found after reweighting characters by their rescaled consistency indices values (best fit), with characters 36 and 46 ordered.

The data matrix was processed with PAUP*4.0b10 under the following search settings: 1000 random stepwise additions with one tree held in memory at any one time (MAXTREES = 1), followed by TBR branch-swapping (MAXTREES = unlimited) on trees in memory. The initial run, with all characters unordered and equally weighted, yielded 25 equally parsimonious trees at 139 steps (CI = 0.4779; RI = 0.694; RC = 0.3395), a strict consensus of which is shown in Figure 11.11A. Reweighting characters by the maximum value of their rescaled consistency indices gives five trees (CI = 0.6722; RI = 0.849; RC = 0.5884). The strict consensus of these (Figure 11.11B) has been used to construct Figure 11.4.

If characters 36 (position of centre of radiation on prearticular) and 46 (enclosure of mandibular canal) are ordered as in Ahlberg and Clack's (1998) analysis, then a PAUP* run with all characters equally weighted gives five most parsimonious trees at 140 steps (CI = 0.4745; RI = 0.7025; RC = 0.3412), the strict consensus of which is illustrated in Figure 11.11C. If characters are reweighted by the maximum value of their rescaled consistency indices, then a single tree is obtained (Figure 11.11D; CI = 0.6772; RI = 0.8641; RC = 0.603).

Chapter 12

The fossil record and molecular clocks: basal radiations within the Neornithes

Gareth J. Dyke

ABSTRACT

The fossil record of the extant clades of birds (Neornithes) is critical to understanding both the timing and pattern of the evolutionary divergences within this major verte-brate group. Interpretations of the fossil record have indicated that this radiation occurred in the aftermath of the Cretaceous–Tertiary (K–T) extinction event. However, the use of 'molecular clocks' to estimate the timing of lineage divergences on the basis of sequence data have instead led to proposals that most of the major lineages of modern birds originated deep in the Cretaceous. Use of the neornithine record to address estimates founded on a molecular timescale is necessary, but remains problematic because of uncertainties surrounding the placement of most fossil taxa within existing phylo-genetic hypotheses for extant clades. Although variance in the relative position of a fossil within a clade will impact on divergence estimates, few attempts have actually been made to distinguish the placements of such taxa with respect to stem- or crown-groups. Here, I present osteological evidence for the phylogenetic placement of some well-preserved fossil neornithines from the early Tertiary and discuss the implica-tions of constraining such taxa to the development of 'molecular clock' hypotheses for the timing of the divergence of modern birds.

Introduction

The debate – molecules and fossils

The more than 10 000 species of modern birds are the living descendants of a large and ancient radiation than can be traced back 150 million years to the earliest bird, *Archaeopteryx*, from the Late Jurassic. The taxonomic diversity and genealogical rela-tionships of the early birds, the origin and refinement of flight, and the origin of avian functional and physiological specialization are just some of the evolutionary issues that have captured the interest of decades of palaeo-ornithological research (for reviews, see Chatterjee 1997; Padian and Chiappe 1998; Feduccia 1999; Chiappe and Dyke 2002). Recently, however, much research interest has been focused on addressing the question of the timing of divergence of the extant lineages of birds (Neornithes *sensu* Cracraft 1988; alternatively 'crown-clade' Aves *sensu* Gauthier 1986; Figure 12.1).

The timing of the divergences of, and within, the Neornithes remains very contro-versial. Debates have revolved around the question of the timing of the appearance

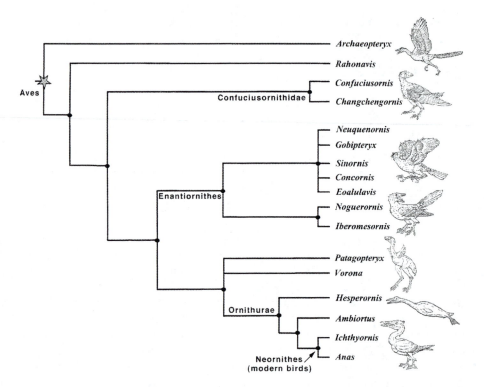

Figure 12.1 Consensus phylogeny (redrawn, with permission, from Chiappe 2001) depicting the
phylogenetic relationships within Aves and showing Neornithes at the crown (see
text for details; for details of competing hypotheses see Chiappe and Dyke 2002).

of these clades (i.e. the extant orders and families; Monroe and Sibley 1990), and the
extent to which clades of Neornithes had diversified by the end-Mesozoic, marked
by the Cretaceous–Tertiary (K–T) extinction event. The classical hypothesis, based
on the published fossil record (Unwin 1993; Hope 2002), and as outlined, for example,
by Olson (1985) and Feduccia (1999), states that a specific modern bird 'morphotype'
was present at low diversity during the Mesozoic – the so-called 'transitional shorebirds' or 'waterbirds' (Feduccia 1999) – and survived the end-Cretaceous extinction.
All the non-neornithine lineages were wiped out as a result of the K–T boundary event
(Feduccia 1995, 1999; Figure 12.1). The second hypothesis, to a large extent based
on lineage divergence times estimated from molecular sequence data and making use
of 'molecular clocks' suggests that the majority, if not all, of the major clades of
Neornithes diverged during the Mesozoic (e.g. Hedges *et al.* 1996; Cooper and
Penny 1997; Rambaut and Bromham 1998; van Tuinen *et al.* 1998, 2000). By use of
different datasets, these works reached the same conclusion – that the major clades
of Neornithes diverged deep in the Cretaceous, largely at odds with the known
fossil record.

The fossil record of Cretaceous bird remains that have been referred to extant
lineages is sparse – only a handful of records are represented by more than isolated
elements (Padian and Chiappe 1998; Chiappe and Dyke 2002). Hence the incompleteness

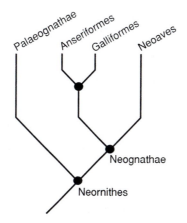

Figure 12.2 Consensus cladogram to illustrate clade relationships at the base of Neornithes (based on various sources, see text for details).

of most of these specimens referred to as Cretaceous neornithines renders few characters useful for phylogenetic analyses (Clarke 1999; Dyke and Mayr 1999; Clarke and Chiappe 2001; Dyke 2001a), and our currently poor understanding of the higher-level phylogenetic relationships of extant lineages (Figure 12.2) further complicates their systematic consideration (Cracraft 1988; Cracraft and Clarke 2001; Livezey and Zusi 2001). In spite of problems, however, this scanty fossil record has been used to hypothesize the existence of a number of extant lineages of birds prior to the end of the Mesozoic (e.g. Pelecaniformes, Charadriiformes, Anseriformes, Galliformes, and Psittaciiformes), either by direct interpretation or by using them for the temporal calibration of molecular phylogenies. None the less, it is imperative that these Cretaceous reports of neornithines are treated with extreme caution (Clarke 1999; Dyke and Mayr 1999; Clarke and Chiappe 2001) – the earliest records of birds that are complete enough to be informative for cladistic analyses (i.e. that present more than a few codeable characters for analysis and as a result are useful for estimating the temporal divergences of the extant lineages), come from rocks that are approximately 55 million years old, about 10 million years after the K–T boundary. Although a few specimens represented by more than single bones do fill this temporal gap (e.g. Vedding-Kristoffersen 2000), these have yet to be considered within phylogenetic analyses. It would not be surprising if future studies of these, or even older specimens, do support their placements within extant lineages, but such work has yet to be undertaken.

How deep?

In spite of inherent problems with the Mesozoic fossil record of the extant clades, the presence of several immediate neornithine outgroups in the latest Cretaceous (e.g. *Ichthyornis* Marsh 1880; Dyke *et al.* 2002; *Limenavis* Clarke and Chiappe 2001; *Apsaravis* Norell and Clarke 2001) implies that the lineage leading to the Neornithes must have diverged *prior* to the end of the Mesozoic (Clarke and Chiappe 2001; Chiappe

Table 12.1 The described fossil records used by Cooper and Penny (1997) to estimate the divergence times of modern bird clades

Taxonomy	Age (Ma)[a]	Reference	Material
Loon (Gaviidae)	70	Olson (1992)	Tibiotarsus
Tropicbird (Phaethontidae)	60	Olson (1994)	Coracoid/humerus
Rhea (Rheidae)	60	Tambussi (1995)	Isolated elements
Ostrich (Struthionidae)	60	Houde & Haubold (1987)	Incomplete skeletons
Penguin (Spheniscidae)	58	Fordyce & Jones (1990)	Incomplete skeletons
Charadriiformes	60	Olson & Parris (1987)	Isolated humeri
Galliformes	50	Crowe & Short (1992)	Complete specimens

[a] Ages of fossils as listed by Cooper and Penny (1997).

and Dyke 2002). However, this inference provides little information regarding temporal divergences *among* extant lineages. Strong debates remain, but these centre around the question of *how deep* the extant lineages of birds (Figure 12.2) can be extended into the Mesozoic (Cracraft 2001; Chiappe and Dyke 2002). If one accepts hypotheses presented by proponents of 'molecular clocks' (e.g. Hedges *et al.* 1996; Cooper and Penny 1997; van Tuinen *et al.* 1998, 2000) then the neornithine clades extend far below the K–T boundary and the known fossil record of these taxa is hopelessly incomplete (Dyke 2001a) – contrasting with the fossil record of other small vertebrates from the Cretaceous (Benton 1999; Fara and Benton 2000).

What use are fossils?

The use of 'molecular clocks' to provide estimates for divergence times within the Neornithes has required the consideration of fossil taxa to provide either (or both) *internal* and *external* calibration points for clades (Cooper and Penny 1997; Table 12.1), even though specimens are incomplete and lack phylogenetic control. The use of internal calibrations for such a 'clock' estimate is preferable since the age of a fossil will provide an estimate *closer* to the true divergence time of the clade(s) in question and hence will require less extrapolation and associated error (van Tuinen and Hedges 2001). But there are problems inherent to previous uses of the fossil record of Neornithes in this way: (1) proposed systematic positions of fossil taxa are often cited uncritically, in the absence of phylogenetic control; and (2) no attempts have been made to distinguish between the placement of fossils with respect to the stem- or crown-groups of the clades in question. The single most important issue pertinent to the use of fossils for the calibration of 'molecular clock' (or other) estimates of divergences is the position of taxa within the phylogeny of the group in question, problematic because the relationship of a given fossil to the crown- or stem-group of a clade can only be known through phylogenetic analysis (Figure 12.3). Although estimates for the divergence times of the major clades of extant birds are becoming more and more numerous in the literature, this important distinction has yet to be made for the vast majority of the fossil record of Neornithes.

In this chapter, I outline osteological evidence for the phylogenetic relationships of a number of well-preserved fossil neornithine taxa within the context of the relationships

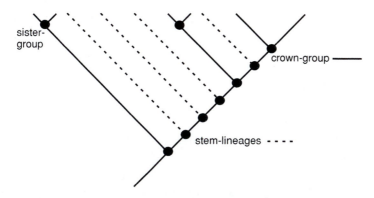

Figure 12.3 The phylogenetic placement of fossils in either the stem-group or crown-group of a particular clade (re-drawn with permission from Magallón and Sanderson 2001, figure 1). The relative phylogenetic placements of fossils imply *minimum ages* for divergences within large clades – the position of fossil taxa relative to either the stem- or crown-group *cannot* be known except through phylogenetic analysis.

of clades at the base of Neornithes. This review draws on a number of recent cladistic analyses based on morphological characters and synopses of phylogenetic relationships within basal Neornithes (e.g. Bledsoe 1988; Cracraft 1988; Livezey 1986, 1997, 1998; Ericson 1997; Lee *et al.* 1997; Cracraft and Clarke 2001; Livezey and Zusi 2001). References to avian taxonomic groups follow the standard checklist of Monroe and Sibley (1990). I will then discuss the implications of these fossil records for addressing the timing of basal neornithine divergences within the context of 'molecular clock' analyses.

Institutional abbreviations

BMNH PAL – The Natural History Museum, London, England (Palaeontology Department Collection); GM – Geiselthal Museum, Halle, Germany.

Fossils and the basal clades of the Neornithes

Although little is known about the phylogenetic interrelationships of the Neornithes based on morphological characters (especially those relevant to the inclusion of fossils), at least the basal clade relationships of extant birds are relatively uncontroversial. A basal split within the group has consistently been recovered through analysis of both molecular and morphological datasets, forming the two major clades Palaeognathae (palaeognaths, comprising the living ratites and tinamou) and Neognathae (the remaining clades of extant birds). Within the more diverse of these two clades, the Neognathae, a second division in the phylogeny is well-supported dividing the sister-taxa Galliformes and Anseriformes (collectively termed Galloanserae; Sibley *et al.* 1988; Groth and Barrowclough 1999) from all other modern birds (this remaining clade is commonly referred to as Neoaves; Sibley *et al.* 1988; Cracraft and Clarke 2001) (Figure 12.2).

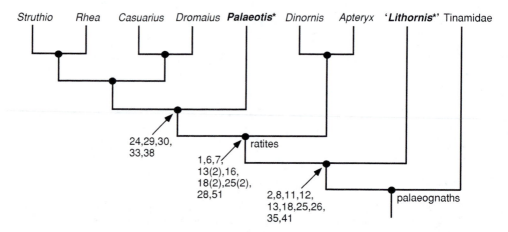

Struthio   Rhea   Casuarius   Dromaius   **Palaeotis***   Dinornis   Apteryx   **'Lithornis*'**   Tinamidae

24,29,30,
33,38

ratites

1,6,7,
13(2),16,
18(2),25(2),
28,51

2,8,11,12,
13,18,25,26,
35,41

palaeognaths

Figure 12.4 Consensus phylogeny depicting relationships among Palaeognathae on the basis of morphological characters and including selected fossil taxa (*Lithornis* and *Palaeotis*; see text for details and sources). Numbers listed at nodes correspond to characters as listed by Lee *et al.* (1997); asterisks indicate fossil taxa scored on the basis of more than one specimen (see text for details).

Palaeognaths

At the base of Neornithes (Cracraft and Clarke 2001; Figure 12.2), the Palaeognathae include the flightless ratites – emu, ostrich, cassowary, and kiwi – as well as the volant tinamous. Although controversy exists regarding precise relationships within palaeognaths, especially within the ratites, most workers agree that tinamous are the basal sister-taxon with respect to the remaining ratite birds (e.g. Cracraft 1974; Bledsoe 1988; Lee *et al.* 1997; Haddrath and Baker 2001; Cooper *et al.* 2001; Figure 12.4). Despite controversy surrounding the results of some molecular analyses, the only recent morphological phylogenetic analysis to include these birds remains that of Lee *et al.* (1997), although a number of their characters were derived from the earlier work of Bledsoe (1988).

The known fossil record of these birds is extensive. Unwin (1993), for example, listed first occurrences for palaeognath families that extend from at least the Lower Eocene to Recent. Furthermore, some workers have postulated the presence of these birds in the Mesozoic (e.g. Kurochkin 1999), but records remain problematic largely as a result of their poor preservation (Hope 2002) and uncertain phylogenetic position (Chiappe and Dyke 2002). So far, little material from the Eocene has been considered within a phylogenetic context, but the members of the 'Lithornithidae' (Houde 1988) and the taxon *Palaeotis weigelti* are particularly poorly understood. Although known under a number of alternative names since the 1840s, Houde's (1988) analysis of the relationships of early Tertiary lithornithids, such as *Lithornis* and *Paracathartes*, led to the proposal that these forms represent part of a flighted basal divergence within the Palaeognathae (significant because most of the extant members of this clade have lost the ability to fly). The affinities of *Palaeotis*, on the other hand, have engendered much debate. Although, originally, this taxon was named by Lambrecht (1928) from a single right tarsometatarsus, *Palaeotis* is currently represented by several partially complete (but extensively crushed) specimens from the Middle

Eocene of Germany. Lambrecht (1928) first classified *Palaeotis* within Gruiformes (in the Otididae, the family that includes the living bustards), but on the basis of subsequently discovered partial specimens, Houde and Haubold (1987) suggested a placement within the Palaeognathae – a 'primitive ratite' within the Struthionidae (ostriches). Evidence for this position included, amongst other characters, lack of the closure of the ilioischiadic foramen (character 24 of Lee *et al.* 1997; hypothesized to be derived within ratites, as noted by Houde and Haubold 1987). This more primitive position for *Palaeotis* within the ratites was rejected by Peters (1988) who considered that this bird should be classified close to the extant rheas (Rheidae) (Figure 12.4).

By use of the character descriptions presented by Lee *et al.* (1997), the known specimens of *Lithornis* from the London Clay Formation were coded (Appendix 12.1) and a phylogenetic analysis was run with the same additional terminals as Lee *et al.* (1997). Although a large number of specimens from the London Clay were referred to *Lithornis* by Houde (1988), only those that consist of more than one element in certain association were included within this analysis (Appendix 12.1). In addition to *Lithornis*, specimens of *Palaeotis weigelti* were also included in the analysis. Codings for this taxon are based on fossil material examined in the Senckenberg Museum, Frankfurt (SMF–ME 1578: a compressed slab comprising a complete articulated individual that was referred to *Palaeotis weigelti* by Peters 1988), and a cast of GM 4362 (a partially complete, but crushed, individual) referred and described by Houde and Haubold (1987). Parsimony analysis of the complete dataset (Appendix 12.1), including a composite 'London Clay *Lithornis*' and *Palaeotis*, resulted in the production of a single MPT (91 steps; CI: 0.758; RI: 0.794). In this tree (Figure 12.4), *Lithornis* is hypothesized to be the sister taxon to the monophyletic ratites. Within this latter grouping, *Palaeotis* is recovered as the sister taxon to two clades, *Casuarius* plus *Dromaius*, and *Struthio* plus *Rhea* (Figure 12.4). Interestingly, inclusion of the fossil taxa effects no changes to the topology for extant taxa proposed by Lee *et al.* (1997, figure 7.2).

The position of the composite London Clay *Lithornis* in a more derived position than tinamous, and as the sister taxon to the other included taxa, is supported in this analysis on the basis of the following derived character-states (numbered as listed by Lee *et al.* 1997, their appendix 1, and as preserved in the specimens considered): 8 (scapula and coracoid fused; seen in BMNH PAL A 5303 and A 5425); 12 (internal tuberosity of humerus knoblike, having a degree of medial protrusion; seen in BMNH PAL A 5204, BMNH PAL A 5303, BMNH PAL A 33138, BMNH PAL A 38934, BMNH PAL A 5425); 13 (deltoid crest raised from the base of the external tuberosity; seen in BMNH PAL A 5204, BMNH PAL A 5303); 18 (pronounced external epicondyle of humerus; seen in BMNH PAL A 5425). Characters that are not preserved in the available London Clay specimens, but that are hypothesized to support the monophyly of this grouping are: 2 (absence of posterior lateral processes on the sternum); 11 (humerus longer than the ulna); 17 (transverse ligamental sulcus shallow; reversed in the moa *Dinornis*); 25 (transverse processes of sacral vertebrae broad and fused); 26 (puboischial bar present); 35 (supratendinal bridge of tibiotarsus absent); 41 (posterior margin of external condyle of tibiotarsus extended laterally).

The position of *Palaeotis weigelti* within the ratites is supported by this analysis on the basis of the following derived characters (listed as given by Lee *et al.* 1997): 24 (obturator process of the ischium fused with the pubis to form a complete obturator foramen); and 30 (anterior metatarsal groove [of tarsometatarsus] deep and narrow for its entire length). Furthermore, and as seen in the Dinornithidae, *Struthio* and *Rhea* (Lee *et al.* 1997: 203), the external and internal cotylar surfaces of the tarsometatarsus are concave in *Palaeotis* (derived state 2 for Lee *et al.* character 29). Additional characters hypothesized to support this grouping, but not preserved in *Palaeotis*, include: 33 (surface of the anterior interarticular area of the cnemial crest of the tibiotarsus mediolaterally compressed); and 38 (internal condyle of tibiotarsus projected anteriorly relative to external condyle in distal view). However, several characters hypothesized on the basis of this analysis to be derived within palaeognaths in general (i.e. Tinamidae and more derived terminals) are also seen primitively in *Palaeotis*. These include the presence of posterior lateral processes on the sternum (character 2) and a trochanteric crest on the femur that extends proximally beyond the level of the iliac facet (character 41).

Galloanserae

Anseriformes

The Order Anseriformes comprises the extant waterfowl, the screamers, ducks, and geese, and has a putative fossil record extending into the latest Cretaceous (e.g. Howard 1955; Olson and Parris 1987; Olson 1999). From the Lower Eocene Green River Formation of North America comes perhaps the most famous of all fossil neornithines, *Presbyornis pervetus*. For many years, *Presbyornis* was thought to be somehow 'intermediate' in its morphology, providing evidence for an evolutionary link between a number of traditional avian orders by the time of the earliest Eocene (e.g. Olson and Feduccia 1980; Feduccia 1995). The recent inclusion of *Presbyornis* within cladistic analyses of the Anseriformes has demonstrated that this taxon can be placed well within the order, closely related to the Anatidae (extant true ducks; Ericson 1997; Livezey 1997, 1998). Because of the incompleteness of the putative Cretaceous records (Hope 2002), *Presbyornis* and another Lower Eocene taxon, *Anatalavis oxfordi* Olson, present the earliest partially complete fossil records for the Anseriformes (Figure 12.5).

On the basis of the analysis presented by Livezey (1997, 1998), I coded the preserved cranial and postcranial morphology of *Anatalavis oxfordi* in light of the proposal made by Olson (1999) that this taxon is basal within Anseriformes, closely related to the extant magpie–goose *Anseranas*. Results of this analysis, the first to consider the relationships of the London Clay taxon in a cladistic context, instead supports a more derived placement for *Anatalavis oxfordi* within Anseriformes, as the sister taxon to *Presbyornis* and Anatidae. Hence, two well-preserved anseriform taxa known from the Lower Eocene have been placed convincingly within the order on the basis of cladistic analysis (Figure 12.5).

This placement for *Anatalavis oxfordi* within the Anseriformes is unambiguous, and on the basis of a number of cranial features, including the presence of a long and ventrally terminating lamina basiparasphenoidale (character 3 of Livezey 1997), this position is recovered by parsimony analysis. *Anatalavis* is the sister taxon of the extant true ducks (Figure 12.5), including *Presbyornis*, because of the presence of a prominent crista fossa parabasalis of the exoccipitale (character 2 of Livezey 1997)

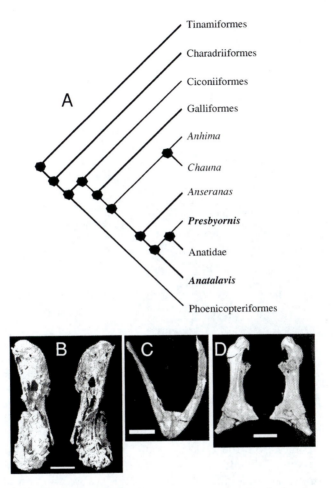

Tinamiformes

Charadriiformes

Ciconiiformes

Galliformes

Anhima

Chauna

Anseranas

Presbyornis

Anatidae

Anatalavis

Phoenicopteriformes

Figure 12.5 (A) Hypothesis for the phylogenetic positions of the lower Eocene taxa *Presbyornis* and *Anatalavis* (boldface) – well-preserved fossil representatives of order Anseriformes (see text for details and sources); (B–D) some of the preserved elements of *Anatalavis oxfordi* Olson (BMNH PAL 5922): (B) skull in lateral views; (C) furcula; (D) coracoid in dorsal and medial views. Scales are 10 mm.

and a prominent processus coronoideus of the mandible (character 18 of Livezey 1997) (Figure 12.5). See Dyke (2001b) for details.

Galliformes

The landfowl of the order Galliformes form a large and cosmopolitan group of about 70 genera (Monroe and Sibley 1990) found on almost all continents across the globe. Again, the described fossil record of these birds is extensive – putative remains of Galliformes have been described from deposits that range in age from the latest Cretaceous to Recent (e.g. Unwin 1993; Hope 2002). However, the majority of this material is incompletely preserved and so one of the oldest *certain* records for the

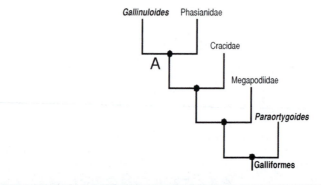

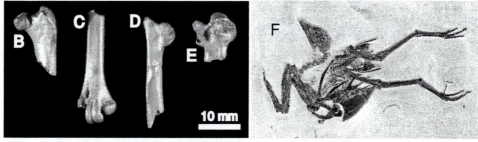

Figure 12.6 (A) Hypothesis for the phylogenetic positions of the Lower Eocene taxa *Gallinuloides* and *Paraortygoides* – well-preserved fossil representatives of order Galliformes (see text for details and sources); (B–E) fossil elements of *Paraortygoides* from the Lower Eocene of England; (F) type specimen of *Gallinuloides*.

order to date comes from the Lower Eocene Green River Formation of the United States. This taxon, *Gallinuloides wyomingensis* (Figure 12.6), was placed within the extinct family Gallinuloididae by Lucas (1900), now also considered to contain a number of somewhat younger and incompletely known taxa (Milne-Edwards 1867–1871; Tordoff and Macdonald 1957; Mayr 2000). Another well-preserved and early fossil representative of these birds is the genus *Paraortygoides* known from a number of specimens from the Lower Eocene London Clay Formation of the UK, and from the Middle Eocene deposit of Messel, Germany (Figure 12.6).

Recently, the first morphological phylogenetic analysis for Galliformes, scored at the level of individual extant genera, was published (Dyke *et al.* 2003). Based on more than 100 osteological characters, it is now possible to investigate the relationships of a number of the better preserved fossil representatives of the order within a phylogenetic context (Figure 12.6). As can be seen from the cladogram in Figure 12.6, the basic relationships of the extant taxa within the order can be summarized with the mound-building megapodes (Megapodiidae) as the most basal taxon, the sister of the cracids (i.e. currassows, guans; Cracidae) and the more familiar phasianoids (i.e. pheasants, partridges, guineafowl; Phasianoidea). Within this latter grouping, a number of the recovered clades are well supported on the basis of osteological evidence, such as the New and Old World quail (Perdicini) and the grouse (Tetraonidae) (see Dyke *et al.* 2003 for details of this analysis).

Including the two Lower Eocene taxa *Gallinuloides* and *Paraortygoides* within the phylogenetic analysis resulted in no changes to original tree topology (not shown), and an increase in tree length of just six steps. However, markedly different placements for the two taxa are suggested – *Paraortygoides* (Figure 12.6) is hypothesized to be a basal (perhaps the most basal) representative of the order, whereas *Gallinuloides* (Figure 12.6) is placed in a more derived position towards the crown of Galliformes (Figure 12.6). This conclusion is of interest because, almost without exception, previous workers have considered *Gallinuloides* to be a member of, or closely related to, the basal galliform clade Cracidae (Tordoff and Macdonald 1957; Brodkorb 1964; Ballmann 1969).

The ages of included fossils – what do these imply?

Although it is possible to view a number of the early fossil representatives of the neornithine clades in a phylogenetic context, in terms of the pattern of the radiation of modern birds, what do the placements of these taxa imply with respect to the two alternative hypotheses outlined above? Although acceptance of a hypothesis in which the bulk of the neornithine radiation occurred in the early Tertiary (Feduccia 1995, 1999) would require taking the known fossil record at face value – the noted lack of Neornithes in the later stages of the Mesozoic would then present evidence for their absence (Cracraft 2001) – it could be that both hypotheses are, to some extent, correct. Perhaps, as has been suggested, some clades diverged deep in the Cretaceous and then were maintained at low diversity until after the K–T boundary (Cooper and Fortey 1998; Cracraft 2001). Workers have used the fossil record and the phylogeny of Neornithes in different ways in order to address this problem. On the basis of some limited phylogenetic placements for fossils and the spatial pattern of extant taxa within phylogenetic topologies, Cracraft (2001) argued that both of these lines of evidence support a Cretaceous divergence for neornithine clades on the Southern Hemisphere continents. On the other hand, Bleiweiss's (1998) stratigraphic gap analysis of the published fossil record of three putatively related clades of Neornithes (Apodiformes, Caprimulgiformes, and Strigiformes) supported the probability of their divergence as late as the Tertiary.

Whatever the outcome of specific debates regarding the taxonomy of individual specimens claimed to be neornithine and from the Mesozoic (Stidham 1998; Dyke and Mayr 1999), the fossil record, when evaluated in a phylogenetic context, does appear to support a pre-Tertiary divergence for at least the more basal clades – Palaeognathae, Galliformes plus Anseriformes. I have argued elsewhere (Dyke 2001a) that the placement of well-represented fossil taxa (such as *Presbyornis* and *Gallinuloides*) from the early Tertiary within crown-groups of the most basal major clades of Neornithes must imply the *divergence* of these clades at an earlier time. Further evidence for this inference simply comes from the proximity of these Tertiary fossils to the K–T boundary (less than 10 million years in some cases) and from the fact that the vast majority of material claimed as neornithine from the Mesozoic has also been placed within more basal clades (Dyke 2001a). Of course *divergence* estimates tell us nothing about *diversity* changes and hence the explosive evolutionary radiation of Neornithes in the aftermath of the K–T boundary extinction is still a strong possibility.

As has been pointed out by Benton (1999) and Brochu (2001), even though techniques for estimating divergence times on the basis of molecular sequence data have become more and more sophisticated, these methods are still reliant on the fossil record to provide a timescale, by use of taxa that are either internal or external with respect to the clade in question. Although problems with the use of fossils for internal calibration are clear (van Tuinen and Hedges 2001), this approach is clearly preferable to the use of distant external events (such as the synapsid–diapsid divergence in the Carboniferous), that is subject to even more sampling bias.

Conclusion

In this chapter I have discussed the phylogenetic evidence for the placement of a number of well-preserved fossil modern birds from the early Tertiary within the most basal clades of Neornithes. Although the fossil record of described taxa from the Cenozoic is vast, it is clear that a degree of phylogenetic control is needed before such evidence can be incorporated into hypotheses that discuss the timing and extent of the neornithine radiation. Fossils that can be dated accurately are important for 'molecular clock' estimates of divergence but are useless in the absence of phylogenetic context. Work on the systematic placements of early neornithines is ongoing and depends to a great extent on improvements in knowledge of the phylogenetic relationships of extant taxa based on morphological characters. This interval of the fossil record is not, however, entirely depauperate – in contrast with described Mesozoic records, a large number of specimens that are well enough preserved to be informatively included within cladistic analyses have been described from the early Tertiary. Although much further work will be required to combine morphological information from these taxa within analyses, some degree of pattern is beginning to emerge.

Acknowledgements

I thank Philip Donoghue and Paul Smith for the invitation to participate in the symposium 'Telling the evolutionary time: molecular clocks and the fossil record' held as part of the 3rd Biennial Meeting of the Systematics Association (London 2001). S. Chapman, L. Claessens, G. Mayr, A. Milner and P. Sweet provided access to specimens, and P. Crab (Natural History Museum Photographic Services) took the photographs used in Figure 12.4. L. Chiappe and S. Magallón kindly allowed reproduction of figures used in their previous papers. This work was funded by the Frank M. Chapman Memorial Fund, American Museum of Natural History (Department of Ornithology).

References

Ballmann, P. (1969) 'Les oiseaux Miocènes de la Grive-Saint-Alban (Isère)', Geobios, 2: 157–204.
Benton, M.J. (1999) 'Early origins of modern birds and mammals: molecules vs. morphology', BioEssays, 21: 1043–51.
Bledsoe, A.H. (1988) 'A phylogenetic analysis of postcranial skeletal characters of ratite birds', Annals of the Carnegie Museum, 57: 73–90.
Bleiweiss, R. (1998) 'Fossil gap analysis supports early Tertiary origin of trophically diverse avian orders', Geology, 26: 323–6.

Brochu, C.A. (2001) 'Progress and future directions in archosaur phylogenetics', *Journal of Paleontology*, 75: 1185–201.

Brodkorb, P. (1964) 'Catalogue of fossil birds, Part 2 (Anseriformes through Galliformes)', *Bulletin of the Florida State Museum Biological Sciences*, 8: 195–335.

Chatterjee, S. (1997) *The Rise of Birds*, Baltimore: Johns Hopkins University Press.

Chiappe, L.M. (2001) 'Phylogenetic relationships among basal birds', in J.A. Gauthier and L.F. Gall (eds) *New Perspectives on the Origin and Evolution of Birds: Proceedings of the International Symposium in Honor of John H. Ostrom*, New Haven: Yale University Press, pp. 125–39.

Chiappe, L.M. and Dyke, G.J. (2002) 'The Mesozoic radiation of birds', *Annual Reviews of Ecology and Systematics*, 33: 91–124.

Clarke, J.A. (1999) 'New information on the type material of *Ichthyornis*: of chimeras, characters and current limits of phylogenetic taxonomy', *Journal of Vertebrate Paleontology*, 19: 38A.

Clarke, J.A. and Chiappe, L.M. (2001) 'A new carinate bird from the Late Cretaceous of Patagonia', *American Museum Novitates*, 3323: 1–23.

Cooper, A. and Fortey, R. (1998) 'Evolutionary explosions and the phylogenetic fuse', *Trends in Ecology and Evolution*, 13: 151–6.

Cooper, A. and Penny, D. (1997) 'Mass survival of birds across the KT boundary: molecular evidence', *Science*, 275: 1109–13.

Cooper, A., Lalueza-Fox, C., Anderson, S., Rambaut, A., Austin, J. and Ward, R. (2001) 'Complete mitochondrial genome sequences of two extinct moas clarify ratite evolution', *Nature*, 409: 704–7.

Cracraft, J. (1974) 'Phylogeny and evolution of the ratite birds', *Ibis*, 116: 494–521.

—— (1988) 'The major clades of birds', in M.J. Benton (ed.) *The Phylogeny and Classification of the Tetrapods, Volume 1: Amphibians, Reptiles, Birds*, Oxford: Systematic Association, pp. 333–55.

—— (2001) 'Avian evolution, Gondwana biogeography and the Cretaceous-Tertiary mass extinction event', *Proceedings of the Royal Society, London*, B268: 459–69.

Cracraft, J. and Clarke, J.A. (2001) 'The basal clades of modern birds', in J.A. Gauthier and L.F. Gall (eds) *New Perspectives on the Origin and Evolution of Birds: Proceedings of the International Symposium in Honor of John H. Ostrom*, New Haven: Yale University Press, pp. 143–56.

Crowe, T.M. and Short, L.L. (1992) 'A new gallinaceous bird from the Oligocene of Nebraska, with comments on the phylogenetic position of the Gallinuloididae', *Science series, Natural History Museum of Los Angeles County*, 36: 179–85.

Dyke, G.J. (2001) 'The evolutionary radiation of modern birds: systematics and patterns of diversification', *Geological Journal*, 36: 305–15.

—— (2001b) 'The fossil waterfowl (Aves, Auseriformes) from the Eocene of England', *American Museum Novitates*, 3354: 1–15.

Dyke, G.J. and Mayr, G. (1999) 'Did parrots exist in the Cretaceous period?', *Nature*, 399: 317–18.

Dyke, G.J., Dortangs, R.W., Jagt, J.W.M., Mulder, E.W.A., Schulp, A.S. and Chiappe, L.M. (2002) 'Europe's last Mesozoic bird', *Naturwissenschaften*, 89: 408–11.

Dyke, G.J., Gulas, B.E. and Crowe, T.M. (2003) 'The suprageneric relationships of galliform birds (Aves, Galliformes): a cladistic analysis of morphological characters', *Zoological Journal of the Linnean Society*, 137: 227–44.

Ericson, P.G.P. (1997) Systematic relationships of the Paleogene family Presbyornithidae (Aves: Anseriformes), *Zoological Journal of the Linnean Society*, 21: 429–83.

Fara, E. and Benton, M.J. (2000) 'The fossil record of Cretaceous tetrapods', *Palaios*, 15: 161–5.

Feduccia, A. (1995) 'Explosive evolution in Tertiary birds and mammals', *Science*, 267: 637–8.

Feduccia, A. (1999) *The Origin and Evolution of Birds*, 2nd edn, New Haven: Yale University Press.

Fordyce, R.E. and Jones, C.M. (1990) 'Penguin history and new fossil material from New Zealand', in L.S. Davis and J.T. Darby (eds) *Penguin Biology*, New York: Academic Press.

Gauthier, J.A. (1986) 'Saurischian monophyly and the origin of birds', *Memoirs of the California Academy of Sciences*, 8: 1–55.

Groth, J.G. and Barrowclough, G.F. (1999) Basal divergences in birds and the phylogenetic utility of the nuclear RAG-1 gene, *Molecular Phylogenetics and Evolution*, 12: 115–23.

Haddrath, O. and Baker, A.J. (2001) 'Complete mitochondrial DNA sequences of extinct birds: ratite phylogenetics and the vicariance biogeography hypothesis', *Proceedings of the Royal Society, London*, B268: 939–45.

Hedges, S.B., Parker, P.H., Sibley, C.G. and Kumar, S. (1996) 'Continental breakup and the ordinal diversification of birds and mammals', *Nature*, 381: 226–9.

Hope, S. (2002) 'The Mesozoic fossil record of Neornithes', in L.M. Chiappe and L.D. Witmer (eds) *Mesozoic Birds: Above the Heads of the Dinosaurs*, Berkeley: University of California Press, in press.

Houde, P. (1988) 'Paleognathous birds from the Tertiary of the northern hemisphere', *Publications of the Nuttall Ornithological Club*, 22: 1–148.

Houde, P. and Haubold, H. (1987) '*Palaeotis weigelti* restudied: a small Middle Eocene ostrich (Aves: Struthioniformes)', *Palaeovertebrata*, 17: 27–42.

Howard, H. (1955) 'A new wading bird from the Eocene of Patagonia', *American Museum Novitates*, 1710: 1–25.

Kurochkin, E.N. (1999) 'The relationships of the Early Cretaceous *Ambiortus* and *Otogornis* (Aves: Ambiortiformes)', *Smithsonian Contributions to Paleobiology*, 89: 275–84.

Lambrecht, K. (1928) '*Palaeotis weigelti* n. g. n. sp., eine fossile Trappe aus der Mitteleozänen Braunkohle des Geiseltales', *J-hall Verband Halle*, 7: 1–11.

Lee, K., Felsenstein, J. and Cracraft, J. (1997) 'The phylogeny of ratite birds: resolving conflicts between molecular and morphological data sets', in D.P. Mindell (ed.) *Avian Molecular Systematics and Evolution*, New York: Academic Press, pp. 173–208.

Livezey, B.C. (1986) 'A phylogenetic analysis of Recent anseriform genera using morphological characters', *Auk*, 103: 737–54.

—— (1997) 'A phylogenetic analysis of basal Anseriformes, the fossil *Presbyornis* and the interordinal relationships of waterfowl', *Zoological Journal of the Linnean Society*, 121: 361–428.

—— (1998) 'Erratum', *Zoological Journal of the Linnean Society*, 124: 397–8.

Livezey, B.C. and Zusi, R.L. (2001) 'Higher-order phylogenetics of modern Aves based on comparative anatomy', *Netherlands Journal of Zoology*, 51: 179–205.

Lucas, F.A. (1900) 'Characters and relations of *Gallinuloides wyomingensis* Eastman, a fossil Gallinaceous bird from the Green River Shales of Wyoming', *Bulletin of the Museum of Comparative Zoology*, 36: 79–84.

Magallón, S. and Sanderson, M.J. (2001) 'Absolute diversification rates in angiosperm clades', *Evolution*, 55: 1762–80.

Marsh, O.C. (1880) *Odontornithes: a Monograph on the Extinct Toothed Birds of North America*, Washington D.C.: Government Printing Office.

Mayr, G. (2000) 'A new basal galliform bird from the Middle Eocene of Messel (Hessen, Germany)', *Senckenbergiana lethaea*, 80: 45–57.

Milne-Edwards, A. (1867–1871) *Recherches Anatomiques et Paléontologiques pour Servir à l'Histoire des Oiseaux Fossiles de la France* (4 volumes), Paris: Victor Masson et Fils.

Monroe, B.L. Jr. and Sibley, C.G. (1990) *A World Checklist of Birds*, New Haven: Yale University Press.

Norell, M.A. and Clarke, J.A. (2001) 'Fossil that fills a critical gap in avian evolution', *Nature*, 409: 181–4.

Olson, S.L. (1985) 'The fossil record of birds', in D.S. Farner, J.R. King and K.C. Parkes (eds) *Avian Biology* (Volume 8), New York: Academic Press, pp. 79–256.

—— (1992) '*Neogaeornis wetzeli* Lambrecht, a Cretaceous loon from Chile (Aves: Gaviidae)', *Journal of Vertebrate Paleontology*, 12: 123–4.

—— (1994) 'A giant *Presbyornis* (Aves: Anseriformes) and other birds from the Paleocene Aquila Formation of Maryland', *Proceedings of the Biological Society of Washington*, 107: 429–35.

—— (1999) 'The anseriform relationships of *Anatalavis* Olson and Parris (Anseranatidae), with a new species from the Lower Eocene London Clay', *Smithsonian Contributions to Paleobiology*, 89: 231–43.

Olson, S.L. and Feduccia, A. (1980) '*Presbyornis* and the origin of the Anseriformes (Aves: Charadriomorphae)', *Smithsonian Contributions to Zoology*, 323: 1–24.

Olson, S.L., and Parris, D.C. (1987) 'The Cretaceous birds of New Jersey', *Smithsonian Contributions to Paleobiology*, 63: 1–22.

Padian, K. and Chiappe, L.M. (1998) 'The origin and early evolution of birds', *Biological Reviews*, 73: 1–42.

Peters, D.S. (1988) 'Ein vollständiges Exemplar von *Palaeotis weigelti* (Aves, Palaeognathae)', *Courier Forschungsinstitut Senckenberg*, 107: 223–33.

Rambaut, A. and Bromham, L. (1998) 'Estimating divergence dates from molecular sequences', *Molecular Biology and Evolution*, 15: 442–8.

Sibley, C.G., Ahlquist, J.E. and Monroe, B.L. Jr. (1988) 'A classification of the living birds of the world based on DNA–DNA hybridization studies', *Auk*, 105: 409–23.

Stidham, T.A. (1998) 'A lower jaw from a Cretaceous parrot', *Nature*, 396: 29–30.

Tambussi, C.P. (1995) 'The fossil Rheiformes from Argentina', *Courier Forschunginsitut Senckenberg*, 181: 121–30.

Tordoff, H.B. and Macdonald, J.R. (1957) 'A new bird (family Cracidae) from the early Oligocene of South Dakota', *Auk*, 74: 174–84.

Unwin, D.M. (1993) 'Aves', in M.J. Benton (ed.) *The Fossil Record II*, London: Chapman & Hall, pp. 717–37.

van Tuinen, M. and Hedges, S.B. (2001) 'Calibration of avian molecular clocks', *Molecular Biology and Evolution*, 18: 206–13.

van Tuinen, M., Sibley, C.G. and Hedges, S.B. (1998) 'Phylogeny and biogeography of ratite birds inferred from DNA sequences of the mitochondrial ribosomal genes', *Molecular Biology and Evolution*, 15: 370–6.

—— (2000) 'The early history of modern birds inferred from DNA sequences of nuclear and mitochondrial ribosomal genes', *Molecular Biology and Evolution*, 17: 451–7.

Vedding-Kristoffersen, A. (2000) 'Lithornithid birds (Aves: Palaeognathae) from the Lower Paleogene of Denmark', *Geologie en Mijnbouw*, 78: 375–81.

Appendix 12.1

Dataset including representative fossils of Palaeognathae (*Lithornis* and *Palaeotis*); '?' denotes 'missing character data'.

```
            00000000011111111112222222222233333333334444444444555555
            12345678901234567890123456789012345678901234567890123456


Outgroup    00000000000000000000000000000000000000000000000000000000

Tinamidae   00000000?0000010000000000000000010000000000000000000000000

Dinornithids 111?111120??????0200000021200001011110001100001000111111

Apteryx     111?111120112?1112000000210000010011100110100010001111111

Casuarius   110101111011211111010010111111111110110111010010110000020

Dromaius    110101111011211111010010111111111110110111010010110000020

Struthio    110001113111101111?110111211121121011111121011011100030

Rhea        111?011131111011111?110111211121121011111121011011000?0

A5204       0?00000?????110?????????????????11???????1110200??0?0?10??

A5303       ???????121?11???????????????????????????????????????????

A33138      0???00?????1?????????????????????????????????????????????

A38935      0???00???????????????????????????????????????????????????

A38934      ???????????1?10??????????????????????????????????????????

A5425       ??0???0121?1?100?1????????????????????????11?????????????

Palaeotis   10??0??1101??1?????0?001??012111?1?20???01?????000???0??
```

[See Appendix 2 of Lee *et al.* (1997: 196) for list of characters.]

Systematics Association Publications

1. Bibliography of key works for the identification of the British fauna and flora, 3rd edition (1967)[†]
Edited by G.J. Kerrich, R.D. Meikie and N. Tebble
2. Function and taxonomic importance (1959)[†]
Edited by A.J. Cain
3. The species concept in palaeontology (1956)[†]
Edited by P.C. Sylvester-Bradley
4. Taxonomy and geography (1962)[†]
Edited by D. Nichols
5. Speciation in the sea (1963)[†]
Edited by J.P. Harding and N. Tebble
6. Phenetic and phylogenetic classification (1964)[†]
Edited by V.H. Heywood and J. McNeill
7. Aspects of Tethyan biogeography (1967)[†]
Edited by C.G. Adams and D.V. Ager
8. The soil ecosystem (1969)[†]
Edited by H. Sheals
9. Organisms and continents through time (1973)[†]
Edited by N.F. Hughes
10. Cladistics: a practical course in systematics (1992)[*]
P.L. Forey, C.J. Humphries, I.J. Kitching, R.W. Scotland, D.J. Siebert and D.M. Williams
11. Cladistics: the theory and practice of parsimony analysis (2nd edition) (1998)[*]
I.J. Kitching, P.L. Forey, C.J. Humphries and D.M. Williams

[*]Published by Oxford University Press for the Systematics Association
[†]Published by the Association (out of print)

Systematics Association Special Volumes

1. The new systematics (1940)
Edited by J.S. Huxley (reprinted 1971)
2. Chemotaxonomy and serotaxonomy (1968)[*]
Edited by J.C. Hawkes
3. Data processing in biology and geology (1971)[*]
Edited by J.L. Cutbill
4. Scanning electron microscopy (1971)[*]
Edited by V.H. Heywood

5. Taxonomy and ecology (1973)*
Edited by V.H. Heywood
6. The changing flora and fauna of Britain (1974)*
Edited by D.L. Hawksworth
7. Biological identification with computers (1975)*
Edited by R.J. Pankhurst
8. Lichenology: progress and problems (1976)*
Edited by D.H. Brown, D.L. Hawksworth and R.H. Bailey
9. Key works to the fauna and flora of the British Isles and northwestern Europe, 4th edition
   (1978)*
Edited by G.J. Kerrich, D.L. Hawksworth and R.W. Sims
10. Modern approaches to the taxonomy of red and brown algae (1978)
Edited by D.E.G. Irvine and J.H. Price
11. Biology and systematics of colonial organisms (1979)*
Edited by C. Larwood and B.R. Rosen
12. The origin of major invertebrate groups (1979)*
Edited by M.R. House
13. Advances in bryozoology (1979)*
Edited by G.P. Larwood and M.B. Abbott
14. Bryophyte systematics (1979)*
Edited by G.C.S. Clarke and J.G. Duckett
15. The terrestrial environment and the origin of land vertebrates (1980)
Edited by A.L. Pachen
16. Chemosystematics: principles and practice (1980)*
Edited by F.A. Bisby, J.G. Vaughan and C.A. Wright
17. The shore environment: methods and ecosystems (2 volumes) (1980)*
Edited by J.H. Price, D.E.C. Irvine and W.F. Farnham
18. The Ammonoidea (1981)*
Edited by M.R. House and J.R. Senior
19. Biosystematics of social insects (1981)*
Edited by P.E. House and J.-L. Clement
20. Genome evolution (1982)*
Edited by G.A. Dover and R.B. Flavell
21. Problems of phylogenetic reconstruction (1982)
Edited by K.A. Joysey and A.E. Friday
22. Concepts in nematode systematics (1983)*
Edited by A.R. Stone, H.M. Platt and L.F. Khalil
23. Evolution, time and space: the emergence of the biosphere (1983)*
Edited by R.W. Sims, J.H. Price and P.E.S. Whalley
24. Protein polymorphism: adaptive and taxonomic significance (1983)*
Edited by G.S. Oxford and D. Rollinson
25. Current concepts in plant taxonomy (1983)*
Edited by V.H. Heywood and D.M. Moore
26. Databases in systematics (1984)*
Edited by R. Allkin and F.A. Bisby
27. Systematics of the green algae (1984)*
Edited by D.E.G. Irvine and D.M. John
28. The origins and relationships of lower invertebrates (1985)‡
Edited by S. Conway Morris, J.D. George, R. Gibson and H.M. Platt
29. Infraspecific classification of wild and cultivated plants (1986)‡
Edited by B.T. Styles

30. Biomineralization in lower plants and animals (1986)‡
Edited by B.S.C. Leadbeater and R. Riding
31. Systematic and taxonomic approaches in palaeobotany (1986)‡
Edited by R.A. Spicer and B.A. Thomas
32. Coevolution and systematics (1986)‡
Edited by A.R. Stone and D.L. Hawksworth
33. Key works to the fauna and flora of the British Isles and northwestern Europe, 5th edition (1988)‡
Edited by R.W. Sims, P. Freeman and D.L. Hawksworth
34. Extinction and survival in the fossil record (1988)‡
Edited by G.P. Larwood
35. The phylogeny and classification of the tetrapods (2 volumes) (1988)‡
Edited by M.J. Benton
36. Prospects in systematics (1988)‡
Edited by J.L. Hawksworth
37. Biosystematics of haematophagous insects (1988)‡
Edited by M.W. Service
38. The chromophyte algae: problems and perspective (1989)‡
Edited by J.C. Green, B.S.C. Leadbeater and W.L. Diver
39. Electrophoretic studies on agricultural pests (1989)‡
Edited by H.D. Loxdale and J. den Hollander
40. Evolution, systematics, and fossil history of the Hamamelidae (2 volumes) (1989)‡
Edited by P.R. Crane and S. Blackmore
41. Scanning electron microscopy in taxonomy and functional morphology (1990)‡
Edited by D. Claugher
42. Major evolutionary radiations (1990)‡
Edited by P.D. Taylor and G.P. Larwood
43. Tropical lichens: their systematics, conservation and ecology (1991)‡
Edited by G.J. Galloway
44. Pollen and spores: patterns and diversification (1991)‡
Edited by S. Blackmore and S.H. Barnes
45. The biology of free-living heterotrophic flagellates (1991)‡
Edited by D.J. Patterson and J. Larsen
46. Plant–animal interactions in the marine benthos (1992)‡
Edited by D.M. John, S.J. Hawkins and J.H. Price
47. The Ammonoidea: environment, ecology and evolutionary change (1993)‡
Edited by M.R. House
48. Designs for a global plant species information system (1993)‡
Edited by F.A. Bisby, G.F. Russell and R.J. Pankhurst
49. Plant galls: organisms, interactions, populations (1994)‡
Edited by M.A.J. Williams
50. Systematics and conservation evaluation (1994)‡
Edited by P.L. Forey, C.J. Humphries and R.I. Vane-Wright
51. The haptophyte algae (1994)‡
Edited by J.C. Green and B.S.C. Leadbeater
52. Models in phylogeny reconstruction (1994)‡
Edited by R. Scotland, D.I. Siebert and D.M. Williams
53. The ecology of agricultural pests: biochemical approaches (1996)**
Edited by W.O.C. Symondson and J.E. Liddell
54. Species: the units of diversity (1997)**
Edited by M.F. Claridge, H.A. Dawah and M.R. Wilson

55. Arthropod relationships (1998)**
Edited by R.A. Fortey and R.H. Thomas
56. Evolutionary relationships among Protozoa (1998)**
Edited by G.H. Coombs, K. Vickerman, M.A. Sleigh and A. Warren
57. Molecular systematics and plant evolution (1999)
Edited by P.M. Hollingsworth, R.M. Bateman and R.J. Gornall
58. Homology and systematics (2000)
Edited by R. Scotland and R.T. Pennington
59. The flagellates: unity, diversity and evolution (2000)
Edited by B.S.C. Leadbeater and J.C. Green
60. Interrelationships of the Platyhelminthes (2001)
Edited by D.T.J. Littlewood and R.A. Bray
61. Major events in early vertebrate evolution (2001)
Edited by P.E. Ahlberg
62. The changing wildlife of Great Britain and Ireland (2001)
Edited by D.L. Hawksworth
63. Brachiopods past and present (2001)
Edited by H. Brunton, L.R.M. Cocks and S.L. Long
64. Morphology, shape and phylogeny (2002)
Edited by N. MacLeod and P.L. Forey
65. Developmental genetics and plant evolution (2002)
Edited by Q.C.B. Cronk, R.M. Bateman and J.A. Hawkins
66. Telling the evolutionary time: molecular clocks and the fossil record (2003)
Edited by P.C.J. Donoghue and M.P. Smith

*Published by Academic Press for the Systematics Association
†Published by the Palaeontological Association in conjunction with Systematics Association
‡Published by the Oxford University Press for the Systematics Association
**Published by Chapman & Hall for the Systematics Association

Index